D0889785

*Functional Analysis
and Boundary-value
Problems: an
Introductory Treatment*

π   Pitman Monographs and
   Surveys in Pure and Applied Mathematics 30

Functional Analysis and Boundary-value Problems: an Introductory Treatment

B. Dayanand Reddy

University of Cape Town

Longman
Scientific &
Technical

Copublished in the United States with
John Wiley & Sons, Inc., New York

Longman Scientific & Technical
Longman Group UK Limited
Longman House, Burnt Mill, Harlow
Essex CM20 2JE, England
and Associated Companies throughout the world.

Copublished in the United States with
John Wiley & Sons, Inc., 605 Third Avenue, New York, NY 10158

First published 1986

AMS Subject Classifications: (main) 46C05, 35J20, 65N30
(subsidiary): 47A05, 35A35
ISSN 0269-3666

British Library Cataloguing in Publication Data

Reddy, B. Dayanand
 Functional analysis and boundary-value
 problems: an introductory treatment.—
 (Pitman monographs and surveys in pure
 and applied mathematics, ISSN 0269-3666;
 30)
 1. Functional analysis
 I. Title   II. Series
 515.7   QA320

ISBN 0-582-98826-8

Library of Congress Cataloging-in-Publication Data

Reddy, B. Dayanand, 1953–
 Functional analysis and boundary-value problems.

 (Pitman monographs and surveys in pure and applied
mathematics, ISSN 0269-3666; 30)
 Bibliography: p.
 Includes index.
 1. Functional analysis.   2. Boundary value problems.
I. Title. II. Series.
QA320.R43 1986   515.7   86-13341
ISBN 0-470-20384-6 (USA only)

Set in 10/12 Times

Printed in Great Britain
at the Bath Press, Avon

Contents

Preface

Modern developments in the physical sciences and engineering are making increasing use of a variety of mathematical techniques that have hitherto been regarded as strictly part of "pure" mathematics. One such area of mathematics is functional analysis, a knowledge of which is essential for a study, for example, of the modern theory of boundary-value problems and their approximation.

For a number of years now I have taught a course in Applied Functional Analysis to senior undergraduates in applied mathematics and physics, and to postgraduate students in engineering. I have made frequent reference to available texts when designing the course, and have also experimented with various ways of teaching the material. It has been my experience that, while students in principle possess the required background to embark on a study of this subject (typically, some advanced calculus and differential equations), they invariably lack the ability to deal comfortably with the degree of abstraction inherent in functional analysis. It is thus necessary that the beginner be led very gently into the subject, with the degree of success depending very much on the extent to which concrete illustrations can be given of abstract ideas.

Also from a pedagogic point of view, I have found after much experimentation that it is preferable to treat in detail a specific application together with those topics in functional analysis which are required for a study of the particular application, rather than to deal with a number of disparate topics in functional analysis together with brief glimpses of a variety of applications, a trend which most current texts on "applied" functional analysis follow. The key here is motivation: students who are interested in a particular application will happily learn the requisite mathematics, whereas a collection of applications in which there is no great interest will produce a correspondingly damp response.

A specific application in which there is nowadays a considerable interest is the use of approximate methods for the solution of boundary-value problems. Among these methods, the finite element method has received by far the greatest attention. Yet there is not a single text available at the moment which deals at an introductory level with functional analysis and its applications to boundary-value problems and the mathematics of finite elements.

This book is intended to fill that gap. It is addressed to a rather wide audience: senior undergraduate and postgraduate students in mathematics, engineering and the physical sciences, as well as those who are no longer students but whose backgrounds are in the more traditional aspects of differential equations and finite elements. Furthermore, the book may also be used by anyone who requires an introduction to functional analysis, for whatever reason.

The prerequisites for reading this book are modest: conventional early undergraduate courses in differential equations, vector analysis and linear algebra will do, though it will be helpful if there has been some experience in the modelling of physical phenomena (which engineers will have had) and in the actual use of finite elements. These latter requirements are by no means essential, though; indeed, one of the aims of this book is to give mathematicians some idea of how the tools of functional analysis can be brought to bear on an area that was once very much the preserve of engineers.

Having said this, I must at the same time beg the indulgence of the mathematical community: analysts may well be perturbed by what they would regard as a lack of rigour in places, and by the absence of certain topics deemed an essential part of any introductory account of functional analysis. These departures from the norm have been deliberate: I have been motivated by a desire to present enough of the subject to make the subsequent material on boundary-value problems intelligible, so that the reader may get on to the material on boundary-value problems as soon as possible without being bogged down or sidetracked by additional material of a peripheral nature. Furthermore, anyone who has absorbed the contents of this book will be in a good position to investigate the more advanced topics.

The book is divided into two parts. Part I, consisting of Chapters 1 to 7, constitutes an introduction to functional analysis, while Part II, Chapters 8 to 11, treats elliptic boundary-value problems and their approximation. Each chapter ends with Bibliographical Remarks which give an indication of where supplementary material on the contents of the chapter may be found. Also at the end of each chapter is a set of exercises: some of these are intended as an aid in consolidating understanding, while others fill in gaps in the text and give an indication of extensions of the theory. The exercises are to be regarded as an integral part of the text.

The book may be used in various ways: if used in a formal course, all the material may be covered sequentially or, in the case of a shorter course, some of the topics and the proofs of some of the longer theorems may be omitted, and filled in in a subsequent course. The same remarks apply if the book is used for self-study. Those who have some background in functional analysis may of course go straight on to Part II, using Part I as a compendium of results.

I am by no means an expert on the subject of this book. I therefore welcome suggestions of any ways in which the presentation of this material can be improved.

Cape Town, South Africa B.D.R.

Acknowledgements

A number of individuals have contributed, directly or indirectly, toward making this book a reality. I am very grateful for having had the privilege of studying and working with John Martin, who with his clear and sharp intellect has helped me to appreciate more fully the rich interaction between mathematics and engineering. I have had many illuminating discussions with David Matravers and George Ellis about the merits of teaching functional analysis to scientists and engineers; to these two colleagues I owe thanks for their willingness to allow me the freedom to implement and test my ideas in the Applied Mathematics curriculum, as well as for their encouragement in this venture. David Matravers and Mike Vorster have provided me with excellent working conditions in the Applied Mathematics and Civil Engineering Departments, respectively.

I should like to express my appreciation to Terry Griffin who has kindly read the whole manuscript, and has made many helpful suggestions.

A number of secretaries have contributed towards the preparation of the manuscript. Val Atkinson, Heather Bain, Anke Burbach, Lee Goddard and Pat Jordaan all typed portions of drafts as well as the final manuscript. Shirley Breed, apart from typing much of the drafts, was also responsible for putting together the final manuscript. To all of these ladies I am most grateful for their cheerful assistance. I should also like to thank my student Helena du Toit for her expert preparation of the line drawings.

Finally, in her gentle but firm criticism of many of my ideas, and in her unfailing moral support, my wife Shaada has contributed in no small way towards bringing this project to fruition. For this, and for much more besides, I am very thankful.

Daya Reddy

Introduction

The usefulness of functional analysis may not be immediately evident to users of mathematics who have not hitherto encountered this branch of mathematics. Indeed, engineers, physicists and other applied mathematicians are often repelled by what they perceive to be an unnecessarily high degree of abstraction inherent in the subject, and the conclusion often reached is that functional analysis cannot possibly be of any use in an area of endeavour in which "concrete" solutions to "concrete" problems are being sought. In some instances this feeling is justified: functional analysis is not a panacea, and there are many kinds of problems for which it is an inappropriate tool, to say the least. Also, examples abound of work that is carried out or written in a highly abstract manner when a simpler approach would have yielded the same results and possibly have thrown more light on the problem.

However, there are many areas in which a knowledge of the subject is indispensable if one hopes to probe deeply into the nature of a problem. In this book we will try eventually to convey some idea of the circumstances under which a student or researcher, equipped with little more than the basics of functional analysis, can gain a great deal of insight into the properties of boundary-value problems and their approximation.

The process of developing mathematical models of a wide range of physical phenomena leads to the description of such phenomena by ordinary or partial differential equations. These equations are posed on a given spatial domain, and are supplemented with boundary conditions, which are equations which must hold on the boundary of the domain. The differential equation together with the boundary conditions and all given other information about the problem (such as the geometry of the domain and various coefficients), called the data, form what is known as a *boundary-value problem* (BVP). Of course the next task is one of finding a solution to the problem.

If a closed-form or exact solution is available then we have essentially completed our task, and information about the system being modelled can be obtained from the solution. But for most BVPs the task of finding a solution can prove fruitless, and we are left with the problem of trying to obtain information about the solution, even though there is little prospect of finding that solution. Here we have two ways of proceeding: we can attempt to gain *qualitative information* about the solution, and we can also seek an *approximate solution* to the problem. These two avenues of enquiry are complementary, and both should ideally be pursued in the absence of an exact solution, though it is generally found that researchers opt for one or the other of these two approaches.

What kind of qualitative information is required? It is generally agreed that above anything else it is necessary to know the answers to the following questions:

Does a solution *exist*?
If so, is the solution *unique*?
Does the solution *depend continuously* on the data?

The first two questions need no clarification; in the case of the third question we are asking whether the solution changes only by a small amount if the data is changed by a small amount (of course, exactly what is meant by "small amount" must be made clear). A problem for which small changes in data cause wild fluctuations in the solution, is clearly unstable. If the answers to all three questions are affirmative, then the problem is said to be *well-posed*. It is of course vital that we know these answers, especially if it is not possible to find the exact solution. In order to study these qualitative properties of BVPs, a knowledge of functional analysis is indispensable.

A variety of methods exist for finding approximate solutions to BVPs. Perhaps the two most well-known methods are the finite difference method and the finite element method. The latter has been a great success, particularly since the availability of high-speed computers, and we will in fact give a self-contained account of the finite element method in Chapter 11. Setting aside details for now, the important point is that either one of these methods will provide an *approximate solution* to the problem. While many practitioners regard this as a suitable point at which to conclude proceedings, it is nevertheless of the utmost importance to obtain some information about the *quality of the approximation*. We need to know whether in some sense our approximation is a good one, and also how it could be improved. There are examples in the recent finite element literature in which a qualitative analysis of this kind has provided an explanation for the erratic behaviour observed in the actual results from a computer program, behaviour which would otherwise have remained baffling. Once again a knowledge of functional analysis is required if we are to pursue this line of enquiry satisfactorily. A considerable part of Chapter 11 will be devoted to the problem of estimating the error due to finite element approximations.

In summary, then, the general (and correct) impression gained is that functional analysis will help us to develop theories that will throw light on the *nature* of solutions to problems; it will not in general provide a means or technique for actually solving a given boundary-value problem. Given that the latter may be impossible to achieve, the need for a knowledge of functional analysis by all those who work with partial differential equations is compelling.

The first seven chapters constitute Part I of this book and provide a systematic development of all the basic functional analysis required for a reasonably in-depth study of boundary-value problems. We start in Chapter 1 with some elementary ideas in set theory; in Chapter 2 we introduce sets of functions which play a central role in future developments. Then in Chapters 3 and 4 we develop the theory of Banach and Hilbert spaces. Linear operator theory is covered in some detail in Chapter 5, while in Chapter 6 the concept of orthonormal bases in Hilbert spaces is developed, culminating in the (generalized) Fourier series theory. Sobolev spaces are an essential ingredient of any serious study of partial differential equations; these spaces of functions are described in Chapter 7, after an account of distribution theory has been given, since Sobolev spaces require distributions for their proper definition.

Part II of the book consists of Chapters 8 to 11, and constitutes a study of boundary-value problems and their approximation. In Chapter 8 we study elliptic boundary-value problems, and in Chapter 9 the theory of variational boundary-value problems is presented. The background to the study of finite elements is provided by Chapter 10, in which approximate solutions of variational boundary-value problems are discussed, and in Chapter 11 we deal with the finite element method itself.

Each chapter is divided into sections, and within each section the equations and theorems are simply numbered 1, 2, ... Then, for example, Theorem 1 of Section 20 is referred to in a subsequent section as Theorem 20.1.

Part I
Basic functional analysis

1

Sets

Functional analysis conventionally takes as its starting point the idea of the existence of collections of mathematical objects, for example, numbers, vectors, or functions. Such collections, which are known as sets, are endowed with additional structure, and when this is done it becomes possible to elaborate on their properties and build up a coherent theory.

Sets are clearly basic to a proper study of mathematics, and for this reason we start the study of functional analysis with some introductory aspects of set theory. After a review of the algebra of sets in Section 1, we go on to take a closer look at sets of numbers and of n-tuples in the remainder of this chapter. In the following chapter we will discuss some important sets whose members are functions.

§1. The algebra of sets

A *set* is any well-defined collection of objects. These objects—in the present context mainly numbers, vectors or functions—are called *members* or *elements* of the set. A set is usually denoted by a capital letter, for example A, and if the object x is a member of A we write

$x \in A$

which is read "x is an element of A" or "x belongs to A". Likewise, the expression

$x \notin A$

reads "x is not an element of A". Various ways of defining sets are given in the following examples.

Examples

(a)   The set A of all positive integers less than or equal to 5 is given by

$$A = \{1, 2, 3, 4, 5\}. \tag{1}$$

Here we have defined A simply by listing its elements. A more concise way of defining the set would be to write

$$A = \{n : n \text{ is an integer}, 1 \leqslant n \leqslant 5\}$$

or, if we denote by Z the set of all integers,

$$A = \{n : n \in Z, 1 \leqslant n \leqslant 5\}. \tag{2}$$

The expression in brackets is read "the set of all n, n being an integer, such that $1 \leqslant n \leqslant 5$".

(b)   The above example was of a *finite set*, that is, a set consisting of a finite number of elements. An example of an *infinite set*, i.e. a set which does not have a finite number of elements, is the set

$$A = \{n : n \in Z, n \geqslant 0\}$$

of all non-negative integers.

(c)   The *empty* or *null* set is the set with no elements, and is denoted by $\varnothing$. For example,

$$\{n : n \in Z, n < 1 \text{ and } n > 1\} = \varnothing. \qquad \square$$

Subsets, equal sets.   If A and B are two sets, we say that A is a *subset* of B if each element of A is also an element of B. This is denoted by

$$A \subset B.$$

According to this definition every set is, of course, a subset of itself, and so in order to distinguish subsets which do not coincide with the set in question, we say that A is a *proper subset* of B if A is indeed a subset of B and, furthermore, B also contains elements which do not belong to A. If it is desirable to indicate that B is a subset of A which is possibly the set A itself, we write

$$A \subseteq B.$$

If A is not a subset of B, this is indicated by

$$A \nsubseteq B.$$

For example, if A is given by (1) or (2), then $C = \{1, 2, 3\}$ is a proper subset of A, but $D \nsubseteq A$ where $D = \{1, 2, 6\}$.

Two sets A and B are equal if they contain exactly the same elements.

When this is the case, we write

$A = B.$

According to the definition of a subset, it is clear that two sets A and B are equal if

$A \subset B$ and $B \subset A.$

For example, if

$A = \{x : x^2 = 4\}, \qquad B = \{2, -2\}$

then we have $A = B$.

Union, intersection, difference.   We assume here and henceforth that all sets under discussion are subsets of a single fixed set called the *universal set*, which is denoted by U.

The *union* of two sets A and B, written $A \cup B$, is the set consisting of all elements that are in A or in B. That is,

$A \cup B = \{x : x \in A \text{ or } x \in B\}.$

This is shown graphically in Figure 1.1, where the universal set is represented by a rectangle and its subsets are points or areas within the rectangle.

The *intersection* of two sets A and B, written $A \cap B$, is the set of all elements that belong to both A and B (Figure 1.2). In other words,

$A \cap B = \{x : x \in A \text{ and } x \in B\}.$

The *difference* of two sets A and B, written $A - B$, is the set of all elements of A that do not belong to B (Figure 1.3):

$A - B = \{x : x \in A, x \notin B\}.$

The *complement* of a set A, denoted by A', is the set of all elements

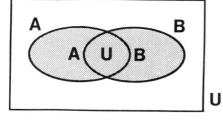

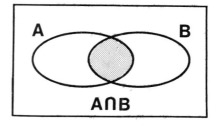

Figure 1.1   Figure 1.2

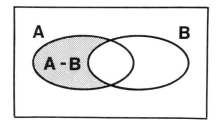

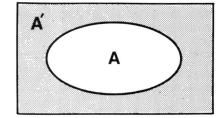

Figure 1.3   Figure 1.4

not in A (Figure 1.4). That is,

$$A' = \{x : x \notin A\}.$$

Example

Let $A = \{x : x \in Z, \ 1 \leq x \leq 10\}$ and $B = \{9, 10, 11, 12\}$. Then

$$A \cup B = \{x : x \in Z, \ 1 \leq x \leq 12\},$$
$$A \cap B = \{9, 10\},$$
$$A - B = \{x : x \in Z, \ 1 \leq x \leq 8\},$$

and

$$A' = \{\ldots -3, -2, -1, 0, 11, 12, 13, \ldots\}$$

(the universal set is taken as Z).   □

Countable sets. It is necessary at times to know whether or not the elements of an infinite set can be "labelled" by positive integers. In other words, we wish to distinguish infinite sets of the form

$$A = \{a_1, a_2, a_3, \ldots\} \tag{3}$$

from those that cannot be labelled as in (3). A set that can be put in one-to-one correspondence (i.e., labelled) with positive integers is called a *countable set*. Of course, any finite set A is countable, for if A has m members we may label them $a_1, a_2, \ldots, a_m$.

Examples

(a)   The set of functions

$$P = \{x^k : k = 1, 2, \ldots\} = \{x, x^2, x^3, \ldots\}$$

is countable.

(b)   As we shall see in the next section, sets like the set of points

$$A = \{x : 0 \leq x \leq 1\}$$

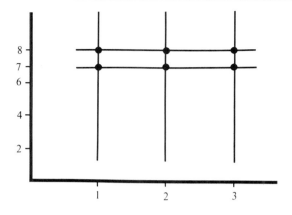

Figure 1.5

on the real line are not countable. That is, the set of real numbers between 0 and 1 cannot be labelled in the form a_1, a_2, ... $\quad\square$

Cartesian products.   Given two sets A and B, their cartesian product $A \times B$ is the set of *all ordered pairs* (a, b) where $a \in A$ and $b \in B$. That is,

$$A \times B = \{(a, b) : a \in A, \ b \in B\}.$$

A simple example of a cartesian product is provided by the case when

$$A = \{1, 2, 3\} \quad \text{and} \quad B = \{7, 8\}: \tag{4}$$

then

$$A \times B = \{(1, 7), (1, 8), (2, 7), (2, 8), (3, 7), (3, 8)\}.$$

If we think of the members of A as lying on a horizontal axis and the members of B as lying on the *vertical* axis of a pair of coordinate axes, then $A \times B$ may be represented by a set of points in the plane, as shown in Figure 1.5.

The idea of a cartesian product may be extended to products of more than two sets. For example, the cartesian product $A_1 \times A_2 \times A_3 \times \ldots A_n$ is defined to be the set of all ordered n-tuples

$$(a_1, a_2, \ldots, a_n) \quad \text{where} \quad a_1 \in A_1, a_2 \in A_2, \ldots, a_n \in A_n.$$

Note that, in general, $B \times A \neq A \times B$. For example, if A and B are as in (4), then

$$B \times A = \{(7, 1), (7, 2), (7, 3), (8, 1), (8, 2), (8, 3)\} \neq A \times B.$$

§2. Sets of numbers

We have already come across the set Z of all integers, defined by

$$Z = \{\ldots, -3, -2, -1, 0, 1, 2, 3, \ldots\}.$$

Occasionally we shall also make use of the set Z^+ of *natural numbers* or positive integers:

$$Z^+ = \{1, 2, 3, \ldots\}.$$

A *rational number* is a number that can be expressed as the ratio of two integers. We denote the set of rational numbers by Q:

$$Q = \{x : x = p/q,\ p \in Z,\ q \in Z,\ q \neq 0\}.$$

An important property of Q, which we record here, is the following: *the set Q of rational numbers is a countably infinite set*. In other words, the rational numbers can be put in one-to-one correspondence with integers. You are asked to show this in Exercise 2.1.

Numbers that do not belong to Q are called *irrational numbers*. For example, $\sqrt{2}$ cannot be written in the form p/q for $p, q \in Z$, and so is irrational. This brings us to R, the set of *real numbers*; R consists of all rational as well as irrational numbers. It is convenient to think of R as represented by an infinitely long line, each point on the line being a real number. Naturally R is an uncountable set.

Figure 1.6

Subsets of R.   Very often we deal not with the whole real line but only a portion of it, called an *interval*. Thus, if a and b are two points on R such that $a \leqslant b$, then we define

(i)   the open interval $(a, b) = \{x : x \in R,\ a < x < b\}$;
(ii)   the closed interval $[a, b] = \{x : x \in R,\ a \leqslant x \leqslant b\}$;
(iii)   the half-open intervals

$$(a, b] = \{x : x \in R,\ a < x \leqslant b\}$$

and

$$[a, b) = \{x : x \in R,\ a \leqslant x < b\}.$$

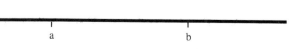

Figure 1.7

With this notation we can also write

$R = (-\infty, \infty)$.

The notions of closed and open sets are extremely important, and we therefore describe now in a little more detail exactly what we mean by open and closed sets in R.

Open sets. Given any point c on the real line, the open interval $(c - \varepsilon, c + \varepsilon) = \{x : c - \varepsilon < x < c + \varepsilon\}$ is called a *neighbourhood of c*; note that we can also write

$(c - \varepsilon, c + \varepsilon) = \{x : |c - x| < \varepsilon\}$.

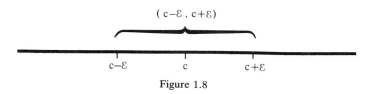

Figure 1.8

Now, let I be any set (or interval) in R; then c is called an *interior* point if we can find a neighbourhood of c, all of whose points belong to I. Finally, a set I in R is called an *open set* if every point of I is an interior point of I.

Examples

(a) The *open interval* (a, b) is an open set: for any point c in (a, b) we can define a neighbourhood lying entirely in (a, b) by choosing ε to be less than $|c - a|$ and $|c - b|$. Thus every point in (a, b) is an interior point.

(b) The closed interval $[a, b]$ is *not* an open set: the points a and b are such that, no matter how small we choose ε, it is impossible to find a neighbourhood of a or b, all of whose points lie in $[a, b]$. Thus a and b are not interior points and so $[a, b]$ is not open. Similar considerations apply to the half-open intervals $[a, b)$ and $(a, b]$; the points a and b respectively, are not interior points.   □

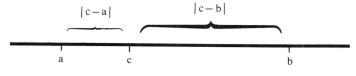

Figure 1.9

Closed sets. In order to define a closed set we define first a *limit point*: let I be a set in R and c a point in R (c does not necessarily belong to I). Then c is called a *limit point* or *accumulation point* of I if *every* neighbourhood of c contains at least one point of I distinct from c. Furthermore, a set $I \subset R$ is a *closed set* if it contains all of its limit points. It is not difficult to show (see Exercise 2.5) that a set I is closed if and only if its complement $I' = R - I$ is *open*.

Finally, we define the *closure $\bar{I}$* of a set $I \subset R$ to be the union of I and all its limit points. Clearly $\bar{I}$ is a closed set.

Examples

(a)   Consider the *open* interval $(0, 1)$: according to the above definition, every point in $(0, 1)$ is a limit point. Furthermore, 0 and 1 are also limit points of $(0, 1)$ since every neighbourhood of these two points contains members of $(0, 1)$. But 0 and 1 do not belong to the set, and so it is not closed. The closure of $(0, 1)$ is $[0, 1]$.

(b)   The *closed interval* $[a, b]$ is a closed set.   □

Supremum and infimum.   Suppose that the set A is a subset of R: if there is a real number p such that $p \geqslant x$ for all points x in A then we call p an *upper bound* of A. Similarly, if there is a real number q such that $q \leqslant x$ for all x in A then q is called a lower bound of A. Note that A can have many (even infinitely many) upper and lower bounds.

Figure 1.10

Now suppose that there is a number m which belongs to A and which also is an upper bound of A; we call m the *maximum* of the set A and write

max $A = m$.

Similarly, if there is a number n which belongs to A and which, furthermore, is a lower bound of A, then this number is called the *minimum* of A and we write

min $A = n$.

Examples

(a) Let A be the closed interval $[0, 1] = \{x : x \in R,\ 0 \leqslant x \leqslant 1\}$. Then any number $a \geqslant 1$ is an upper bound, any number $b \leqslant 0$ is a lower bound, and

$$\max A = 1, \qquad \min A = 0.$$

(b) Let $A = (0, 1) = \{x : x \in R,\ 0 \leqslant x \leqslant 1\}$. In this case A has no maximum or minimum, although 0 is a lower bound and 1 an upper bound of A. Note that 0 and 1 do not belong to A. □

The above examples illustrate once again the essential difference between closed and open intervals: closed intervals have minima and maxima while open intervals do not. Still, we would like to be able to express the fact that a set such as $(0, 1)$ has a *least upper bound*, which is the smallest of all its upper bounds, and a *greatest lower bound*, which is the largest of all its lower bounds.

In general the *supremum* or *least upper bound* of a set A is a number p' which is an upper bound of A, and which satisfies $p' \leqslant p$ for all upper bounds p. When p' exists, we write

$$p' = \sup A.$$

Similarly, the *infimum* or *greatest lower bound* of A is a number q' which is a lower bound of A, and which satisfies $q' \geqslant q$ for all lower bounds q. We normally write this as

$$q' = \inf A.$$

We note that when $\max A$ exists then clearly

$$\max A = \sup A.$$

Similarly, if $\min A$ exists then

$$\min A = \inf A.$$

Examples

(a) Let $A = (0, 1] = \{x : 0 < x \leqslant 1\}$. Then $\max A = \sup A = 1$, and $\inf A = 0$ although $\min A$ does not exist.
(b) Let A be the positive real line

$$R^+ = \{x : x \in R,\ x \geqslant 0\}.$$

Then $\inf R^+ = \min R^+ = 0$ while $\sup R^+$ does not exist, since R^+ is not bounded above. □

§3. Subsets of R^n

In the previous section we dealt in some detail with subsets of the real
line or intervals: here we extend some of the concepts to higher-
dimensional regions. We start with a description of R^2, which is defined
by

$$R^2 = R \times R$$

so that members of R^2 are *ordered pairs of real numbers*:

$$R^2 = \{(x, y) : x \in R, y \in R\}.$$

Just as R is represented geometrically by a line, so may R^2 be thought of
as a *plane* extending indefinitely in all directions. If we use the notation

$$\mathbf{x} \equiv (x, y)$$

to denote a typical member of R^2 then clearly $\mathbf{x}$ represents a point in the
plane with coordinates (x, y), as shown in Figure 1.11.

The above situation is easily generalized to higher dimensions: for
example, the set $R^3 \equiv R \times R \times R$ is the set of all ordered triples (x, y, z)
of real numbers, that is,

$$R^3 = \{\mathbf{x} = (x, y, z) : x, y, z \in R\}.$$

As for R^2, we simply represent a typical member of R^3 by $\mathbf{x} = (x, y, z)$.
Geometrically we can regard R^3 as being synonymous with three-

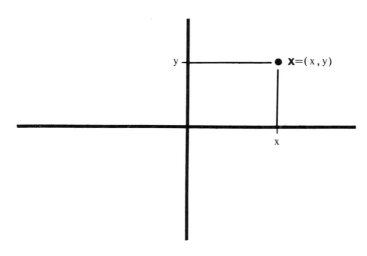

Figure 1.11

dimensional space, any $\mathbf{x} \in R^3$ being a point in this space with coordinates x, y and z.

Generally, we define R^n to be the set of all *ordered n-tuples* of real numbers:

$$R^n = \{\mathbf{x} = (x_1, x_2, \ldots, x_n) : x_i \in R, \ i = 1, \ldots, n\}.$$

When working in two or three dimensions it is convenient to use the alternative notations

$\mathbf{x} = (x, y)$ or $\mathbf{x} = (x_1, x_2)$ in R^2,

$\mathbf{x} = (x, y, z)$ or $\mathbf{x} = (x_1, x_2, x_3)$ in R^3,

depending on circumstances.

Open sets in R^n. The generalization to higher dimensions of the interval on the real line is the *domain* in R^n; in order to describe exactly what we mean by a domain, however, it is necessary first of all to define an open set in R^n. Recall that a neighbourhood of a point c in R is an open interval of points x satisfying $|x - c| < \varepsilon$; now we can read this inequality as "the distance from x to c is less than ε", and so it follows that all we need in order to extend the idea of a neighbourhood to R^n is a means of measuring distance. In R^2 the distance between two points $\mathbf{x}$ and $\mathbf{y}$ is defined by

$$|\mathbf{x} - \mathbf{y}| = \surd((x_1 - y_1)^2 + (x_2 - y_2)^2)$$

where (x_1, x_2) are the coordinates of $\mathbf{x}$ and (y_1, y_2) the coordinates of $\mathbf{y}$; hence, we can define a neighbourhood of a point $\mathbf{c}$ in R^2 to be the set of points that are a distance less than ε away from $\mathbf{c}$, for some $\varepsilon > 0$: that is, if we denote a neighbourhood of $\mathbf{c}$ by $N(\mathbf{c}; \varepsilon)$, then

$$N(\mathbf{c}; \varepsilon) = \{\mathbf{x} : \mathbf{x} \in R^2, \ |\mathbf{x} - \mathbf{c}| < \varepsilon\},$$

and ε is called, for obvious reasons, the *radius* of the neighbourhood (Figure 1.12).

We immediately generalize to R^n and define a neighbourhood of a point $\mathbf{c}$ in R^n to be the set

$$N(\mathbf{c}; \varepsilon) = \{\mathbf{x} : \mathbf{x} \in R^n, \ |\mathbf{x} - \mathbf{c}| < \varepsilon\}$$

where ε is the radius of the neighbourhood and the distance $|\mathbf{x} - \mathbf{c}|$ from $\mathbf{x}$ to $\mathbf{c}$ is given by

$$|\mathbf{x} - \mathbf{c}| = \surd((x_1 - c_1)^2 + (x_2 - c_2)^2 + \cdots + (x_n - c_n)^2).$$

Now that we have at our disposal the concept of a neighbourhood in R^n, we can define open sets in R^n simply by modifying appropriately the

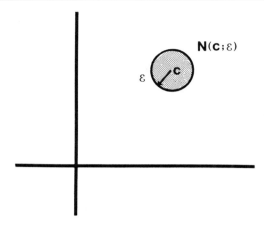

Figure 1.12

definition given in Section 2 for subsets of R; specifically, $\mathbf{c} \in R^n$ is called an *interior point* of a set Ω of points in R^n if we can always find a neighbourhood of $\mathbf{c}$, all of whose points belong to Ω. The situation is depicted in Figure 1.13 for the case $n = 2$; the set Ω is defined to be the set of all points lying *inside* the curve Γ. We can define a neighbourhood $N(\mathbf{c}; \varepsilon)$ lying entirely inside Ω by choosing ε to be less than d, the shortest distance from $\mathbf{c}$ to the boundary Γ.

Finally, a subset Ω of R^n is an *open set* if every point of Ω is an interior point.

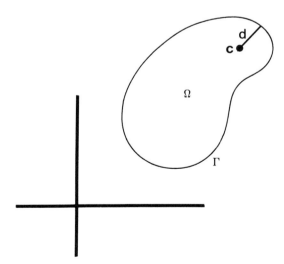

Figure 1.13

Example

The set

$$\Omega = \{\mathbf{x} : \mathbf{x} = (x_1, x_2) \in R^2,\ 0 < x_1 < 1 \text{ and } 0 < x_2 < 1\}$$

is an open set; a neighbourhood $N(\mathbf{c}; \varepsilon)$ of any point $\mathbf{c}$ that lies entirely in Ω may be defined by choosing ε to be less than any of the distances α, β, γ, δ shown in Figure 1.14. The boundary Γ of the set Ω is the set of points lying on the unit square; if we define $\bar{\Omega}$ to be the set of points inside and on the boundary, i.e.

$$\bar{\Omega} = \Omega \cup \Gamma,$$

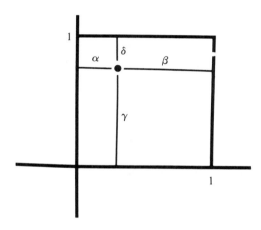

Figure 1.14

then $\bar{\Omega}$ is easily shown not to be open.   □

Closed sets in R^n. As with open sets, closed sets in R^n are defined in much the same way as their counterparts in R. Specifically, if Ω is a subset of R^n and $\mathbf{c}$ is a point in R^n (not necessarily in Ω, though), then $\mathbf{c}$ is called a limit point of Ω if *every* neighbourhood of $\mathbf{c}$ contains at least one point of Ω distinct from $\mathbf{c}$. Then, if the set Ω contains all of its limit points it is called a closed set.

Example

The unit square $\bar{\Omega} = \Omega \cup \Gamma = \{\mathbf{x} : \mathbf{x} \in R^2,\ 0 \leqslant x_1 \leqslant 1,\ 0 \leqslant x_2 \leqslant 1\}$ is closed (see previous example); however Ω is not closed since all the points lying on Γ are limit points of Ω but do not belong to Ω.   □

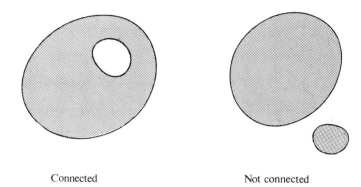

Connected   Not connected

Figure 1.15

Domain in R^n. We now describe the kinds of sets in R^n with which we shall be most concerned. First, we define a *connected set* Ω in R^n to be a set which has the property that *every* pair of points in Ω can be connected by a curve which lies entirely in Ω. Examples of connected and disconnected sets are shown in Figure 1.15.

We define next a *domain* in R^n to be an *open connected* set in R^n. Our interest will be exclusively confined to domains in R, R^2 and occasionally in R^3; in the case of R^2 and R^3 the boundary Γ (that is, the curve (in R^2) or surface (in R^3) within which all points of the domain lie) is assumed to be sufficiently smooth, in the sense that it possesses no cusps or suchlike singularities. Examples of admissible and inadmissible domains are shown in Figure 1.16.

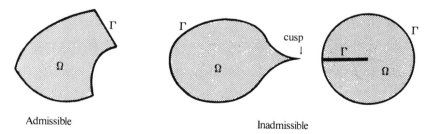

Admissible   Inadmissible

Figure 1.16

Bibliographical remarks

Further details on the topics discussed in this chapter can be found in most textbooks bearing titles such as *Real Analysis* or *Advanced Calculus*. Particularly readable accounts are available in the texts by Apostol [2], Binmore [7] and Simmons [45].

Exercises

1.1   Let $A = \{x \in Z: x^2 - x - 6 = 0\}$
      and $B = \{x \in Z: x^2 < 10\}$.
      List the elements of A and B. What are $A \cup B$, $A \cap B$, $A \cap Z^+$ and
      $A - Z^+$?

1.2   Let $A = \{1, 2\}$, $B = \{7, 8\}$, $C = \{9, 1\}$. Find $B \times (A \cup C)$ and
      $(A \cap C) \times B$.

1.3   Show that

      $$A \cap (B \cup C) = (A \cap B) \cup (A \cap C),$$
      $$A \cup (B \cap C) = (A \cup B) \cap (A \cup C).$$

      Illustrate these identities.

2.1   Show that the set Q of rational numbers is countable.
      [Hint: Set up a table of the form

1/1	1/2	1/3	$\ldots$
2/1	2/2	2/3	$\ldots$
3/1	$\ldots$		]

2.2   Show that $\sqrt{2}$ is irrational. [Hint: assume that $\sqrt{2}$ is rational and
      obtain a contradiction.]

2.3   Find all the limit points of the following subsets of R:

      (i) $[a, b]$;   (ii) Q;   (iii) $(0, 1) \cup \{2\}$.

2.4   Which of the sets in Exercise 2.3 are (a) open; (b) closed?

2.5   Show that a set $I \subset R$ is closed if and only if its complement is open.

2.6   Find max A, min A, sup A and inf A when

      (i) $A = \{1/n : n = 1, 2, 3, \ldots\}$;   (ii) $A = \{x : 0 < x^2 < 1\}$.

2.7   Show that the supremum has the following properties:
      (a)   if $I \subset R$ and α is any positive real number, then

      $$\sup_{x \in I} \alpha x = \alpha \sup_{x \in I} x;$$

(b)   if $I \subset R$ and α is any real number, then

$$\sup_{x \in I} (\alpha + x) = \alpha + \sup_{x \in I} x.$$

3.1   Determine the limit points of the following sets and establish which of these sets are open, closed or neither:

(a) $\Omega = \{x : x \in R^2,\ 0 \leqslant x \leqslant 1,\ 0 < y \leqslant 1\}$;

(b) $\Omega = \{x : x \in R^3,\ x^2 + y^2 + z^2 < a^2,\ z > 0\}$.

3.2   The *diameter* dia (Ω) of a set in R^n is defined by

$$\text{dia}\,(\Omega) = \sup\,\{|x - y|,\ x, y \in \Omega\}.$$

Find dia (Ω) for the sets in Exercise 3.1.

2

The spaces $C^m(\Omega)$ and $L_p(\Omega)$

In due course we are going to endow sets with particular properties and on the basis of these assumed properties construct a theory for special kinds of sets such as Hilbert spaces. In the development of this theory it is not necessary to appeal to the precise character of a set: the basic axioms and the theorems that follow from these axioms apply equally to sets whose members are numbers or matrices or functions. Before embarking on the task of describing this general framework, however, we first introduce two important examples of sets, or *spaces* (as they are usually called when endowed with additional properties) of functions, namely the spaces of continuous functions and the L_p spaces of functions whose pth powers are integrable. With these at our disposal it will be possible in subsequent chapters to illustrate aspects of the general theory, using as special examples sets such as R or R^n, as well as the spaces $C^m(\Omega)$ and $L_p(\Omega)$.

In Section 4 the concept of continuity is introduced, and the space $C^m(\Omega)$ of m-times continuously differentiable functions is defined. This set of course excludes many well-behaved functions which may not be continuous but which are nevertheless integrable. In the remaining sections of this chapter we discuss a framework in which it is possible to define spaces of integrable functions.

§4. Continuous functions

Continuous functions occur in great abundance in applied mathematics. This is not surprising, since many natural phenomena which are modelled mathematically involve quantities which may be represented (perhaps approximately) by continuous functions. Our aim in this section is to describe, rather informally at first, the concept of continuity, and subsequently to arrive at a mathematically suitable definition of a continuous function.

23

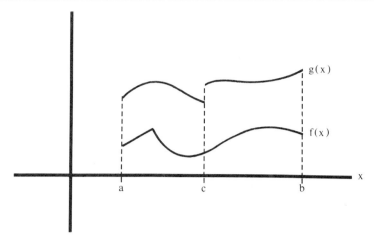

Figure 2.1

We begin by considering two arbitrary functions $f(x)$ and $g(x)$ that are defined on the interval $a \leqslant x \leqslant b$ or $[a, b]$, as shown in Figure 2.1. The function $f(x)$ is continuous, by which we mean that it is possible to draw the graph of $f(x)$ without lifting one's pen. On the other hand, the function $g(x)$ is discontinuous—its graph has a break at $x = c$. Roughly speaking, then, a continuous function is one whose graph is an uninterrupted curve.

Another type of discontinuous function is one that is *unbounded* at some point. For example, the function $h(x) = 1/x^2$ in Figure 2.2 is not

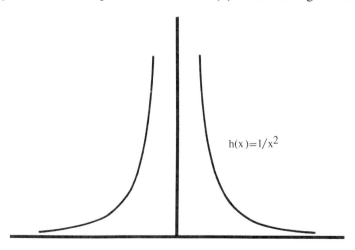

Figure 2.2

continuous at $x = 0$ since $h(x)$ "tends to infinity" as x approaches zero. Again, it is not possible to represent $h(x)$ by an uninterrupted curve.

The above examples give a qualitative idea of what a continuous function looks like. But for subsequent work we need a definition of continuity which agrees with our intuition and, furthermore, which is capable of being applied to any function whatsoever. The following is a suitable definition.

Continuous functions of one variable. Let $f(x)$ be a function on an interval I (open or closed) of the real line. Then $f(x)$ is continuous at a point $x_0 \in I$ if, *given* any positive number ε, no matter how small, it is possible to find a positive number δ (depending on ε and x_0) such that

$$|f(x) - f(x_0)| < \varepsilon \quad \text{whenever} \quad |x - x_0| < \delta. \tag{1}$$

If (1) holds at all points in I then we simply say that f is *continuous*. Generally the number δ will vary from point to point for a given ε, but if it so happens that δ depends only on ε and not on x, we say that f is *uniformly continuous*.

The above definition of continuity has the advantage of a very simple geometrical meaning. To see this, consider the graph of the function $f(x)$ shown in Figure 2.3. Choose a positive number ε and draw lines parallel to the x axis at heights $f(x_0) + \varepsilon$ and $f(x_0) - \varepsilon$. Then f is continuous at x_0 if we can find a positive number δ such that the graph of $f(x)$ lies inside the horizontal band formed by $f(x_0) \pm \varepsilon$ (i.e., $|f(x) - f(x_0)| < \varepsilon$) for all

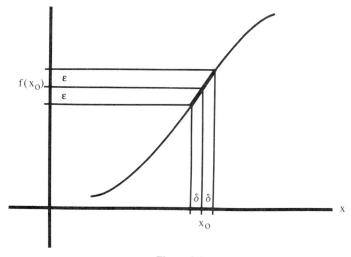

Figure 2.3

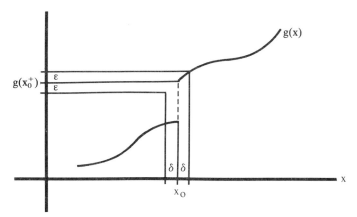

Figure 2.4

values of x in the vertical band $|x - x_0| < \delta$. In other words, the whole portion of the graph lying in the vertical band is contained in the horizontal band. For the function $f(x)$ shown in Figure 2.3 this is clearly possible at any point x_0 lying in I, no matter how small a value of ε is chosen. Thus $f(x)$ is continuous. On the other hand, the function $g(x)$ in Figure 2.4 is continuous at all points in I *except* at $x = x_0$: no matter how small we make δ, the vertical band will always include a portion of the graph that lies outside the horizontal band.

Examples

(a) $f(x) = x^2$ is continuous on $[0, 1]$ since

$$|f(x) - f(x_0)| = |x^2 - x_0^2| = |(x - x_0)(x + x_0)|$$
$$< 2\,|x - x_0| \quad \text{for} \quad x, x_0 \in [0, 1].$$

If $|x - x_0| < \delta$ then $|f(x) - f(x_0)| < 2\delta$. Hence, if ε is given, it suffices to take $\delta = \varepsilon/2$ to guarantee that $|f(x) - f(x_0)| < \varepsilon$ whenever $|x - x_0| < \delta$. Since δ does not depend on x_0, f is also uniformly continuous.

(b) $f(x) = 1/x$ is continuous but not uniformly continuous on the half-open interval $(0, 1]$. To show that f is continuous, we have

$$|f(x) - f(x_0)| = \frac{|x_0 - x|}{|x|\,|x_0|}.$$

For any x in the interval $|x - x_0| < \delta$ we have $x > x_0 - \delta$ (remember that x and x_0 are always non-negative since we are working in $(0, 1]$). Hence,

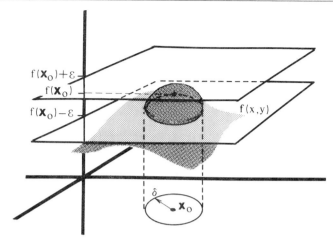

Figure 2.5

whenever $|x - x_0| < \delta$ we have

$$|f(x) - f(x_0)| < \frac{\delta}{x_0(x_0 - \delta)}.$$

Thus if ε is given, we choose $\delta = \varepsilon x_0^2/(1 + \varepsilon x_0)$ (this is found by setting $\delta/x_0(x_0 - \delta) = \varepsilon$). Then f is continuous for any x_0 in $(0, 1]$. But f is not uniformly continuous since δ depends on both ε and x_0. ☐

For functions of more than one variable the above ideas are easily extended. For example, consider a function $f(\mathbf{x}) \equiv f(x, y)$ of two variables defined on an open subset Ω of R^2, as shown in Figure 2.5. To check continuity at a point $\mathbf{x}_0 = (x_0, y_0)$ in Ω we choose a positive number ε and construct a pair of horizontal *planes* at heights $f(\mathbf{x}_0) + \varepsilon$ and $f(\mathbf{x}_0) - \varepsilon$ above the xy plane. Then $f(\mathbf{x})$ is continuous at $\mathbf{x}_0$ if it is always possible to construct a *cylinder* of radius δ (i.e., the set of points $\mathbf{x}$ for which $|\mathbf{x} - \mathbf{x}_0| < \delta$), this radius depending on ε, such that that part of the surface lying within the cylinder is contained in the horizontal band $|f(\mathbf{x}) - f(\mathbf{x}_0)| < \varepsilon$. This is but a special case of the general definition of continuity defined for functions of any number of variables, which we now state.

Continuity in R^n. A function $f(\mathbf{x})$ defined on an open subset Ω of R^n is continuous at a point $\mathbf{x}_0$ in Ω if, for every positive number ε, no matter how small, it is possible to find a positive number δ (depending on ε and

$\mathbf{x}_0)$ such that

$$|f(\mathbf{x}) - f(\mathbf{x}_0)| < \varepsilon \quad \text{whenever} \quad |\mathbf{x} - \mathbf{x}_0| < \delta. \tag{2}$$

If (2) holds for every $\mathbf{x}_0$ in Ω then f is said to be continuous on Ω. Furthermore, if δ does not depend on $\mathbf{x}_0$ then f is uniformly continuous.

The space $C(\Omega)$. For any domain Ω in R^n the collection of all continuous functions defined on Ω forms a set or space which is denoted by $C(\Omega)$. That is,

$$C(\Omega) = \{u : u(\mathbf{x}) \text{ is continuous on } \Omega\}.$$

For functions defined on a subset $\Omega = (a, b)$ of the real line, we simply write $C(a, b)$. The space of functions which are continuous on the closed set $\bar{\Omega} = \Omega \cup \Gamma$ (Ω and its boundary Γ) is denoted by $C(\bar{\Omega})$, and by $C[a, b]$ for functions on the closed interval (there is a difference: for example, $u(x) = 1/x$ is continuous on $(0, 1)$ but not on $[0, 1]$).

The spaces $C^m(\Omega)$, $C^\infty(\Omega)$. Among all the continuous functions defined on a subset Ω of R^n, some have the property that their first derivatives and possibly some derivatives of higher order are also continuous functions. We define $C^m(\Omega)$ to be the set or space of functions which, together with all their derivatives up to and including those of order m, are continuous. That is,

$$C^m(a, b) = \{u : u, u', \dots, u^{(m)} \text{ are all continuous functions}\}$$

for $\Omega = (a, b) \subset R$, and

$$C^m(\Omega) = \{u : u, \partial u/\partial x, \partial u/\partial y, \dots, \partial^m u/\partial x^k \partial y^{m-k} \ (k = 0, \dots, m)$$
$$\text{are all continuous functions}\}$$

for $\Omega \subset R^2$.

We define $C^\infty(\Omega)$ to be the space of functions *all* of whose derivatives are continuous on Ω. Clearly the inclusions $C^\infty(\Omega) \subset \cdots \subset C^m(\Omega) \subset C^{m-1}(\Omega) \subset \cdots \subset C(\Omega)$ hold.

Examples

(a)   The function

$$u(x) = \begin{cases} 0, & -1 \le x < 0, \\ x^2, & 0 \le x \le 1, \end{cases}$$

belongs to $C^1[-1, 1]$ since u and du/dx are both continuous, but d^2u/dx^2 is discontinuous (Figure 2.6).

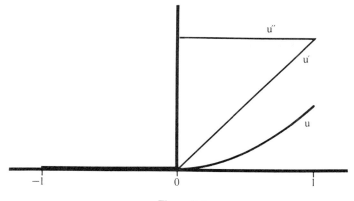

Figure 2.6

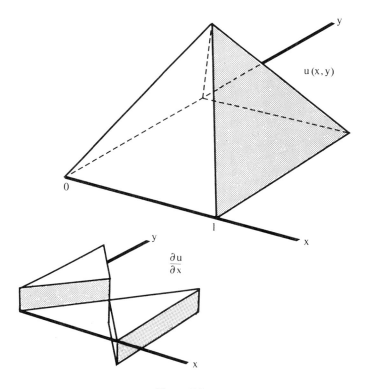

Figure 2.7

(b) $u(x) = \sin x$ belongs to $C^\infty(-\infty, \infty)$ since u and all of its derivatives are continuous on the whole real line.

(c) The pyramid-shaped function shown in Figure 2.7 is a member of $C(\Omega)$ where $\Omega = (0, 1) \times (0, 1)$ is the unit square in R^2, since $\partial u/\partial x$ and $\partial u/\partial y$ are not continuous functions.   □

We conclude this section with an important result which will be useful later.

Theorem 1. *If* u *is a continuous function defined on a closed, bounded subset* $\bar{\Omega}$ *of* R^n*, then* u *is bounded and* u *achieves its supremum and infimum on* $\bar{\Omega}$.

Remark.   Theorem 1 says that $|u(\mathbf{x})| \leqslant K$ for some constant $K > 0$ and for all $\mathbf{x}$ in $\bar{\Omega}$; furthermore, $\sup |u(\mathbf{x})| = \max |u(\mathbf{x})|$ and $\inf |u(\mathbf{x})| = \min |u(\mathbf{x})|$, for $\mathbf{x} \in \bar{\Omega}$, and these quantities are naturally bounded.

Examples

(a) Consider the function $u(x) = \sin x$ defined on $[0, 2\pi]$, which is closed. The supremum of $u(x)$ is 1 which is achieved at $x = \pi/2$, while the infimum is -1 which is achieved at $x = 3\pi/2$.

(b) Let $u(x) = 1/x^2$ on $[0, 1]$; the interval is closed but u is not continuous at the origin, so the theorem cannot be applied. As a matter of interest, $\inf u = 1$ (at $x = 1$) but $\sup u$ does not exist.

(c) Let $u(x) = x^2$ on $(0, 1)$; then $\sup u = 1$ and $\inf u = 0$. The function u achieves its supremum and infimum outside the set $(0, 1)$, which is open.
                                                                              □

§5. Measure of sets in R^n

Many functions which occur in practical applications are not continuous, and cannot therefore be accommodated in one of the spaces $C^m(\Omega)$. Perhaps the simplest example of such a function is the step function

$$H(x) = \begin{cases} 0, & x \leqslant 0, \\ 1, & x > 0. \end{cases}$$

Though functions like $H(x)$ are not continuous, they do nevertheless possess the important property that they are *integrable*, that is, their integrals exist. For example, the integral of $H(x)$ is the ramp function $R(x)$ shown in Figure 2.8; clearly, $R(x) \in C(-\infty, \infty)$.

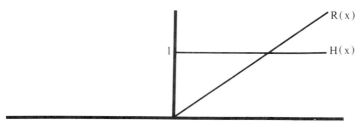

Figure 2.8

Our aim is to set up a space or set of functions which may be integrated, in a sense to be described. There are, however, certain preliminary concepts such as the measure of a set which we need to discuss first. After dealing with measure, we discuss Riemann and Lebesgue integrations and then go on to describe the very important space $L_p(\Omega)$ of integrable functions.

The concept of measure of a set Ω in R^n is simply a means of quantifying the "size" of Ω. For example, if $n = 1$ and Ω is the closed interval $[a, b]$, then by the measure of $[a, b]$ we understand the length $|b - a|$. Similarly, if $\Omega \subset R^2$ then the measure of Ω is its area. Generally, given any set Ω in R^n we write $\mu(\Omega)$ for the measure of Ω.

In R^n, an m-dimensional region is said to be *degenerate* and has zero measure when $m < n$. For example, if $n = 1$ then any isolated point is zero-dimensional and has zero length, if $n = 2$ any curve in the plane is one-dimensional and has zero area, and so on. In particular, the *boundary* Γ of any set Ω in R^n has zero measure in R^n (Figure 2.9).

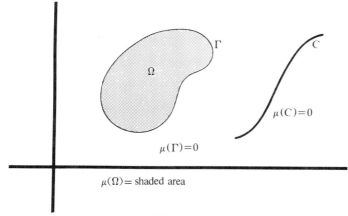

Figure 2.9

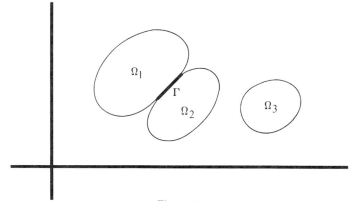

Figure 2.10

Furthermore, any *countable* set has zero measure. Recall from Section 1 that a set is countable if it is finite or denumerable; it follows that if Z and Q are the sets of integers and rational numbers in R, then

$$\mu(Z) = \mu(Q) = 0.$$

We record here also the following intuitively obvious result : if Ω_1 and Ω_2 are *disjoint*, i.e. if

$$\Omega_1 \cap \Omega_2 = \varnothing,$$

or if Ω_1 and Ω_2 are non-overlapping, i.e. if $\bar{\Omega}_1 \cap \bar{\Omega}_2$ is a set of zero measure, then

$$\mu(\Omega_1 \cup \Omega_2) = \mu(\Omega_1) + \mu(\Omega_2). \tag{1}$$

In Figure 2.10, Ω_1 and Ω_2 have a common boundary Γ, so that

$$\bar{\Omega}_1 \cap \bar{\Omega}_2 = \Gamma.$$

But $\mu(\Gamma) = 0$, since Γ is degenerate, hence (1) holds. Indeed, (1) holds for any combination of the sets Ω_1, Ω_2 and Ω_3 shown in Figure 2.10, and

$$\mu(\Omega_1 \cup \Omega_2 \cup \Omega_3) = \mu(\Omega_1) + \mu(\Omega_2) + \mu(\Omega_3).$$

Now consider any set Ω in R^n. We may subdivide Ω into a finite number of non-overlapping subsets Ω_1, Ω_2, ..., Ω_k, the resulting collection $\{\Omega_1 \ldots \Omega_k\}$ being called a *partition* of Ω. For example, if $n = 1$ and $\Omega = I = [a, b]$, we may obtain a partition of I by specifying

$$a = x_0 < x_1 < x_2 < \cdots < x_k = b$$

so that the collection $\{I_1, \ldots, I_k\}$, where

$$I_r = [x_{r-1}, x_r],$$

Figure 2.11

forms a partition of I. Clearly $I_{r-1} \cap I_r = \{x_r\}$ and

$$\mu(I_1 \cup I_2 \cup \cdots \cup I_k) = \sum_{r=1}^{k} \mu(I_r).$$

Similarly, if $n = 2$ and Ω is the rectangle $[a, b] \times [c, d]$, then we obtain a partition of Ω by specifying

$$a = x_0 < x_1 < \cdots < x_k = b,$$
$$c = y_0 < y_1 < \cdots < y_k = d,$$

and by defining rectangular subsets $\Omega_1, \Omega_2, \ldots, \Omega_m$ as shown in Figure 2.12.

We are now in a position to discuss integration of functions.

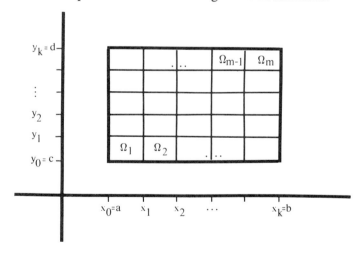

Figure 2.12

§6. Integration and the space $L_p(\Omega)$

Let $u(\mathbf{x})$ be a function defined on a domain Ω in R^n. Construct a partition $P = \{\Omega_1, \ldots, \Omega_k\}$ of Ω and evaluate $u(\mathbf{x}_r)$, $r = 1, \ldots, k$, where $\mathbf{x}_r$ is a point in Ω_r (Figure 2.13). Then we define the Riemann

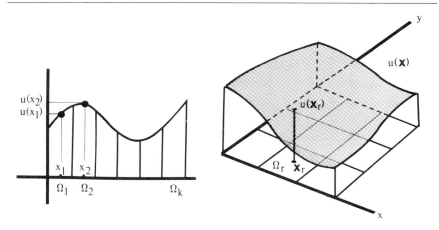

Figure 2.13

integral of u to be

$$\lim_{k\to\infty} \sum_{r=1}^{k} u(\mathbf{x}_r)\mu(\Omega_r) \tag{1}$$

where the limit is taken so that the maximum value of $\mu(\Omega_r)$ over all subsets approaches zero as $k\to\infty$. In other words, as the partition is progressively refined the quantity (1) approaches a limit. We denote this limit by

$$\int_\Omega \cdots \int u(\mathbf{x}) \, dx_1 \, dx_2 \cdots dx_n \qquad \text{or, more briefly,} \qquad \int_\Omega u(\mathbf{x}) \, dx. \tag{2}$$

When $n = 1$ we simply write, for $\Omega = [a, b]$,

$$\int_a^b u(x) \, dx$$

while for $n = 2$ we write

$$\iint_\Omega u(\mathbf{x}) \, dx \quad \text{or} \quad \iint_\Omega u(x, y) \, dx \, dy \quad \text{or simply} \quad \int_\Omega u(\mathbf{x}) \, dx.$$

The Riemann integral is the integral used in everyday applications, and it is generally adequate for most purposes. However, it also suffers from certain deficiencies; for one, it is possible to set up sequences of Riemann-integrable functions whose limits are not Riemann-integrable (sequences are covered in Section 10). Another serious drawback is that there are certain "nasty" functions which we are unable to deal with using the Riemann integral; for example, the function

$$u(x) = \begin{cases} 1, & x \text{ is rational,} \\ 0, & x \text{ is irrational,} \end{cases} \quad x \in [0, 1] \tag{3}$$

is not Riemann-integrable. With the more general *Lebesgue* integral we avoid these problems: the Lebesgue integral is able to handle functions like (3) and, furthermore, gives the same result as the Riemann integral if the function is Riemann-integrable. Also, limits of Lebesgue-integrable functions are always Lebesgue-integrable. While it might seem rather pedantic to abandon the Riemann integral for the above reasons—after all, how often are we required to integrate something like the function defined in (3)?—we will find later on that spaces of Lebesgue-integrable functions possess properties which allow us to classify them as Banach spaces or Hilbert spaces, and to draw on the vast reservoir of results for such spaces. From a practical point of view, Riemann and Lebesgue integrals coincide when the former exists, so all we have done is to broaden the class of functions that can be integrated. We now give a brief account of Lebesgue integration.

Let u be a function defined on a subset Ω of R^n with measure $\mu(\Omega)$. Now suppose that the set A of all those $\mathbf{x}$ in Ω for which $u(\mathbf{x}) < m$ is measurable (i.e. A can be assigned a measure), for any number $m > 0$. Then we say that $u(\mathbf{x})$ is *measurable* on Ω (Figure 2.14). From our point of view, all functions encountered in subsequent applications will be assumed measurable.

We now describe, first for functions of a single variable, the idea of Lebesgue integration. Consider a function $u(x)$ defined on a closed interval $[a, b]$. Let m be any lower bound of u and M any upper bound, and construct a partition of $[m, M]$, say

$$I_r = [y_{r-1}, y_r], \qquad m = y_0 < y_1 < \cdots < y_k = M.$$

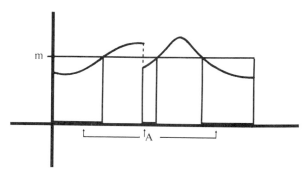

Figure 2.14

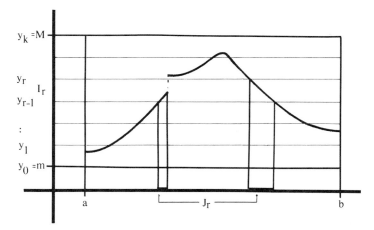

Figure 2.15

Next denote by J_r the set of points x in $[a, b]$ for which $u(x)$ lies in I_r, as shown in Figure 2.15. Then, roughly speaking, we define the Lebesgue integral of $u(x)$ between $x = a$ and $x = b$ by

$$\int_a^b u(x) \, dx = \lim_{k \to \infty} \sum_{i=1}^k y_i \mu(J_i),$$

it being understood that the intervals I_r making up the partition of $[m, M]$ become progressively smaller as $k \to \infty$.

As mentioned earlier, functions which are Riemann-integrable are also Lebesgue-integrable, and the two integrals coincide. Indeed, for well-behaved functions such as the one shown in Figure 2.15, it is clear that the Lebesgue integral, like the Riemann integral, amounts to the area under the graph of the function. However, as we indicated earlier, there are Lebesgue-integrable functions which are not Riemann-integrable; we demonstrate such a case in the following example.

Example

Let

$$u(x) = \begin{cases} 2, & x \text{ is rational,} \\ 1, & x \text{ is irrational,} \end{cases}$$

for $a < x < b$. This function is not Riemann-integrable, but if we partition the interval $[y_0, y_n]$ as shown in Figure 2.16, then $u(x) = 2$ is contained in one such interval, and $u(x) = 1$ in another. But the set J_1 of points x in $[a, b]$ for which $u(x) = 2$ consists of all rational numbers; this is a

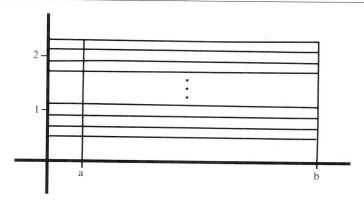

Figure 2.16

countable set, and so has measure *zero*. On the other hand, the measure of the set J_2 of points x in $[a, b]$ for which $u(x) = 1$ may be found from

$$\mu([0, 1]) = \mu(J_1) + \mu(J_2)$$

or

$$b - a = 0 + \mu(J_2)$$

since J_1 and J_2 are disjoint, with their union equal to $[a, b]$. Hence

$$\int_a^b u(x) \, dx = \lim_{k \to \infty} \sum_{i=1}^k y_i \mu(J_i) = 1 \cdot (b - a) = b - a. \qquad \square$$

The concept of Lebesgue integration is readily extended to functions on subsets of R^n for any value of n, and the arguments closely parallel those we have given for functions of a single real variable. Furthermore, all the usual properties of Riemann integrals extend to Lebesgue integrals; we summarize without proof some of these properties.

Theorem 1. *Let $u(x)$ and $v(x)$ be Lebesgue-integrable functions on $\Omega \subset R^n$. Then*

(i) $\int_\Omega [\alpha u(\mathbf{x}) + \beta v(\mathbf{x})] \, dx = \alpha \int_\Omega u(\mathbf{x}) \, dx + \beta \int_\Omega v(\mathbf{x}) \, dx$ *for   constants* α, β;

(ii) *if $u(\mathbf{x}) \leq v(\mathbf{x})$ for all $\mathbf{x} \in \Omega$, then $\int_\Omega u(\mathbf{x}) \, dx \leq \int_\Omega v(\mathbf{x}) \, dx$;*

(iii) $\left| \int_\Omega u(\mathbf{x}) \, dx \right| \leq \int_\Omega |u(\mathbf{x})| \, dx$;

(iv) *if $u(\mathbf{x})$ is bounded above and below by numbers m and M, then*

$$m\mu(\Omega) \leq \int_\Omega u(\mathbf{x}) \, dx \leq M\mu(\Omega).$$

From a practical point of view, the functions with which we shall be dealing are all Riemann-integrable. For technical reasons that will become apparent later, though, it is preferable simply to regard all integrals as Lebesgue integrals.

The space $L_p(\Omega)$.   Let p be a real number with $p \geqslant 1$. A function $u(\mathbf{x})$ defined on a subset Ω of R^n is said to belong to $L_p(\Omega)$ if u is measurable and if the (Lebesgue) integral

$$\int_\Omega |u(\mathbf{x})|^p \, d\mathbf{x}$$

is finite. Of course, every bounded continuous function on Ω belongs to $L_p(\Omega)$, but there are many other functions which have this property, as we show below.

Examples

(a)   The step function $H(x)$ defined by

$$H(x) = \begin{cases} 0, & -1 < x < 0 \\ 1, & 0 \leqslant x < 1 \end{cases},$$

belongs to $L_p(-1, 1)$ for any $p \geqslant 1$ since

$$\int_{-1}^{1} |H(x)|^p \, dx = \int_0^1 (1)^p \, dx = 1 < \infty.$$

(b)   The function $u(x) = x^{-1/3}$ belongs to $L_p(0, 1)$ for any $p < 3$, since

$$\int_0^1 |u(x)|^p \, dx = \int_0^1 x^{-p/3} \, dx = \frac{3}{3-p} [x^{(3-p)/3}]_0^1$$

which is finite for $p < 3$.   □

Some results which are frequently useful are embodied in the following theorem.

Theorem 2.   (i) $L_p(\Omega) \subset L_{p'}(\Omega)$ *for $p \geqslant p'$;*
(ii)   *If $u \in L_p(\Omega)$ then the integrals*

$$\int_\Omega |u(\mathbf{x})| \, d\mathbf{x} \qquad and \qquad \int_\Omega u(\mathbf{x}) \, d\mathbf{x}$$

are finite;
(iii)   *If $u, v \in L_2(\Omega)$ then the integral*

$$\int_\Omega u(\mathbf{x})v(\mathbf{x}) \, d\mathbf{x}$$

is finite.

Proof. The proof of (i) relies on the inequality

$$\left[\int_\Omega |u(\mathbf{x})|^{p'}\, dx\right]^{1/p'} \leqslant \left[\int_\Omega |u(\mathbf{x})|^p\, dx\right]^{1/p} \tag{4}$$

which holds if $p \geqslant p'$ (see Exercise 9.7 later for a derivation of (4)). If u belongs to $L_p(\Omega)$ then the integral on the right is finite, and hence so is the integral on the left. Thus $u \in L_{p'}(\Omega)$ also, which proves (i).

Part (ii) is a trivial consequence of (i): set $p' = 1$; then we have, for $u \in L_p(\Omega)$,

$$\int_\Omega u(\mathbf{x})\, dx \leqslant \int_\Omega |u(\mathbf{x})|\, dx \leqslant \left(\int_\Omega |u(\mathbf{x})|^p\, dx\right)^{1/p} < \infty.$$

Part (iii) is a result of the inequality

$$\int_\Omega u(\mathbf{x})v(\mathbf{x})\, dx \leqslant \sqrt{\left(\int_\Omega |u(\mathbf{x})|^2\, dx\right)} \sqrt{\left(\int_\Omega |v(\mathbf{x})|^2\, dx\right)}$$

(see Exercise 6.2). ∎

In order to deal meaningfully with the space $L_p(\Omega)$ we must first remove a source of ambiguity. Suppose that $f(x)$ and $g(x)$ are two measurable functions which differ at only a *finite* number of points (Figure 2.17): clearly,

$$\int_a^b |f(x)|^p\, dx = \int_a^b |g(x)|^p\, dx,$$

since f and g differ only on a set of measure zero. We say that f and g are

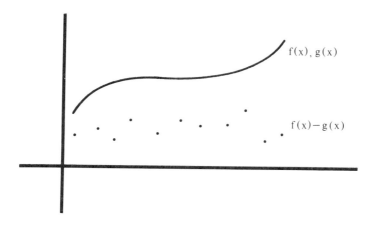

Figure 2.17

equal *almost everywhere* (abbreviated a.e.) and write

$$f = g \quad \text{(a.e)}. \tag{5}$$

We shall always take it as understood that for any $u \in L_p(\Omega)$ there are infinitely many functions which differ from u only on a set of measure zero, and when working in L_p we regard these functions as being identical with u, in the sense of (5).

The space $L_\infty(\Omega)$. If we let $p \to \infty$ then we may define the space $L_\infty(\Omega)$ to be the space of all measurable functions on Ω which are *bounded* almost everywhere on Ω (that is, except possibly on subsets of zero measure):

$$L_\infty(\Omega) = \{u : |u(\mathbf{x})| \leq k \text{ a.e. on } \Omega\}.$$

Clearly, $L_\infty(\Omega)$ is a subset of $L_p(\Omega)$ for all $p \geq 1$, since any $u \in L_\infty(\Omega)$ satisfies

$$\int_\Omega |u(x)|^p \, dx \leq \int_\Omega k^p \, dx < \infty,$$

so that $u \in L_p(\Omega)$ also.

Example

The function

$$u(x) = \begin{cases} x^2, & 0 \leq x < 1, \quad x \neq \tfrac{1}{2} \\ \to \infty, & x = \tfrac{1}{2}, \end{cases}$$

is bounded a.e. on $(0, 1)$ since $u(x) \to \infty$ only on a set of measure zero (the point $x = \tfrac{1}{2}$). □

It is interesting to note that while we have

$$L_\infty(\Omega) \subset \cdots \subset L_p(\Omega) \subset \cdots \subset L_1(\Omega),$$

the space $C(\Omega)$ of continuous functions is not a subset of any of the L_p spaces. For example, the function $u(x) = x^{-1}$ belongs to $C(0, 1)$ but not to $L_\infty(0, 1)$ since it is not bounded. But the space $BC(\Omega)$ of *bounded continuous* functions, defined by

$$BC(\Omega) = \{u \in C(\Omega) : |u(\mathbf{x})| < \infty \text{ for all } \mathbf{x} \in \Omega\}$$

is a subset of $L_\infty(\Omega)$.

Figure 2.18 shows schematically how the spaces $C^m(\Omega)$ and $L_p(\Omega)$ are related.

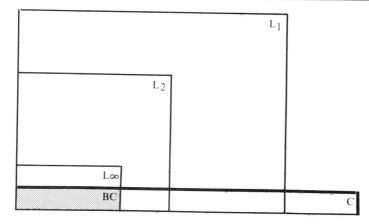

Figure 2.18

Bibliographical remarks

A good treatment of the concept of continuity may be found in Apostol [2] and Binmore [7]. The treatment of measure and integration given in Sections 5 and 6 is deliberately superficial, since a more complete discussion would take us too far off course. What has been covered is sufficient background, though, for what follows. There exist many readable accounts of the Lebesgue theory, notable examples being Kolmogorov and Fomin [24], Reed and Simon [38], Smirnov [46], Naylor and Sell [29], and Roman [40]; these should be consulted for a more comprehensive account of the theory.

Exercises

4.1 Discuss the continuity of the following functions after sketching their graphs:

  (a) $u(x) = x/(x^2 - 1)$, $-\infty < x < \infty$

  (b) $u(x) = \begin{cases} (x - |x|)/x, & x < 0, \\ 2, & x = 0. \end{cases}$

4.2 Show that a polynomial of degree n defined on an interval $[a, b]$ is continuous.

4.3 Show that $u(x) = x^{1/2}$ is continuous in $[0, \infty)$.

4.4 Show that $f(\mathbf{x}) = x^2 + 2y$ is continuous at any $\mathbf{x} \in R^2$.

4.5 Show that $f(x) = 1/x$ is uniformly continuous on $[a, b]$ where $b > a > 0$.

4.6 If f is continuous at a point x_0 in $[a, b]$ with $f(x_0) > 0$, show that there must be a neighbourhood $(x_0 - h, x_0 + h)$ about x_0 in which f is positive.

4.7 Prove Bolzano's theorem: *if $f(x)$ is a continuous function on $[a, b]$ with $f(a)f(b) < 0$ (i.e., $f(a)$ and $f(b)$ have different signs), there is at least one point c in $[a, b]$ such that $f(c) = 0$.* [Use the result in Exercise 4.6.]

4.8 To which spaces $C^m(\bar{\Omega})$ do the following functions belong?

(a) $u(x) = \begin{cases} 0, & -1 \leqslant x < 0, \\ x(1 + x), & 0 \leqslant x \leqslant 1 \end{cases}$

(b) $u(\mathbf{x}) = (\sin x)(1 - y), \qquad (x, y) \in [0, \pi] \times [0, 1]$.

(c) $u''(x) = \begin{cases} 0, & 0 \leqslant x \leqslant \frac{1}{2} \\ 1, & 0 < x < 1 \end{cases}$

6.1 Verify that the function $u(x) = x^{-1/a} (a > 0)$ belongs to $L_p(0, 1)$ if $a > p$. Does u belong to $L_\infty(0, 1)$?

6.2 Prove Theorem 6.2(iii). [Hint: consider

$$\int_\Omega [u(\mathbf{x}) - \alpha v(\mathbf{x})]^2 \, dx > 0, \qquad \text{for any } \alpha \in R.$$

Expand and then choose

$$\alpha = \int_\Omega u(\mathbf{x})v(\mathbf{x}) \, dx \Big/ \int_\Omega [v(\mathbf{x})]^2 \, dx.]$$

3

Linear spaces

From elementary courses in vector algebra and analysis we know that the idea of a vector as a directed line segment is not sufficient for us to build up a non-trivial theory, let alone use it in concrete applications. Additional structure has to be added: we agree to add together vectors using the parallelogram law, and we define various forms of multiplication of vectors, for example the scalar (dot) product and the vector (cross) product. Once these properties have been adopted, it becomes possible to construct a fairly sophisticated theory.

The same is true of sets in general. A set without structure is sterile, and not of much use from the point of view of the analyst. The question of what kinds of properties to assume is generally answered by looking at the properties of simple sets like R or the set of vectors, and by generalizing accordingly. This process of generalization will be a recurrent theme in the next few chapters; in this chapter we begin the process by defining in Section 7 a linear space which is, broadly speaking, a set whose members behave like vectors. In Sections 8 and 9 we show how properties such as "length", "distance" and "scalar product" can be defined for any set.

§7. Linear spaces and subspaces

We are familiar with the idea of a *set* being a collection of objects, all of which have a specified property. In most applications, though, it is useful to be able to add together multiples of members of a set and to have the assurance that the result of such an operation will yield something that is also a member of that set. This is the essence of a linear space.

Suppose we generalize from the behaviour of vectors, starting by first reviewing some familiar properties of the set V of all vectors. Given

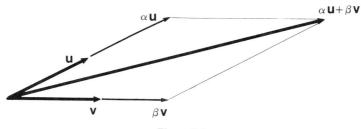

Figure 3.1

vectors **u**, **v**, **w** and real numbers α, β we know that:

1. $\alpha\mathbf{u} + \beta\mathbf{v}$ is also a vector (sums of multiples of vectors are also vectors) (Figure 3.1);

2. $\mathbf{u} + \mathbf{v} = \mathbf{v} + \mathbf{u}$, $\mathbf{u} + (\mathbf{v} + \mathbf{w}) = (\mathbf{u} + \mathbf{v}) + \mathbf{w}$ (i.e. when adding vectors together, the result does not depend on the order in which the addition is carried out);

3.   there is a vector **0** called the *zero vector* which has the property that $\mathbf{u} + \mathbf{0} = \mathbf{u}$ for all vectors **u**;

4.   there is a vector $-\mathbf{u}$ called the *negative* of **u**, which has the property that $\mathbf{u} + (-\mathbf{u}) = \mathbf{0}$ (we normally write this as $\mathbf{u} - \mathbf{u} = \mathbf{0}$). This in turn defines *subtraction*: by the difference $\mathbf{u} - \mathbf{v}$ we mean the vector $\mathbf{u} + (-\mathbf{v})$, as in Figure 3.2;

5. $(\alpha\beta)\mathbf{u} = \alpha(\beta\mathbf{u})$;

6. $(\alpha + \beta)\mathbf{u} = \alpha\mathbf{u} + \beta\mathbf{u}$, $\alpha(\mathbf{u} + \mathbf{v}) = \alpha\mathbf{u} + \alpha\mathbf{v}$;

7. $1 \cdot \mathbf{u} = \mathbf{u}$ (this, with 6, tells us that $\mathbf{u} = (1 + 0)\mathbf{u} = 1 \cdot \mathbf{u} = 1 \cdot \mathbf{u} + 0 \cdot \mathbf{u} = \mathbf{u}$ so that $0 \cdot \mathbf{u} = 0$.

Now all of these properties of vectors are readily generalized to any set, and this is what we do next.

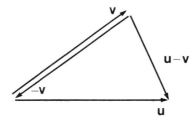

Figure 3.2

Linear space. A set U is called a linear space if it has an operation " + " called addition, an operation of multiplication by a scalar (real number), and satisfies the following *axioms*:

LS1. For all $u, v \in U$ and $\alpha, \beta \in R$, $\alpha u + \beta v$ is also a member of U.
LS2. $u + v = v + u$ and $u + (v + w) = (u + v) + w$ for all $u, v, w \in U$.
LS3. There is an element 0 of U called the zero element which has the property that

$$u + 0 = u \qquad \text{for all} \qquad u \in U.$$

LS4. For every $u \in U$ there is an element $-u$ which satisfies $u + (-u) = 0$. Then by the difference $u - v$ we understand $u + (-v)$.
LS5. $(\alpha\beta)u = \alpha(\beta u)$ for all $\alpha, \beta \in R$ and $u \in U$.
LS6. $(\alpha + \beta)u = \alpha u + \beta u$, $\alpha(u + v) = \alpha u + \alpha v$ for all $\alpha, \beta \in R$ and $u, v \in U$.
LS7. $1 \cdot u = u$.

Examples

(a) The set V of vectors is a linear space: indeed, V served as a model when setting up the axioms of a linear space.

(b) The set R^n of n-tuples is a linear space, with addition defined by

$$\mathbf{u} + \mathbf{v} = (u_1, u_2, \ldots, u_n) + (v_1, v_2, \ldots, v_n)$$
$$= (u_1 + v_1, u_2 + v_2, \ldots, u_n + v_n) \quad \text{for } \mathbf{u}, \mathbf{v} \in R^n$$

and scalar multiplication by

$$\alpha\mathbf{u} = \alpha(u_1, u_2, \ldots, u_n) = (\alpha u_1, \alpha u_2, \ldots, \alpha u_n) \quad \text{for } \alpha \in R.$$

The zero element is $\mathbf{0} = (0, \ldots, 0)$ and the element $-\mathbf{u}$ is $-\mathbf{u} = (-u_1, \ldots, -u_n)$.

(c) The set $C^m(\Omega)$ is a linear space. For, if u and v are two m-times continuously differentiable functions, then so is the function $\alpha u + \beta v$ defined by $(\alpha u + \beta v)(\mathbf{x}) = \alpha u(\mathbf{x}) + \beta v(\mathbf{x})$. The zero element is simply the zero function and the function $-u$ is the function satisfying $-u(\mathbf{x}) = -1 \cdot u(\mathbf{x})$.

(d) The space $L_p(\Omega)$ is a linear space for $1 \leqslant p < \infty$; this follows from the *Minkowski inequality for integrals*

$$\left[\int_\Omega |u \pm v|^p \, dx \right]^{1/p} \leqslant \left[\int_\Omega |u|^p \, dx \right]^{1/p} + \left[\int_\Omega |v|^p \, dx \right]^{1/p} \tag{1}$$

which is proved in Exercise 7.6. If we replace u by αu and v by αv in (1) then we see that

$$\int_\Omega |\alpha u + \beta v|^p \, dx < \infty,$$

since the terms on the right-hand side are bounded. Hence $\alpha u + \beta v \in L_p(\Omega)$.

The space $L_\infty(\Omega)$ is likewise a linear space, as is readily verified.

(e)   The set U defined by

$$U = \{u : u(\mathbf{x}) \in C(\Omega),\ u(\mathbf{x}) \geq 0,\ \mathbf{x} \in \Omega\}$$

is *not* a linear space since, for example, $\alpha u(\mathbf{x})$ is not a member of U for negative values of α.   □

Since all linear spaces are sets, it is natural to enquire whether subsets of linear spaces are also linear spaces. This is not always true, but in those cases in which it is true we give the subset a special name.

Subspace.   A subspace V of a linear space U is a subset of U which is also a linear space.

Examples

(a)   Consider the linear space R^3; all points of the form $(x, y, 0)$ form a subspace of R^3—the xy plane—since sums of multiples of points in the xy plane also lie in this plane. Indeed, the set of points of any plane *passing through the origin* is a subspace of R^3 (Figure 3.3).

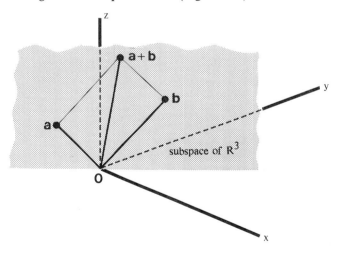

Figure 3.3

(b)   The set $P_3[0, 1]$ of polynomials of degree ≤ 3 forms a subset of C $[0, 1]$ and constitutes a subspace: for any polynomials $p(x)$, $q(x) \in$

$P_3[0, 1]$,

$$\alpha p(x) + \beta q(x)$$

is also a polynomial of degree $\leqslant 3$, and therefore belongs to $P_3[0, 1]$.

(c)   The set $BC(\Omega)$ of bounded continuous functions forms a subspace of $L_p(\Omega)$ (see Section 6) for $1 \leqslant p \leqslant \infty$.   □

Sum of subspaces.   Given two subspaces V, W of a linear space U, we define the *sum* of V and W, denoted $V + W$, to be the set of all members of U of the form $v + w$ with $v \in V$ and $w \in W$. That is,

$$V + W = \{u \in U : u = v + w \quad \text{for} \quad v \in V, \ w \in W\}.$$

The set $V + W$ is also a subspace of U since, for any u and $\bar{u}$ in $V + W$, we have

$$\alpha u + \beta \bar{u} = \alpha(v + w) + \beta(\bar{v} + \bar{w}) \quad (\text{since } u = v + w, \ \bar{u} = \bar{v} + \bar{w})$$

$$= \underbrace{(\alpha v + \beta \bar{v})}_{\in V} + \underbrace{(\alpha w + \beta \bar{w})}_{\in W},$$

so that $\alpha u + \beta \bar{u}$ is also in $V + W$.

Example

Let $U = R^3$ and $V = \{\mathbf{x} \in R^3;\ \mathbf{x} = (\alpha, 0, 0)\}$, $W = \{\mathbf{x} \in R^3 : \mathbf{x} = (0, \beta, 0)\}$. Then the sum of V and W is the subspace of U consisting of all points $\mathbf{x} = (\alpha, \beta, 0)$ (Figure 3.4).   □

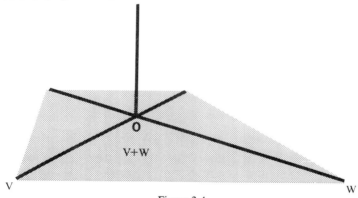

Figure 3.4

Direct sum of subspaces.   If V and W are subspaces of a linear space U, then U is said to be the *direct sum* of $V + W$ if $U = V + W$ and $V \cap W = \{0\}$. We denote the direct sum by $V \oplus W$.

Example

Let $U = R^3$, $V = \{\mathbf{x} \in R^3 : \mathbf{x} = (\alpha,\, \beta,\, 0)\}$, $W = \{\mathbf{x} \in R^3 : \mathbf{x} = (0,\, \beta,\, \gamma)\}$ and $X = \{\mathbf{x} \in R^3 : \mathbf{x} = (0,\, 0,\, \gamma)\}$. Then clearly we have

$$U = V + W \quad \text{and} \quad U = V + X.$$

But $V \cap W = \{\mathbf{x} : \mathbf{x} = (0,\, \beta,\, 0)\} \neq \{\mathbf{0}\}$ and $V \cap X = \{\mathbf{0}\}$. Hence we may write $U = V \oplus X$ (Figure 3.5). □

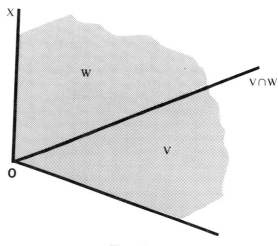

Figure 3.5

The question of when an arbitrary member of u of a linear space U has a *unique* representation $u = v + w$ for $v \in V$, $w \in W$ is easily resolved, as we show in the next result.

Theorem 1. *Let U be a linear space. Then $U = V \oplus W$ if and only if every $u \in U$ has the unique representation*

$$u = v + w$$

for some $v \in V$, $w \in W$.

The proof of this theorem is left as an exercise (Exercise 7.4).

§8. Inner product

We are now in a position to generalize to linear spaces many of the fundamental concepts of vector analysis, and we start with the concept of

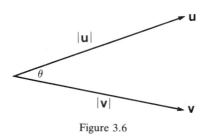

Figure 3.6

an *inner product* or *scalar* product. Recall from vector analysis that the inner or scalar product $\mathbf{u} \cdot \mathbf{v}$ of two vectors $\mathbf{u}$ and $\mathbf{v}$ is a real number given by

$$\mathbf{u} \cdot \mathbf{v} = |\mathbf{u}|\,|\mathbf{v}| \cos \theta$$

where θ is the angle between $\mathbf{u}$ and $\mathbf{v}$. The inner product has the following properties: $\mathbf{u} \cdot \mathbf{v} = \mathbf{v} \cdot \mathbf{u}$ (symmetry); $(\alpha \mathbf{u} + \beta \mathbf{v}) \cdot \mathbf{w} = \alpha \mathbf{u} \cdot \mathbf{w} + \beta \mathbf{v} \cdot \mathbf{w}$ (linearity); and $\mathbf{u} \cdot \mathbf{u} \geqslant 0$ with $\mathbf{u} \cdot \mathbf{u} = 0$ iff $\mathbf{u} = 0$ (positive-definiteness). Still in the context of vectors, the scalar product in turn provides us with a means of measuring the *length* or *norm* of a vector, which is defined by $|\mathbf{u}| = (\mathbf{u} \cdot \mathbf{u})^{1/2}$. Finally, when equipped with the inner product operation it is also possible to measure the distance between two points $\mathbf{x}$ and $\mathbf{y}$ in R^3: if this is denoted by $d(\mathbf{x}, \mathbf{y})$, then

$$d(\mathbf{x}, \mathbf{y}) = \surd((\mathbf{y} - \mathbf{x}) \cdot (\mathbf{y} - \mathbf{x}))$$
$$= |\mathbf{y} - \mathbf{x}|$$
$$= \surd((y_1 - x_1)^2 + (y_2 - x_2)^2 + (y_3 - x_3)^2).$$

The function $d(\ldots)$, being a device for measuring distances between points, is called the *metric*.

The concepts of inner product, norm and metric are defined in much the same way for arbitrary linear spaces. In this section we deal with the inner product.

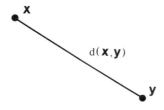

Figure 3.7

Inner product.   Let U be a linear space: then the inner product (u, v) of $u, v \in U$ is a real-valued operation which satisfies the following *axioms*:

IP1. $(u, v) = (v, u)$ (symmetry)

IP2. $(\alpha u + \beta v, w) = \alpha(u, w) + \beta(v, w)$ (linearity)

IP3. $(u, u) \geqslant 0$ and $(u, u) = 0$ iff $u = 0$ (positive-definiteness),

for all $u, v, w \in U$ and $\alpha, \beta \in R$. Thus we have defined an inner product to be an operation that has the same properties as the dot or scalar product in respect of vectors.

Examples

(a)   Let $U = R^3$: then the conventional scalar product defined by

$$(\mathbf{x}, \mathbf{y}) = \mathbf{x} \cdot \mathbf{y} = x_1 y_1 + x_2 y_2 + x_3 y_3$$

for $\mathbf{x} = (x_1, x_2, x_3)$ and $\mathbf{y} = (y_1, y_2, y_3)$ satisfies the inner product axioms.

(b)   Let U be the space $L_2(a, b)$ of square-integrable functions defined on the interval (a, b). An inner product for $L_2(a, b)$ may be defined by

$$(u, v) \equiv \int_a^b u(x)v(x)\,\mathrm{d}x \quad \text{for} \quad u, v \in L_2(a, b). \tag{1}$$

We have

$$(v, u) = \int_a^b v(x)u(x)\,\mathrm{d}x = (u, v)$$

so that axiom IP1 is satisfied. Secondly,

$$(\alpha u + \beta v, w) = \int_a^b (\alpha u(x) + \beta(x))w(x)\,\mathrm{d}x$$

$$= \alpha \int_a^b u(x)w(x)\,\mathrm{d}x + \beta \int_a^b v(x)w(x)\,\mathrm{d}x$$

$$= \alpha(u, w) + \beta(v, w)$$

and so axiom IP2 is satisfied. Finally,

$$(u, u) = \int_a^b u^2(x)\,\mathrm{d}x$$

which is clearly positive since it is the integral of a positive function. The only function $u(x)$ for which this integral vanishes is $u(x) = 0$ (a.e.), and so axiom IP3 is satisfied. Hence (1) defines an inner product on $L_2(a, b)$. $\qquad\square$

Inner product space.   A linear space U on which an inner product has been defined is called an inner product space.

Orthogonality.   Recall from vector analysis the concept of orthogonality: two vectors $\mathbf{u}$ and $\mathbf{v}$ are orthogonal if $\mathbf{u} \cdot \mathbf{v} = 0$, that is, if they are at right angles to each other. Since we have at our disposal the concept of an inner product, it is a very simple matter to define orthogonality in *any* inner product space: two members u, v of an inner product space U are orthogonal if

$$(u, v) = 0. \tag{2}$$

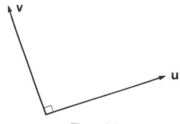

Figure 3.8

Example

Consider the functions $u(x) = \sin x$, $v(x) = \cos x$, with u, $v \in L_2(-\pi, \pi)$. Making use of the inner product (1) we find that

$$(u, v) = \int_{-\pi}^{\pi} \sin x \cos x \, dx = 0$$

and so u and v are orthogonal.   □

The Schwarz inequality.   We return for a moment to vectors and observe another property of the dot product. Recall that

$$\mathbf{u} \cdot \mathbf{v} = |\mathbf{u}| \, |\mathbf{v}| \cos \theta$$

or

$$\mathbf{u} \cdot \mathbf{v} = (\mathbf{u} \cdot \mathbf{u})^{1/2}(\mathbf{v} \cdot \mathbf{v})^{1/2} \cos \theta. \tag{3}$$

But $|\cos \theta| \leqslant 1$, hence we have

$$|\mathbf{u} \cdot \mathbf{v}| \leqslant (\mathbf{u} \cdot \mathbf{u})^{1/2}(\mathbf{v} \cdot \mathbf{v})^{1/2}. \tag{4}$$

The generalization of (4) to any inner product space U is called the *Schwarz inequality*, which states that

$$|(u, v)| \leqslant (u, u)^{1/2}(v, v)^{1/2} \qquad \text{for all } u, v \in U. \tag{5}$$

The proof follows from the observation that

$$(u - \alpha v, u - \alpha v) \geqslant 0$$

for an arbitrary real number α, using axiom IP3. From the linearity and symmetry of the inner product this becomes

$$(u, u) - 2\alpha(u, v) + \alpha^2(v, v) \geqslant 0.$$

Now α is arbitrary: and so if we choose α to be equal to $(u, v)/(v, v)$, then

$$(u, u) - (2(u, v)^2/(v, v)) + ((u, v)^2/(v, v)) \geqslant 0$$

or, rearranging terms and multiplying throughout by (v, v),

$$(u, v)^2 \leqslant (u, u)(v, v).$$

The result then follows upon taking the square root of both sides.

§9. Norm, metric

At the beginning of Section 8 we introduced the concept of a norm by drawing analogies with lengths of vectors. As with the definition of an inner product, we start from scratch with an arbitrary linear space U and *define* a *norm* $\|\cdot\|$ to be a real-valued operation that satisfies the following *axioms*:

N1. $\|u\| \geqslant 0$ and $\|u\| = 0$ iff $u = 0$ (positive-definiteness),

N2. $\|\alpha u\| = |\alpha| \, \|u\|$ (linearity),

N3. $\|u + v\| \leqslant \|u\| + \|v\|$ (triangle inequality),

for all $u, v \in U$, $\alpha \in R$. The third axiom relates to the case of two vectors **u** and **v**: it is well known (Figure 3.9) that $\|\mathbf{u} + \mathbf{v}\| \leqslant \|\mathbf{u}\| + \|\mathbf{v}\|$ for this case.

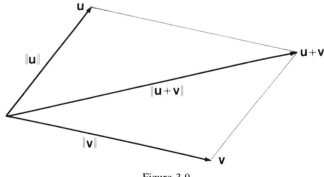

Figure 3.9

Examples

(a) Let $U = R^3$: then the usual norm defined on R^3 is

$$\|\mathbf{x}\| = (x_1^2 + x_2^2 + x_3^2)^{1/2} \quad \text{for} \quad \mathbf{x} = (x_1, x_2, x_3).$$

The extension to R^n is obvious.

(b) The quantity $\|\cdot\|_p$ defined by

$$\|\mathbf{x}\|_p = \left[\sum_{i=1}^{n} |x_i|^p \right]^{1/p} \quad (1 \le p < \infty)$$

is a norm on R^n. Axioms N1 and N2 are seen by inspection to hold, and the triangle inequality is a consequence of the *Minkowski inequality for sums*

$$\left[\sum_{i=1}^{n} |x_i \pm y_i|^p \right]^{1/p} \le \left[\sum_{i=1}^{n} |x_i|^p \right]^{1/p} + \left[\sum_{i=1}^{n} |y_i|^p \right]^{1/p} \tag{1}$$

for $1 \le p < \infty$. The proof of this inequality is outlined in Exercise 9.6. If we let $p \to \infty$ then the quantity $\|\cdot\|_\infty$ defined on R^n by

$$\|\mathbf{x}\|_\infty = \max_{1 \le i \le n} |x_i|$$

is also a norm on R^n.

(c) Let $U = L_p(a, b)$ with $1 \le p < \infty$: the L_p-norm is defined by

$$\|u\|_{L_p} = \left[\int_a^b |u(x)|^p \, dx \right]^{1/p}, \quad u \in L_p(a, b). \tag{2}$$

Axioms N1 and N2 are easily shown to hold, and the triangle inequality follows from the Minkowski inequality for integrals (7.1) which can be written as

$$\|u \pm v\|_{L_p} \le \|u\|_{L_p} + \|v\|_{L_p}.$$

(d) The space $L_\infty(a, b)$ of bounded measurable functions is a normed space, with norm $\|\cdot\|_\infty$ defined by

$$\|u\|_\infty = \sup |u(x)|, \tag{3}$$

the supremum being taken over all subsets of (a, b) with non-zero measure. The first two norm axioms obviously hold, while the triangle inequality follows from (Figure 3.10)

$$\begin{aligned} \|u + v\|_\infty &= \sup |u(x) + v(x)| \\ &\le \sup \left(|u(x)| + |v(x)| \right) \\ &\le \sup |u(x)| + \sup |v(x)| = \|u\|_\infty + \|v\|_\infty. \end{aligned}$$ □

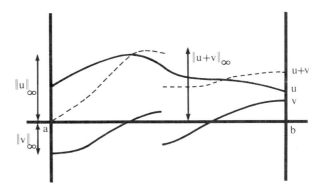

Figure 3.10

Now recall from Section 6 that the space of bounded continuous functions $BC(a, b)$ is a subset of $L_\infty(a, b)$; hence $BC(a, b)$ with the sup-norm $\|\cdot\|_\infty$ is a normed space, as is $C[a, b]$. (Theorem 6.1 guarantees that continuous functions defined on the closed interval $[a, b]$ are bounded.)

In a similar way we can show that $L_\infty(\Omega)$, $BC(\Omega)$ and $C(\bar{\Omega})$ are normed spaces with the norm $\|u\|_\infty = \sup |u(\mathbf{x})|$, for $\Omega \subset R^n$.

Normed space. A linear space U with a norm $\|\cdot\|$ defined on it is called a normed space.

Norm generated by inner product. The norm is a primitive concept which does not require for its definition the existence of an inner product. Indeed, a norm is any operation $\|\cdot\|$ which satisfies N1–N3. But if we happen to have available an inner product, then we can define a norm by

$$\|u\| = (u, u)^{1/2}, \tag{4}$$

and we say that $\|\cdot\|$ in (4) is the norm generated by the inner product. The analogy with vectors in R^3 is clear: given the scalar (inner) product, the norm or length of a vector $\mathbf{u}$ is given by

$$\|\cdot\| = (\mathbf{u} \cdot \mathbf{u})^{1/2}.$$

The question now arises: is $\|u\|$ defined in (4) really a norm? That is, does it satisfy all of the norm axioms? The answer, of course, is yes: from the properties IP1–IP3 (Section 8) of the inner product, $(u, u)^{1/2} > 0$ and $(u, u)^{1/2} = 0$ iff $u = 0$. Thus axiom N1 holds. Secondly,

$$\|\alpha u\|^2 = (\alpha u, \alpha u)$$
$$= \alpha^2 (u, u) = \alpha^2 \|u\|^2$$

using the linearity of the inner product. Thus

$$\|\alpha u\| = |\alpha| \, \|u\|$$

and so axiom N2 is satisfied. Finally, from (4),

$$\|u + v\|^2 = (u + v, u + v)$$
$$= (u, u) + 2(u, v) + (v, v)$$
$$= \|u\|^2 + 2(u, v) + \|v\|^2$$
$$\leqslant \|u\|^2 + 2\,|(u, v)| + \|v\|^2$$
$$\leqslant \|u\|^2 + 2\,\|u\|\,\|v\| + \|v\|^2 = (\|u\| + \|v\|)^2$$

where we have used the Schwarz inequality (8.5) together with (4). Thus, taking the square root of both sides we find that the triangle inequality is satisfied and so the quantity $(.\,,.)^{1/2}$ does indeed define a norm.

We record here an alternative way of writing the Schwarz inequality for an inner product space U: (8.5) with (4) yields the inequality

$$|(u, v)| < \|u\| \, \|v\|, \qquad u, v \in U \tag{5}$$

where the norm in (5) is that generated by the inner product on U.

While the above discussion shows that it is true that *every inner product space is automatically a normed space,* the converse is not true; that is, given a norm, it is not generally possible to define a corresponding inner product. Suppose then, that we are given a norm, and would like to know whether there is an inner product from which this norm can be derived. A very simple way of carrying out this check is by appealing to the parallelogram law: if U is an *inner product space* with norm $\|u\| = (u, u)^{1/2}$, then the identity

$$\|u + v\|^2 + \|u - v\|^2 = 2(\|u\|^2 + \|v\|^2) \tag{6}$$

holds for all $u, v \in U$. The reason for calling this identity the parallelogram law is evident from its interpretation, in Figure 3.11, when U is the

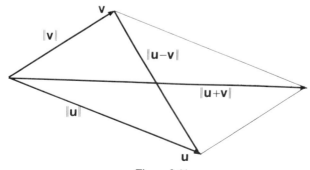

Figure 3.11

set of vectors: from the cosine rule, $\|\mathbf{u} - \mathbf{v}\|^2 = \|\mathbf{u}\|^2 + \|\mathbf{v}\|^2 - 2\|\mathbf{u}\|\|\mathbf{v}\|\cos\theta$ and $\|\mathbf{u} + \mathbf{v}\|^2 = \|\mathbf{u}\|^2 + \|\mathbf{v}\|^2 - 2\|\mathbf{u}\|\|\mathbf{v}\|\cos(180° - \theta)$; adding, we obtain (6). The result (6) is easily proved for an arbitrary inner product space (see Exercise 9.2). What concerns us here, though, is the following conclusion: for any normed space V, *if the norm does not satisfy the parallelogram law then there is no inner product which generates this norm.* When this is so, the space V is not an inner product space.

Example

Consider $C[a, b]$ with the sup-norm $\|\cdot\|_\infty$. Then, with $b > a > 0$ and

$$u(x) = 1, \qquad v(x) = x$$

we have

$$\|u + v\|_\infty = \sup |u(x) + v(x)| = \sup |1 + x| = 1 + b,$$
$$\|u - v\|_\infty = \sup |1 - x| = 1 - a.$$

Thus

$$\|u + v\|_\infty^2 + \|u - v\|_\infty^2 = (1 + b)^2 + (1 - a)^2.$$

But

$$\|u\|_\infty = \sup |1| = 1, \qquad \|v\|_\infty = \sup |x| = b$$

and so $\|u\|_\infty^2 + \|v\|_\infty^2 = 1 + b^2$.

The parallelogram law does not hold, and so $C[a, b]$ with the sup-norm is not an inner product space. In the same way we can show that $C(\bar{\Omega})$ with the sup-norm is also not an inner product space. $\qquad\square$

Later on we shall be dealing almost exclusively with Hilbert spaces, which are inner product spaces with additional properties. It is important, however, to be aware of spaces such as $C(\bar{\Omega})$ with the sup-norm; though not an inner product space, this is a very important normed space.

Metric spaces. The final geometrical property that we wish to extend to linear spaces in general is that of the metric. The motivation comes from the situation in R^3, in which the distance $d(\mathbf{x}, \mathbf{y})$ between two points $\mathbf{x}$ and $\mathbf{y}$ is given by

$$d(\mathbf{x}, \mathbf{y}) = \sqrt{((x_1 - y_1)^2 + (x_2 - y_2)^2 + (x_3 - y_3)^2)}. \tag{7}$$

A further property in R^3 is the triangle inequality (Figure 3.12): the length of one side of a triangle is less than or equal to the sum of the lengths of the other two sides, that is,

$$d(\mathbf{x}, \mathbf{z}) \leqslant d(\mathbf{x}, \mathbf{y}) + d(\mathbf{y}, \mathbf{z}). \tag{8}$$

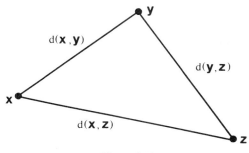

Figure 3.12

As with the concepts of inner product and norm, we define a quantity $d(.,.)$ which has similar properties to (7) and (8) in R^3.

Metric.   Let U be a *set*. A real number $d(u, v)$ is called a metric on U if:

M1. $d(u, v) \geqslant 0$ and $d(u, v) = 0$ iff $u = v$,

M2. $d(u, v) = d(v, u)$,

M3. $d(u, v) \leqslant d(u, w) + d(w, v)$,

for all $u, v, w \in U$.

Note that U need not be a linear space in order for it to have a metric.

Metric space.   A set of U with a metric $d(.,.)$ defined on it is called a metric space.

Metric generated by a norm.   Metric spaces do not have much structure, and are consequently of limited use in our subsequent studies. Instead of defining a metric from scratch, we shall always be working with normed (including inner product) spaces, and will define the corresponding metric by

$$d(u, v) = \|u - v\| . \tag{9}$$

That (9) does indeed satisfy the metric axioms is easily shown, and left as an exercise.

Bibliographical remarks

Linear spaces are also called vector spaces. The concept of a linear space is a purely algebraic one, requiring as it does only a set of rules for combining elements and multiplying them by scalars. Further details on linear spaces may be found in textbooks on linear algebra, for example Hoffman and Kunze [21] and Strang [47], as well as in the book by Halmos [19].

Good accounts of metric, normed and inner product spaces may be found in Kolmogorov and Fomin [23], Kreyszig [25], Naylor and Sell [29], Oden [32], Roman [40], [41] and Smirnov [46].

Exercises

7.1   Which of the following are linear spaces?

(a)   The set of $m \times n$ matrices;

(b)   The set of points $U = \{\mathbf{x} : \mathbf{x} = (x_1, x_2) \in R^2, \ x_2 \geqslant 0\}$ (i.e. the upper half plane);

(c)   The set of solutions of the differential equation

$$a(x)\frac{d^2u}{dx^2} + b(x)\frac{du}{dx} + c(x)u + d(x) = 0, \qquad 0 < x < 1.$$

(d)   The set of solutions of the differential equation

$$a(x)\frac{d^2u}{dx^2} + b(x)\frac{du}{dx} + c(x)u + d(x) = 0, \qquad 0 < x < 1.$$

7.2   Consider the linear space R^2 of ordered pairs. Which of the following subsets of R^2 are subspaces?

(a) $V = \{\mathbf{x} = (x, y) : x = 0\}$;

(b) $V = \{\mathbf{x} = (x, y) : x + y = 1\}$.

7.3   The set $C[a, b]$ is a linear space. Which of the following subsets are subspaces?

(a) $V = \{u \in C[a, b] : u(a) = u(b) = 0\}$;

(b) $V = \{u \in C[a, b] : u(a) = u(b) = 1\}$;

(c) $V = \left\{u \in C[a, b] : \int_a^b u(x)\, dx = 0\right\}$.

7.4   Prove Theorem 7.1.

7.5   Let $U = C[0, 1]$, $V = \{v \in C[0, 1] : v(x) = v(-x)\}$ (the set of even functions) and $W = \{w \in C[0, 1] : w(x) = -w(-x)\}$ (the set of odd functions). Verify that $U = V \oplus W$.

7.6 The purpose of this exercise is to prove the Minkowski inequality for integrals

$$\left[\int_{\Omega}|u \pm v|^p \, dx\right]^{1/p} = \left[\int_{\Omega}|u|^p \, dx\right]^{1/p} + \left[\int_{\Omega}|v|^p \, dx\right]^{1/p}.$$

(a) Show that $\alpha\beta \leqslant (\alpha^p/p) + (\beta^q/q)$ where $1/p + 1/q = 1$ and $\alpha, \beta \in R$.

[Consider the sketch in Figure 3.13: show that area $A = \alpha^p/p$, area $B = \beta^q/q$.]

Set $\alpha = \dfrac{u(\mathbf{x})}{\left[\int_{\Omega}|u(\mathbf{x})|^p \, dx\right]^{1/p}}$ and $\beta = \dfrac{v(\mathbf{x})}{\left[\int_{\Omega}|v(\mathbf{x})|^q \, dx\right]^{1/q}}$, integrate and

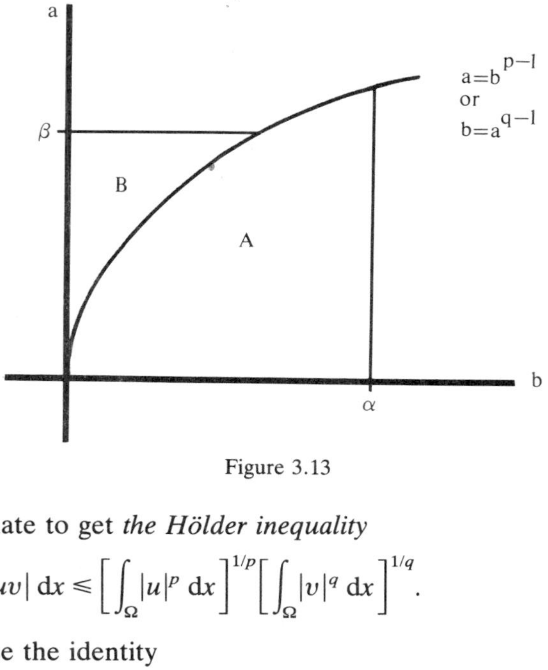

Figure 3.13

manipulate to get *the Hölder inequality*

$$\int_{\Omega}|uv| \, dx \leqslant \left[\int_{\Omega}|u|^p \, dx\right]^{1/p}\left[\int_{\Omega}|v|^q \, dx\right]^{1/q}. \tag{i}$$

(b) Use the identity

$$(|\alpha| + |\beta|)^p = (|\alpha| + |\beta|)^{p-1}|\alpha| + (|\alpha| + |\beta|)^{p-1}|\beta|$$

to obtain the Minkowski inequality.

8.1 Consider the space $C^m[0, 1]$ with inner product $(. , .)_m$ defined by

$$(u, v)_m = \int_0^1 (uv + u'v' + \cdots + u^{(m)}v^{(m)}) \, dx.$$

Given $u(x) = x^3$, $v(x) = 1 - (3x^2/2)$, show that u and v are orthogonal with respect to the inner product $(.\,,.)_0$. Are they orthogonal with respect to $(.\,,.)_1$?

Verify the Schwarz inequality using the inner product $(.\,,.)_2$.

9.1   For an inner product space U show that

$$\bigl|\,\|u\| - \|v\|\,\bigr| \leqslant \|u - v\| \qquad \text{for all } u,\, v \in U.$$

9.2   Prove the parallelogram law

$$\|u + v\|^2 + \|u - v\|^2 = 2(\|u\|^2 + \|v\|^2) \qquad \text{for } u,\, v \in U,$$

where U is an inner product space.

9.3   Show that the quantity

$$\|u\| = \left[\int_a^b \left(\frac{du}{dx}\right)^2 dx \right]^{1/2}, \qquad u \in U$$

satisfies the norm axioms for the case where U is the space

$$U = \{u : u \in C^1[a,\, b], \qquad u(a) = u(b) = 0\}.$$

9.4   If $(u,\, w) = (v,\, w)$ for all $w \in U$, show that $u = v$.

9.5   A subset V of a linear space U is said to be *convex* if, for every $u,\, v \in V$, $\alpha u + (1 - \alpha)v$ is also in V, where $0 \leqslant \alpha \leqslant 1$ (Figure 3.14). Show that the closed ball $B = \{u \in U : \|u\| \leqslant 1\}$ is convex. What does B look like when $U = C(0,\, 1)$ with the sup-norm?

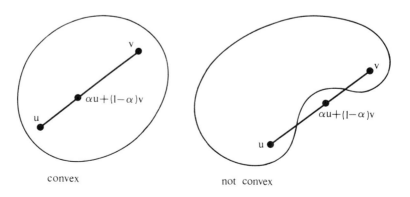

convex   not convex

Figure 3.14

9.6 The purpose of this exercise is to show that

$$\|\mathbf{x}\|_p = [|x_1|^p + \cdots + |x_n|^p]^{1/p}$$

defines a norm on R^n, for $1 \leqslant p < \infty$. In Exercise 7.6(a) set

$$\alpha = \frac{|x_i|}{\|\mathbf{x}\|_p}, \qquad \beta = \frac{|y_i|}{\|\mathbf{y}\|_q},$$

sum over 1 to n, and manipulate to get the *Hölder inequality for sums*

$$\sum_{i=1}^n |x_i y_i| \leqslant \|\mathbf{x}\|_p \|\mathbf{y}\|_q .$$

Use the identity in Exercise 7.6(b) to prove the *Minkowski inequality for sums*

$$\left[\sum_{i=1}^n |x_i \pm y_i|^p\right]^{1/p} \leqslant \|\mathbf{x}\|_p + \|\mathbf{y}\|_p,$$

which confirms that $\|\cdot\|_p$ is a norm for R^n.

9.7 The aim of this exercise is to show that

$$\|u\|_{L_r} \leqslant C \|u\|_{L_p} \tag{ii}$$

for $p \geqslant r \geqslant 1$, so that if $u \in L_p(\Omega)$ then $u \in L_r(\Omega)$ also. First, let p, q, r be real numbers such that

$$\frac{1}{p} + \frac{1}{q} = \frac{1}{r} \qquad \text{or} \qquad \frac{1}{(p/r)} + \frac{1}{(q/r)} = 1. \tag{iii}$$

Replace u by u^r and v by v^r in Hölder's inequality ((i) above) and use (iii) to obtain the generalization

$$\left[\int |uv|^r\right]^{1/r} \leqslant \|u\|_{L_p} \|v\|_{L_q} \tag{iv}$$

of Hölder's inequality. Then use (iv) to obtain (ii).

9.8 Let u and v be non-zero elements in an inner product space U. Show that

$$\|u + v\| = \|u\| + \|v\|$$

if and only if $v = \alpha u$ for some real number $\alpha > 0$.

4

Properties of normed spaces

Some of the most important properties of normed and inner product spaces are those of a topological nature, that is, those that are based on the concept of distances between points. In order to discuss these properties, we need to start by looking at convergence of sequences. This is done in Section 10, where it is shown how the definition of convergence of a sequence in a normed space is a natural extension of the definition in R. In Section 11 we focus attention on sequences in spaces of functions; these are a special case which will occur so often in the future that it is worth devoting some time to the elucidation of their characteristics.

The notion of completeness pervades all of functional analysis; we describe completeness in Section 12, and then show in Section 13 how completeness of a space is related to the closedness of that space.

Finally, in Section 14, we draw on the additional properties of Hilbert spaces: we show that every Hilbert space H can be split into two subspaces V and $V^\perp$ which are such that $(v, w) = 0$ for $v \in V$ and $w \in V^\perp$, and that every member u of H can be represented uniquely as $u = v + w$ for some $v \in V$ and $w \in V^\perp$. This is the projection theorem, which will also feature later on.

§10. Sequences

The concept of convergence of sequences of functions is of central importance to subsequent developments. We start by defining what is meant by a sequence and then show how the concept of convergence of a sequence of real numbers is generalized in such a way that one is able to talk about convergence in any normed space.

Sequences. Let U be *any* set; a sequence $\{u_1, u_2, \ldots, u_n, \ldots\}$ in U is a set of elements of U which have a definite order of occurrence. If the sequence has a *finite* number of elements, it is called a *finite sequence*; otherwise it is called an *infinite sequence*.

Most of the time we shall be dealing with infinite sequences, and we shall generally use the notation $\{u_n\}_{n=1}^{\infty}$ or simply $\{u_n\}$ to denote the infinite sequence $\{u_1, u_2, \ldots, u_n, \ldots\}$.

Example

Sequences may be described either by displaying them or by giving a general formula for determining them. For example, the sequence

$$\{1, 0, \tfrac{1}{3}, 0, \tfrac{1}{5}, 0, \ldots\} \subset [0, 1]$$

is defined by actually displaying the first few elements; alternatively, it could be defined by stating that the nth term u_n of the sequence is

$$u_n = \begin{cases} 0 & \text{for even } n, \\ \dfrac{1}{n} & \text{for odd } n. \end{cases}$$

Another example of a sequence is the sequence of *functions* described by (Figure 4.1)

$$\{u_n\}_{n=1}^{\infty} \subset C[a, b],$$
$$u_n(x) = n(x - a).$$ □

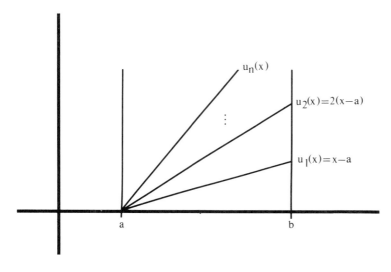

Figure 4.1

Ultimately what interests us most about sequences is the way in which they behave as n gets larger and larger, which brings us to the next topic, namely that of convergence of sequences.

Convergence of sequences.   Consider the sequence of real numbers $\{u_n\} = \{1/n\}_{n=1}^{\infty}$. As n gets larger, the term u_n gets closer to zero, and we say that the sequence *converges to* 0. On the other hand, the sequence $\{2^n\} = \{2, 4, 8, \ldots\}$ increases indefinitely: no matter what number α we specify, it will always be possible to choose a value of n such that 2^n will be greater than α. We say that the sequence $\{2^n\}$ *diverges.*

The above examples behave in a fairly obvious way, and so they could be discussed without recourse to a rigorous definition of convergence. But later on we will need to discuss convergence of arbitrary sequences in any *normed space,* and so we must pin down some of these ideas. Let U be a subset of a normed space V, and $\{u_n\}$ a sequence in U. Let u belong to U, and form the sequence of *real numbers* $\{\|u_1 - u\|, \|u_2 - u\|, \ldots, \|u_n - u\|, \ldots\}$. If the number $\|u_n - u\|$ gets smaller as n gets larger, we agree to call the sequence convergent. Another way of stating this is as follows: pick *any number* $\varepsilon > 0$. Then $\{u_n\}$ converges to some element $u \in U$ if we can always make $\|u_n - u\|$ smaller than ε simply by choosing n large enough, larger than some number N, say (Figure 4.2). We are now ready to give a precise definition of convergence.

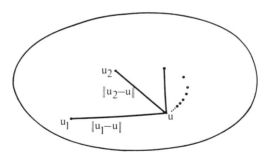

Figure 4.2

Convergence of a sequence in a normed space.   A sequence $\{u_n\}$ in a subset U of a normed space is *convergent* if there is a $u \in U$ for which, given *any* $\varepsilon > 0$, a number N can be found such that

$$\|u_n - u\| < \varepsilon \qquad \text{for all } n > N. \tag{1}$$

If this is the case, we write $u_n \to u$ (u_n converges to u); u is called the *limit* of the sequence. Yet another way of stating (1) informally is

$$\lim_{n\to\infty} \|u_n - u\| = 0 \quad \text{or} \quad \lim_{n\to\infty} u_n = u, \tag{2}$$

which is read "the limit as n tends to ∞ of u_n, is u". Note, however, that by (2) we mean (1).

Example

Consider the sequence $\{\alpha_n\} = \{(3n^2 - 1)/(n^2 - 5n)\}_{n=6}^{\infty}$. As n gets very large we would expect this sequence to approach the limit 3 (since the terms $3n^2$ and n^2 dominate the numerator and denominator, respectively). We check by asking whether a number N can be found such that

$$\|\alpha_n - 3\| = |\alpha_n - 3| = \frac{15n - 1}{n^2 - 5n} < \varepsilon$$

whenever $n > N$, for any $\varepsilon > 0$. This is equivalent to seeking N such that

$$\varepsilon(n^2 - 5n) > 15n - 1 \quad \text{or} \quad \varepsilon n^2 - 5(\varepsilon + 3)n + 1 > 0 \tag{3}$$

for $n > N$. Denote the left-hand side of (3) by $f(n)$ and treat n as a real number: then the graph of $f(n)$ has the shape shown in Figure 4.3 with roots n_1 and n_2 as shown. If we choose $N = n_2$, then clearly $f(n) > 0$ for $n > N$, or $|\alpha_n - 3| < \varepsilon$ for $n > N$, so that $\alpha_n \to 3$. ☐

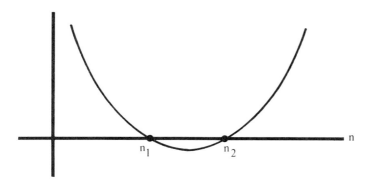

Figure 4.3

§11. Convergence of sequences of functions

When discussing convergence of sequences in normed spaces whose members are functions, it is particularly important to specify which norm is being used, as convergence with respect to one norm does not

necessarily imply convergence with respect to another. We are acquainted so far with two types of norms when dealing with spaces of
functions: the sup-norm (9.3) and the L_p-norm (9.2). As we shall see in
this section, the type of convergence associated with the sup-norm
(namely, uniform convergence) implies convergence in the L_p-norm, but
not vice versa. We begin with a discussion of *pointwise* and *uniform*
convergence.

Suppose that we know that a sequence $\{u_n(\mathbf{x})\}$ of continuous functions
converges to a limit *for each* $\mathbf{x} \in \Omega$. This implies the following: if we *fix* $\mathbf{x}$,
then the sequence of real numbers $u_n(\mathbf{x})$ $(n = 1, 2, \ldots)$ converges to a
real number $u(\mathbf{x})$, say. In other words, for every $\varepsilon > 0$ there exists a
number N such that

$$|u_n(\mathbf{x}) - u(\mathbf{x})| < \varepsilon \qquad \text{whenever } n > N. \tag{1}$$

Of course N will depend on $\mathbf{x}$ and on the number ε. If we now move to
another value of $\mathbf{x}$ the statement (1) may not be true for the same N, a
situation which is obviously not desirable. However, if we can find a
number N independent of x such that (1) holds *for all* $\mathbf{x} \in \Omega$ then we say
that u_n *converges uniformly* to u on Ω. We now define these concepts
formally for the more general case of functions defined on a subset of R^n.

Pointwise and uniform convergence.   A sequence $\{u_n\}$ of functions
defined on an open subset Ω of R^n converges *pointwise* to $u(\mathbf{x})$ if for
every $\varepsilon > 0$ there exists a number N depending on $\mathbf{x}$ and ε such that (1)
holds. If N does not depend on the value of $\mathbf{x}$, then u_n converges
uniformly to u on Ω and we write $\lim_{n \to \infty} u_n = u$ (uniformly).

Uniform convergence has a very simple geometrical interpretation
which is illustrated in Figure 4.4 for the case $\Omega = [a, b]$: according to the
definition, for any given ε all the functions $u_n(x)$, $u_{n+1}(x)$, $\ldots$ lie in the
"tube" of height 2ε situated symmetrically about the limit function $u(x)$,
for n greater than a number N which of course depends on ε, but *not*
on x.

Now that uniform convergence has been defined, one might ask how it
is related to the formal definition (10.1) of convergence in terms of a
norm. To answer this question, consider a sequence $\{u_n\}$ of functions
which belong to the normed space $C[a, b]$ with the norm

$$\|u\|_\infty = \sup |u(x)|, \qquad x \in [a, b].$$

Suppose that this sequence is convergent in the sup-norm; that is, given
any $\varepsilon > 0$ it is possible to find a number N such that

$$\|u_n - u\|_\infty = \sup |u_n(x) - u(x)| < \varepsilon \tag{2}$$

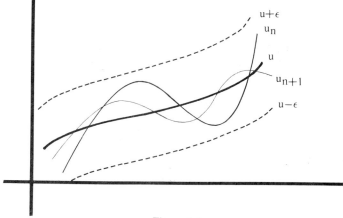

Figure 4.4

whenever $n > N$, for $x \in [a, b]$. But since $|u_n(x) - u(x)| \leqslant \sup |u_n(x) - u(x)|$, it follows that (1) also holds. In other words, *convergence in the sup-norm implies uniform convergence.* Conversely, suppose that $\{u_n\}$ is a uniformly convergent sequence, so that (1) holds. Then ε is an upper bound for $|u_n(x) - u(x)|$, for any x in $[a, b]$. But this implies that the *least upper bound* or *supremum* of $|u_n(x) - u(x)|$ must be less than ε, so that

$$\|u_n - u\|_\infty \equiv \sup |u_n(x) - u(x)| < \varepsilon \qquad \text{for } n > N, \ x \in [a, b]$$

or alternatively,

$$\lim_{n \to \infty} [\sup |u_n(x) - u(x)|] = 0.$$

That is, *uniform convergence implies convergence in the sup-norm.* This useful result can be proved in much the same way for functions defined on domains Ω in R^n and so we simply record the general result.

Theorem 1. *A sequence of functions $\{u_n\}$, where $u_n \in C(\bar{\Omega})$ and Ω is a bounded subset of R^n, converges uniformly if and only if*

$$\lim_{n \to \infty} [\sup |u_n(\mathbf{x}) - u(\mathbf{x})|] = 0, \qquad \text{for } \mathbf{x} \in \bar{\Omega}. \tag{3}$$

Examples

(a) Let $u_n = x^n$, defined on $[0, 1]$. This sequence converges *pointwise* to 0 for $0 \leqslant x < 1$, and to 1 at $x = 1$. If we set $u(x) = 0$, $0 \leqslant x < 1$ and $u(x) = 1$ for $x = 1$, then

$$\sup |u_n(x) - u(x)| = 1 \qquad \text{for all } n,$$

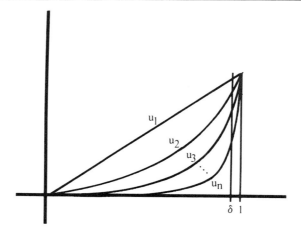

Figure 4.5

this supremum being attained at a value of x "infinitesimally" close to $x = 1$ (Figure 4.5). Hence the sequence does not converge uniformly on $[0, 1]$. However it does converge uniformly to zero on $[0, \delta]$, where $0 < \delta < 1$, since $\sup |u_n(x) - u(x)| = \delta^n$ which $\to 0$ as $n \to \infty$.

(b) Consider the sequence $\{u_n(x) = n^2 x(1 - x)^n\}$; the larger n is, the larger and the closer to the y-axis is the maximum value of $u_n(x)$. For each fixed $x \in [0, 1]$ the sequence converges to zero; but as n increases, the supremum of $|u_n(x) - u(x)| = |u_n(x)|$ also increases (Figure 4.6).

Condition (3) cannot be satisfied, and so we do not have uniform convergence. But convergence is uniform on any interval $[\delta, 1]$ where

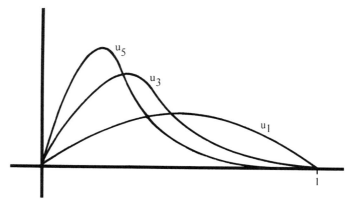

Figure 4.6

$0 < \delta < 1$; indeed, as $n \to \infty$ the undesirable behaviour of the maximum value of u_n will not affect matters in $[\delta, 1]$. $\qquad\qquad\qquad$ □

L_p-*convergence.* We continue our discussion of convergence of sequences of functions, and move on to the larger normed space $L_p(\Omega)$ with the usual L_p-norm with $1 \leq p < \infty$. The definition (10.1) says that a sequence $\{u_n\} \subset L_p(\Omega)$ converges in the L_p-norm to an element $u \in L_p(\Omega)$ if for any given $\varepsilon > 0$ it is possible to find a number N such that

$$\|u_n - u\|_{L_p} < \varepsilon \qquad \text{whenever } n > N, \tag{4}$$

or

$$\left[\int_{\Omega} |u_n(\mathbf{x}) - u(\mathbf{x})|^p \, dx \right]^{1/p} < \varepsilon \qquad \text{whenever } n > N, \tag{5}$$

or

$$\lim_{n \to \infty} \int_{\Omega} |u_n(\mathbf{x}) - u(\mathbf{x})|^p \, dx = 0. \tag{6}$$

This type of convergence is referred to as L_p-*convergence*, and in the case $p = 2$ it is referred to as *convergence in the mean*. It is important to note that while uniform convergence implies L_p-convergence, the converse is not true (see Exercise 11.4). The relationship between uniform, L_p- and pointwise convergence is shown in Figure 4.7.

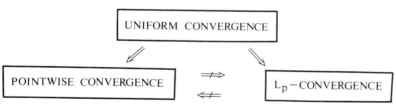

Figure 4.7

Example

Let $\{u_n\} = \{(1 + nx)^{-1}\}$, $n = 1, 2, \ldots$. This sequence converges to 0 on $[0, 1]$ in the L_2 norm since

$$\int_0^1 [u_n(x) - 0]^2 \, dx = \int_0^1 (1 + nx)^{-2} \, dx = (1 + n)^{-1}$$

which tends to zero as $n \to \infty$. It can be shown that $u_n \to 0$ in the L_p-norm for any $p > 1$. $\qquad\qquad\qquad$ □

§12. Completeness

As we have seen, convergent sequences all have the property that the distance between successive members of a sequence, measured by means of some appropriate norm, becomes progressively smaller, and the sequence approaches a definite limit which is, however, a member of the normed space concerned. Unfortunately, the situation is not always so clear-cut: it is always possible to set up sequences in some spaces which do have the property that the distance between successive members becomes progressively smaller, but it is found that the limit that is approached is not a member of the space. For example, suppose we take a look at the half-open interval $(0, 1]$ with the norm $\|\cdot\| = |\cdot|$, and consider the sequence $\{u_n\} = \{1/n\}_{n=1}^{\infty}$. This sequence behaves in all respects like a convergent sequence, and converges to 0, but 0 is not in the space $(0, 1]$!

This behaviour is undesirable for a number of reasons, and we shall always make a strong distinction between spaces in which sequences that behave like convergent sequences do in fact converge to a limit, and those in which the limits of such sequences are "missing". In order to proceed with the discussion, we first define a sequence that has the property that the distance between successive members decreases—this is called a *Cauchy sequence*.

Cauchy sequence.   A sequence $\{u_n\}$ in a subset U of a normed space V is called a Cauchy sequence if

$$\lim_{m,n \to \infty} \|u_m - u_n\| = 0$$

or, more formally, if there exists a number N such that

$$\|u_m - u_n\| < \varepsilon \qquad \text{whenever } m, n > N$$

for any given $\varepsilon > 0$.

Every convergent sequence is a Cauchy sequence (see Exercise 12.5) but, as we discussed, not every Cauchy sequence is convergent for the simple reason that, while the members may try to converge to a limit, the limit may not be part of the space. When this is so, then we say that the space is *incomplete*. The situation may be remedied, however, by adding to the space those elements that are the limits of Cauchy sequences but which were not originally in the space. This process is called *completion* of the space, which is then said to be *complete*. We first define formally a complete space, and then give some simple but important examples of complete spaces.

Complete space. A subset U of a normed space V is complete if *every* Cauchy sequence in U converges to an element of U.

Examples

(a) The set R of real numbers with the norm $\|\cdot\| = |\cdot|$ is complete, as is any *closed* interval of R.

(b) The set R^n with any of the norms $\|\cdot\|_p$ defined by $\|\mathbf{x}\|_p = [\sum_{i=1}^{n} |x_i|^p]^{1/p}$ for $1 \leqslant p < \infty$ is complete, as is R^n with the norm $\|\cdot\|_\infty$ defined by $\|\mathbf{x}\|_\infty = \max_{1 \leqslant i \leqslant n} |x_i|$. This follows from the completeness of R (see Exercise 12.4).

(c) The space $C[0, 1]$ with the integral norm $\|u\|^2 = \int_0^1 u^2 \, dx$ is *not* complete. To see this, consider the sequence $\{u_n\}$, where

$$u_n = \begin{cases} 0, & 0 \leqslant x < \frac{1}{2}, \\ (x - \frac{1}{2})^{1/n}, & \frac{1}{2} \leqslant x \leqslant 1, \end{cases}$$

(see Figure 4.8). It is readily verified that $\{u_n\}$ is a Cauchy sequence; however, its limit $u(x)$ is the *discontinuous* function

$$u(x) = \begin{cases} 0, & 0 < x < \frac{1}{2}, \\ 1, & \frac{1}{2} \leqslant x \leqslant 1. \end{cases}$$

Hence $C[0, 1]$ (and in general $C[a, b]$) is not complete in the L_2-norm, and indeed it is not complete in the L_p-norm for any p such that $1 \leqslant p < \infty$. In a similar way we may show that $C(\bar{\Omega})$ with the L_p norm $\|u\| = [\int_\Omega |u|^p \, dx]^{1/p}$ is not complete, for $1 \leqslant p < \infty$.

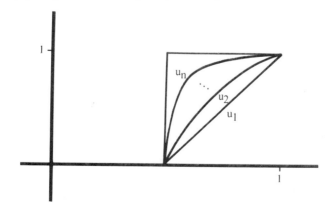

Figure 4.8

(d)   The space $C[a, b]$ with the norm $\|u\|_\infty = \sup |u(x)|$, $x \in [a, b]$, *is complete*. We may show this by demonstrating first of all that every Cauchy sequence converges uniformly to some function $u(x)$, and that this limiting function is necessarily continuous (see Exercise 12.3). Similarly, $C(\bar{\Omega})$ with the norm $\|u\|_\infty = \sup |u(\mathbf{x})|$, $\mathbf{x} \in \bar{\Omega}$, is complete. This example and the previous one demonstrate that completeness of a space depends crucially on the choice of norm.

(e)   The space $L_2(\Omega)$ with the usual L_2-norm is complete. We do not give the proof though, as this would take us too far afield. It is important to note that the notion of *Lebesgue* integration is essential for the completeness of L_2; the space R_2 of functions which are Riemann-square integrable is *not* complete, since it is possible to construct Cauchy sequences of Riemann-integrable functions whose limits are not Riemann-integrable. Generally, for any $p \geqslant 1$ the space $L_p(\Omega)$ with the L_p-norm is complete (this holds for $L_\infty(\Omega)$ as well).   □

Completeness is an extremely important property, because complete spaces possess many useful characteristics that are absent from incomplete spaces. Fortunately, most of the spaces of functions with which we will be working are complete. Normed and inner product spaces that are complete have special names, which are introduced here.

Banach and Hilbert spaces.   A complete normed space is called a *Banach* space; a complete inner product space is called a *Hilbert* space.

Since every inner product defines a norm, every Hilbert space is a Banach space.

Examples

(a) R^n with the norm $\|\cdot\|_p$ defined in Example (b) above, $1 \leqslant p \leqslant \infty$, is a Banach space, while R^n with the Euclidean norm $\|\cdot\|_2$ is a Hilbert space.

(b)   The space $C[a, b]$ with the norm $\|\cdot\|_\infty$ is a Banach space, as are the spaces $L_p(a, b)$ with the L_p-norm. The space $L_2(a, b)$ with the L_2-norm is a Hilbert space.   □

§13.   Open and closed sets, completion

The notion of completeness, which we have just met, is an example of a *topological* property of a normed space. Topological properties of a space are those which are based on the concept of distances between points. Naturally we may use a norm to measure distances between points, although the broader concept of a topological space does not require a norm for its definition; a normed space is just a special case of a topological space.

In this section we are going to discuss a few more topological concepts which are necessary for a proper understanding of later material. Common to all of these concepts—and indeed to topological considerations in general—is the idea of an *open set*, the definition of which in turn depends on the idea of a neighbourhood.

Neighbourhood. Let U be a normed space and let u_0 be an arbitrary point in U. The set

$$N(u_0, \varepsilon) = \{u \in U : \|u - u_0\| < \varepsilon\}$$

is called an *open neighbourhood* of u_0 with radius ε, where $\varepsilon > 0$ (Figure 4.9). Similarly, the set

$$\bar{N}(u_0, \varepsilon) = \{u \in U : \|u - u_0\| \leq \varepsilon\}$$

is called a *closed neighbourhood* of u_0 with radius ε.

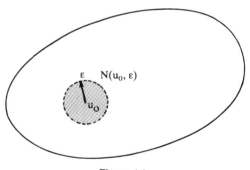

Figure 4.9

Examples

(a) We have already met neighbourhoods of points in R^n (see Section 3); there, the norm used is of course the Euclidean norm.

(b) Consider the space $C[0, 1]$ with the sup-norm $\|u\|_\infty = \sup |u(x)|$. An open neighbourhood of the function $u_0(x)$ of radius ε is the set of all continuous functions on $[0, 1]$ for which (Figure 4.10)

$$\|u - u_0\|_\infty = \sup |u(x) - u_0(x)| < \varepsilon, \qquad x \in [0, 1].$$

(c) In the normed space $L_2(0, 1)$ with the usual L_2-norm, an open

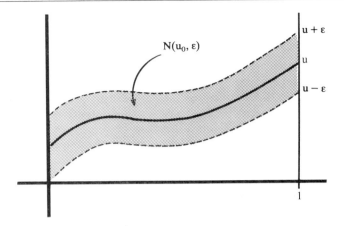

Figure 4.10

neighbourhood of $u_0(x)$ is the set of all functions $u \in L_2(0, 1)$ for which

$$\|u - u_0\|_{L_2} = \left[\int_0^1 [u(x_0) - u(x)]^2 \, \mathrm{d}x \right]^{1/2} < \varepsilon.$$

This open neighbourhood is unfortunately not easy to represent graphically.   □

As can be seen from the definition and the above examples, the idea of an open neighbourhood is generalized in an almost trivial way from the concept in R^n. This is a common feature of functional analysis in normed spaces, and we shall come across it many more times. Indeed, we proceed now to the idea of open and closed subsets in normed spaces, and essentially generalize what was said in Section 3.

Open and closed sets. A subset V of a normed space U is an *open set* if, for every point v in V, there is an open neighbourhood $N(v, \varepsilon)$ of v which lies entirely in V. A point v_0 in V is a *limit point* or *point of accumulation* if every open neighbourhood of v_0, no matter how small, also contains at least one point v in V. Finally, V is a *closed* set if it contains all of its limit points. We also define the *closure* $\bar{V}$ of an open set V to be the union of V and all of its limit points.

Examples

(a) To start with, it may be worth having another look at the situation in R^n (Section 3).

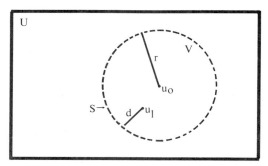

Figure 4.11

(b)   The set $\{u : u \in U, \|u - u_0\| < r\}$ where U is any normed space, is called the *open ball* with centre u_0 and radius r, and is an open set. Indeed, for any point u_1 in V the open neighbourhood $N(u_1, \varepsilon)$ lies entirely in V provided ε is less than d, the shortest distance from u_1 to the boundary of V (Figure 4.11). More formally, set

$$S = \{u \in U, \|u - u_0\| = r\};$$

then

$$d = \inf \|u_1 - u\|, \qquad u \in S$$

and we require $\varepsilon < d$.

(c)   The set $\bar{V} = \{u \in U, \|u - u_0\| \leqslant r\} = V \cup S$, V and S being as in the previous example, is called the *closed ball* of radius r, and is a closed set since, for any $v \in \bar{V}$, the neighbourhood $N(v, \varepsilon)$ contains points of $\bar{V}$ other than v. Hence every member of V is a limit point. Furthermore, for any point u_1 not in $\bar{V}$ we can always construct a neighbourhood $N(u_1, \varepsilon)$ which contains no point of $\bar{V}$. Indeed, let $l = \inf \|u_1 - v\|$, $v \in \bar{V}$; then we simply choose $\varepsilon < l$. Thus $\bar{V}$ contains all its limit points and is closed (Figure 4.12).

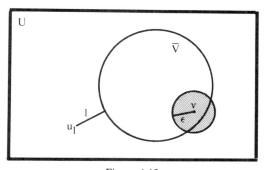

Figure 4.12

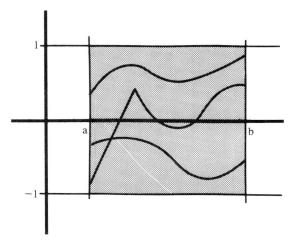

Figure 4.13

(d) Consider the space $C[a, b]$ with the sup-norm, and let $V = \{u \in C[a, b], |u(x)| \leqslant 1\}$. Then V is closed, since V is in fact the closed ball of radius 1, centred at $u(x) = 0$, as can be seen in Figure 4.13. □

There is yet another way of characterizing closed sets, namely by looking at the limits of convergent sequences. This characterization is described in the following theorem.

Theorem 1. *A subset V of a normed space U is closed if and only if every convergent sequence of points in V has its limit in V.*

Proof. First assume that V is closed; we want to show that a convergent sequence $\{u_n\} \subset V$ has its limit in V. Let u be the limit of $\{u_n\}$, i.e., given any $\varepsilon > 0$, there exists a number N such that

$$\|u_n - u\| < \varepsilon \qquad \text{whenever } n > N.$$

In other words, for every $\varepsilon > 0$ the neighbourhood $N(u, \varepsilon)$ contains at least one member of V (i.e., from the sequence $\{u_n\}$) distinct from u. Thus u is a point of accumulation and, since V is closed, $u \in V$.

Conversely, assume that every convergent sequence has its limit in V. Let u_0 be a point in $\bar{V}$; then there is at least one point u_n, say, in the neighbourhood $N(u_0, 1/n)$. This holds for all values of n, so that $\lim_{n \to \infty} u_n = u_0$ (Figure 4.14). But since every convergent sequence has its limit in V by assumption, it follows that $u_0 \in V$ and so $V = \bar{V}$, i.e. V is closed. ■

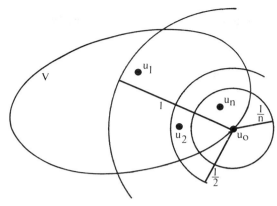

Figure 4.14

It appears from the foregoing result that closed sets, like complete spaces, have no "holes" in them, in the sense that convergent sequences have their limits in the sets. On the other hand, it would seem that open sets have the same deficiencies as incomplete spaces; for example, the interval $(0, 1) \subset R$ is an open set and is also incomplete, whereas $[0, 1]$ is both closed and complete. Still, it is not clear under what circumstances a closed subset of normed space is complete; this we clarify in the next result.

Theorem 2. *Let U be a complete normed space and let V be a subset of U. Then V is complete if and only if V is closed in U.*

Proof. This is not too difficult, and is left as an exercise (see Exercise 13.6). ∎

Note that U is always a closed set in U so that, according to the above theorem, U is closed if and only if it is complete. In other words, *completeness implies closedness and vice versa.*

Example

Let $U = C[0, 1]$ and let $V = P[0, 1]$, the set of all polynomials. First, we recall that $C[0, 1]$ is complete in the sup-norm $\|\cdot\|_\infty$. Now $P[0, 1]$ is a subspace of $C[0, 1]$, but P is *not closed* in C. To see this, consider the fact that $u(x) = e^x$ is a limit point of P: for any $\varepsilon > 0$ we can always find at least one polynomial $p(x) \in P$ lying in a neighbourhood of u; indeed, given any $\varepsilon > 0$, it is possible to find a polynomial $p(x) = 1 + t + t^2/2! +$

$\ldots + t^n/n!$ such that

$$\|u - p\|_\infty = \sup |e^x - p(x)| < \varepsilon$$

for sufficiently large n. But the limit point e^x does not belong to $P[0, 1]$ and so, since $P[0, 1]$ does not contain all its limit points, it is not closed.   □

Earlier we described in a vague fashion how an incomplete space U may be made complete by adding to it those limits of Cauchy sequences that were not originally in U. The resulting space $\bar{U}$, say, is then called the completion of U. We conclude this section by recording some properties of the completion $\bar{U}$ of an incomplete space U, but first we need to define a few more topological concepts.

Dense sets.   A subset V of a normed space U is said to be *dense in U* if the closure of V is U, that is, $\bar{V} = U$. This definition implies of course that *every* member of U is either a member of V or a limit point of V, so that a neighbourhood $N(u_0, \varepsilon)$ of any point u_0 in U contains at least one member of V. This in turn leads to an alternative definition of a dense set: V is dense in U if and only if there are points in U arbitrarily close to points in V, or, given any point $u_0 \in U$ and any number $\varepsilon > 0$, it is possible to find a point $v \in V$ such that $\|u_0 - v\| < \varepsilon$.

Examples

(a)   The set Q of rational numbers is dense in R; the closure $\bar{Q}$ $(=R)$ of Q consists of all rational and irrational numbers.

(b)   *The Weierstrass theorem* states that, for any $u \in L_2[a, b]$ and for every $\varepsilon > 0$, it is possible to find a polynomial $p \in P[a, b]$, $p(x) = a_0 + a_1 x + \ldots$, such that

$$\|u - p\|_{L_2} < \varepsilon.$$

That is, every square-integrable function can be approximated arbitrarily closely by a polynomial. It follows that $P[a, b]$ is dense in $L_2[a, b]$.   □

Separable space.   A normed space U is said to be separable if it contains a *dense* set V that is *countable*. Recall from Section 2 that a countable set is one whose elements can be put in one-to-one correspondence with integers. It follows then that a separable space is one with the property that for each $\varepsilon > 0$ and for each u in U there is a member v_n, say, of a countable set $\{v_1, v_2, \ldots\}$, with $\|u - v_n\| < \varepsilon$.

Example

We have seen that the set of rationals Q is dense in R; since Q is countable, it follows that R is separable. In the same way, R^n is separable: a countable dense subset is the set Q^n of n-tuples of rational numbers.   □

We return now to the idea of completion of a space. Recall from Theorem 2 that an arbitrary subset V of a Banach space (i.e., a complete normed space) U is itself complete if and only if V is closed. It follows that the *closure* $\bar{V}$ of V is *complete*; furthermore, according to the definition of a dense set, V is *dense* in $\bar{V}$. What we have described is of course a way of completing an incomplete space V. In future, by the completion $\tilde{V}$ of a space V, we shall always mean the *closure* of V, so that $\tilde{V} = \bar{V}$. This definition obviously relies on the fact that V must be a subset of a complete space U.

Recall from Section 12 that the space $C(\Omega)$ with the norm $\|u\|_{L_p} = [\int_\Omega |u(\mathbf{x})|^p \, dx]^{1/p}$ is not complete. The completion of this space is actually $L_p(\Omega)$, which is obtained by adding to $C(\Omega)$ those limits of Cauchy sequences (for example, the function $u(x)$ in Example (c) on page 71) that are not in $C(\Omega)$. Thus $C(\Omega)$ is dense in $L_p(\Omega)$. In Section 26 we will show how the result for $p = 2$ is obtained as a special case of a much broader result, the essence of which is that the space $C^m(\Omega)$ is dense in the Sobolev space $H^m(\Omega)$ of functions which, together with their derivatives of order $\leqslant m$, are square-integrable. Our result quoted above is for the special case $m = 0$. Similarly, $L_\infty(\Omega)$ is the closure or completion of $C(\Omega)$ in the L_∞-norm.

§14. Orthogonal complements in Hilbert spaces

In this section we exploit the concept of orthogonality in Hilbert spaces and present results that are generalizations of well-known geometrical situations in R^3. Let U be an *inner product space* and let V be any subspace of U; we define the *orthogonal complement* $V^\perp$ of V to be the set

$$V^\perp = \{w \in U : (w, v) = 0 \qquad \text{for all } v \in V\};$$

that is, $V^\perp$ consists of all those members of U that are orthogonal to every member of V. If w belongs to $V^\perp$ we say that w is orthogonal to V and write $w \perp V$. Since $(v, v) = 0$ implies that $v = 0$, it is clear that the only member of both V and $V^\perp$ is the zero element: $V \cap V^\perp = \{0\}$.

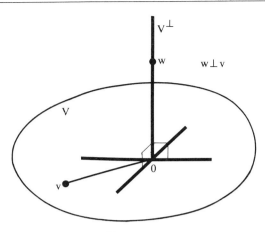

Figure 4.15

Our main aim in this section is to show that if H is any *Hilbert* space and M is a *closed subspace* of H, then $H = M \oplus M^{\perp}$: that is, every $u \in H$ has the unique representation

$$u = v + w \qquad \text{for } v \in M \text{ and } w \in M^{\perp};$$

(recall our discussion of direct sums in Section 7). Before doing so, however, we need to prove another intuitively obvious result which is embodied in the following theorem.

Theorem 1. (i) *Let U be a Hilbert space and let V be a closed subspace of U. Then for every $u \in U$ it is possible to find a unique member v_0 in V such that*

$$d = \|u - v_0\| = \inf \|u - v\|, \qquad v \in V.$$

Moreover, $(u - v_0, v) = 0$ for all $v \in V$, that is $(u - v_0) \perp V$.

(ii) *If W is also a closed subspace of U with $W \subset V$ and $W \neq V$, then there is a member $\tilde{v}$ in V such that $\tilde{v} \neq 0$ and $\tilde{v} \perp W$.*

Remark. Part (i) of the theorem says that, provided V is a closed subspace, we can always find a unique point v_0 in V which is closer to u than any other point in V. Furthermore, this point may be found by "dropping a perpendicular" from u on to V. Part (ii) is illustrated in Figure 4.16 for $U = R^3$.

Proof. (i) We start with a sequence $\{v_n\}$ in V such that

$$d_n = \|u - v_n\| \qquad \text{converges to } d. \tag{1}$$

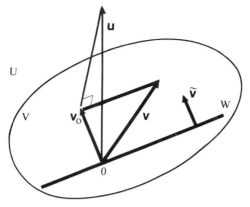

Figure 4.16

If we can show that v_n is a Cauchy sequence then it follows from the completeness of V (see Theorem 13.2) that the limit v_0, say, of the sequence will lie in V. To show that v_n is Cauchy, consider

$$\|v_n - v_m\|^2 = \|(v_n - u) - (v_m - u)\|^2$$
$$= -\|(v_n - u) + (v_m - u)\|^2 + 2(\|v_n - u\|^2 + \|v_m - u\|^2)$$

using the parallelogram law, Exercise 9.2. Now

$$\|(v_n - u) + (v_m - u)\| = \|v_n + v_m - 2u\| = 2 \left\|\tfrac{1}{2}(v_n + v_m) - u\right\| \geqslant 2d$$

and hence

$$\|v_n - v_m\|^2 \leqslant -(2d)^2 + 2(d_n^2 + d_m^2) \to 0 \quad \text{as } m, n \to \infty,$$

from (1). Hence v_n is a Cauchy sequence, and $v_n \to v_0$ in V. But since v_0 is in V we must have

$$\|u - v_0\| \geqslant d;$$

furthermore,

$$\|u - v_0\| = \|u - v_n + v_n - v_0\| \leqslant \|u - v_n\| + \|v_n - v_0\|$$
$$= d_n + \|v_n - v_0\| \to d.$$

Hence $\|u - v_0\| = d$. The proof that v_0 is unique is left as an exercise (see Exercise 14.1).

To show that $(u - v_0, v) = 0$ for all $v \in V$, consider any point $v_0 + \alpha v$ in V: clearly,

$$d^2 \leqslant \|u - (v_0 + \alpha v)\|^2 = \|u - v_0\|^2 - 2\alpha(v, u - v_0) + \alpha^2 \|v\|^2. \qquad (2)$$

Now suppose that $u - v_0$ is not orthogonal to v, and that $(u - v_0, v) = \beta \neq 0$. Also, take α to be equal to $\beta/\|v\|^2$ (recall that α is arbitrary): then we obtain from (2)

$$d^2 \leq \|u - v_0\|^2 - \beta^2/\|v\|^2; \tag{3}$$

but the right-hand side of (3) is less than d^2 since $\|u - v_0\|^2 = d^2$ and $\beta^2/\|v\|^2 > 0$. This leads to a contradiction, and so we must have $\beta = 0$.

(ii) Choose $\bar{v} \in V$ such that $\bar{v} \notin W$, and let $\bar{w}$ be the point in W closest to $\bar{v}$ (we are applying (i) to W). Then $\tilde{v} = \bar{v} - \bar{w}$ is such a point. ■

We are now ready to prove the following theorem.

Theorem 2 (The projection theorem).   *Let V be a closed subspace of Hilbert space H. Then every $u \in H$ can be uniquely written in the form*

$$u = v + w, \qquad v \in V, \qquad w \in V^\perp;$$

that is, $H = V \oplus V^\perp$.

Proof.   First of all, V is complete by Theorem 13.2. Secondly, according to Theorem 1, for each $u \in H$ there is a unique $\tilde{v}$ in V such that $u - \tilde{v}$ is orthogonal to V. That is,

$$u - \tilde{v} = w \quad \text{or} \quad u = \tilde{v} + w, \qquad w \in V^\perp. \tag{4}$$

To prove that there is only one such w, suppose that (4) holds for two elements w_1 and w_2 in $V^\perp$. Then

$$u = \tilde{v} + w_1, \qquad u = \tilde{v} + w_2$$

and so $w_1 - w_2 = 0$ or $w_1 = w_2$. This proves the theorem. ■

We conclude this section with a result that will prove useful later on.

Theorem 3.   *Let V be a subspace of a Hilbert space H. Then $V^\perp = \{0\}$ if and only if V is dense in H.*

In order to prove Theorem 3 we introduce the following property of orthogonal complements.

Lemma 1. $V^{\perp\perp} = \bar{V}$, *where $\bar{V}$ denotes the closure of V and $V^{\perp\perp} = (V^\perp)^\perp$.*

Proof of Lemma 1.   First, we note that $V \subset V^{\perp\perp}$, since if v is any point in V, then $v \perp V^\perp$ so that v also belongs to $V^{\perp\perp}$. Furthermore, since $V^{\perp\perp}$ is closed (see Exercise 14.2), clearly $\bar{V} \subset V^{\perp\perp}$. All that remains is to show that $\bar{V} = V^{\perp\perp}$. Suppose that $\bar{V} \neq V^{\perp\perp}$; from Theorem 1(ii), there is a non-zero point $w \in V^{\perp\perp}$ such that $w \perp \bar{V}$. Since $V \subset \bar{V}$ this means that $w \perp V$ also, so that $w \in V^{\perp}$. But $w \in V^{\perp} \cap V^{\perp\perp}$ implies that $w = 0$, a contradiction. Hence $V^{\perp\perp} = \bar{V}$.   ■

Proof of Theorem 3.   First assume that V is dense in H: then for any $u \in H$ we can always find a point v, say, in V such that $\|v - u\| < \varepsilon$ for any $\varepsilon > 0$. In particular, if $u \in V^{\perp}$ then

$$\varepsilon^2 > \|v - u\|^2 = (v - u, v - u) = \|v\|^2 + \|u\|^2 \geq \|u\|^2.$$

Since ε is arbitrary we must have $u = 0$. Conversely, assume that $V^{\perp} = \{0\}$: then $V^{\perp\perp} = \{0\}^{\perp} = H$, so that from Lemma 1, $\bar{V} = H$, that is, V is dense in H.   ■

Example

Recall from Section 13 that $C(\Omega)$ is dense in $L_2(\Omega)$ with the L_2-norm; it follows that if $u \in L_2(\Omega)$ and $(u, v) = 0$ for all $v \in C(\Omega)$ then $u = 0$ according to Theorem 3. That is,

$$\int_{\Omega} u(\mathbf{x})v(\mathbf{x})\,\mathrm{d}x = 0 \qquad \forall\, v \in C(\Omega) \Rightarrow u(\mathbf{x}) = 0.\qquad \square$$

Bibliographical remarks

The results covered in this chapter are usually given a detailed treatment in books on functional analysis. Good references include Binmore [8], Kreyszig [25], Oden [32], Naylor and Sell [29], Roman [40] and Smirnov [46]; these texts all contain a wealth of information on normed and inner product spaces.

The treatment of completion given in Section 13 is a simplified version of the true story. In general, by the completion of a normed space U with norm $\|\cdot\|_U$ is meant any complete space U^* with norm $\|\cdot\|_{U^*}$, which has the property that U^* has a dense subset V that is isometric to U (two spaces U and V are isometric if a mapping f from U to V exists such that $\|u - v\|_U = \|f(u) - f(v)\|_V$).

We have restricted the definition to the special case where $U = V$, the norm $\|\cdot\|_{U^*}$ being the one used (e.g. $U^* = L_2$, $V = U = C$, $\|\cdot\|_{U^*} = \|\cdot\|_{L_2}$), and in which the mapping f is just the identity mapping (see Chapter 5 for details on maps or operators). The general procedure, though, allows one to find a completion even when the incomplete space is not given as a subset of a complete normed space. The references mentioned earlier all describe the general approach, but for our purposes the description given in Section 13 suffices.

Exercises

10.1   Write down the first few terms of the following sequences:

(i) $\{(-1)^n/n!\}$;   (ii) $\{\frac{1}{2}(1-(-1)^n)\}$;   (iii) $\{3n^2/(5n^2-6)\}$.

10.2   Determine which of the following sequences are convergent, and find their limits:

(a) $\dfrac{4-2n-3n^2}{2n^2+n}$;   (b) $\dfrac{(-1)^n n^4}{2+n^4}$;   (c) $\dfrac{n}{1+n}$.

10.3   The sequence $\{(3n+2)/(n-1)\}$ converges to 3. Find the smallest integer N such that

$$|((3n+2)/(n-1))-3| < \varepsilon \qquad \text{whenever } n > N,$$

for the case $\varepsilon = 0.001$.

10.4   A sequence $\{u_n\}$ is *bounded* if there are constants m_1 and m_2 such that $m_1 \leqslant u_n \leqslant m_2$ for all n. Also, $\{u_n\}$ is *monotone increasing* if $u_{n+1} \geqslant u_n$ for all n, and *monotone decreasing* if $u_{n+1} \leqslant u_n$ for all n. Show that every bounded monotone (increasing or decreasing) sequence converges.

10.5   Let U be an inner product space and suppose that $\{u_n\}$ and $\{v_n\}$ are convergent sequences in U, convergence being defined via the norm generated by the inner product on U. Show that $(u_n, v_n) \to (u, v)$, and deduce that $(u_n, v) \to (u, v)$, and that $\|u_n\| \to \|u\|$.

11.1   Determine intervals on which the following sequences of functions converge pointwise:

(a) $u_n(x) = x^n$;   (b) $u_n(x) = 1/(1+n^2 x^2)$.

11.2   Show that the following sequences converge pointwise to 0 on the intervals given, but that they do not converge in the mean:

(a) $u_n(x) = \begin{cases} 0, & 0 \leqslant x \leqslant 1/n, \\ n, & 1/n < x < 2/n, \\ 0, & 2/n \leqslant x \leqslant 1; \end{cases}$

(b) $u_n(x) = n^{3/2} x e^{-n^2 x^2}$ on $[-1, 1]$.

11.3   Does the sequence $u_n(x) = nx/(1 + n^2 x^2)$, $n = 1, 2, \ldots$ converge uniformly in $[0, 1]$? in $(a, 1]$ (for $1 > a > 0$)?

11.4   Show that uniform convergence of a sequence of functions implies L_p-convergence. Give an example to show that the converse does not hold.

12.1   Show that the sequence $u_n(x) = x^{1/n}$ is a Cauchy sequence in $L_2(0, 1)$.

12.2   Consider the sequence $u_n(x) = x^n$ in $C[0, 1]$. Is this a Cauchy sequence with respect to the L_1-norm?

12.3   The purpose of this exercise is to show that $C[a, b]$ is complete with respect to the sup-norm. Let $u_n(x)$ be a Cauchy sequence; show that $u_n(x_0)$ is a Cauchy sequence of real numbers for every fixed x_0 in $[a, b]$ and deduce that $u_n(x_0)$ converges to a number $u(x_0)$, say. Next, show that $u_n(x)$ converges uniformly to the function $u(x)$. Finally, since $u_n \to u$ uniformly, we have

$$|u_n(x) - u(x)| < \varepsilon \qquad \text{for all } n > N;$$

use the triangle inequality to show that

$$|u(x) - u(y)| \leqslant 2\varepsilon + |u_n(x) - u_n(y)|$$

and deduce from this result and the continuity of u_n (see Section 4) that u is continuous.

12.4   Show that R^n with the norm $\|\cdot\|_p$ $(1 \leqslant p \leqslant \infty)$ is complete.

12.5   Show that every convergent sequence is a Cauchy sequence.

12.6   Consider $C[0, 1]$ with the L_2-norm. Show that the sequence $\{u_n\}$ is a Cauchy sequence, where $u_n(x)$ is as shown in Figure 4.17.
     Next, show that if u_n converges to $u(x)$, then we should have

$$u(x) = \begin{cases} 0, & 0 \leqslant x < \tfrac{1}{2}, \\ 1, & \tfrac{1}{2} < x \leqslant 1, \end{cases}$$

so that $C[0, 1]$ with the L_2-norm is not complete.

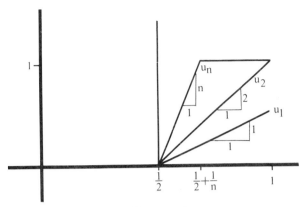

Figure 4.17

13.1 Show that the function

$$u(x) = \begin{cases} -1, & -1 \leqslant x < 0, \\ +1, & 0 \leqslant x \leqslant 1, \end{cases}$$

is a limit point of $C[-1, 1]$ with respect to the L_2-norm.

13.2 Consider the space $C[a, b]$ with the sup-norm, and let V be the subset consisting of functions v satisfying $v(0) = 0$ and $|v(x)| < 1$. Is the function $u(x) = 1$ a limit point of V?

13.3 Sketch the open ball with centre 0 and radius 1 in the space R^2 with the norm

$$\|\mathbf{x}\|_p = [|x_1|^p + |x_2|^p]^{1/p},$$

for the cases $p = 1, 2$ and ∞.

13.4 Find the smallest value of r such that the function $v(x) = \cos 2\pi x$ lies in the closed ball with centre u_0 and radius r in the space $C[0, 1]$ with the sup-norm, where $u_0(x) = \sin 2\pi x$.

13.5 Show that a set V in a normed space U is closed if and only if its complement $V' = U - V$ is open.

13.6 Prove Theorem 13.2.

13.7 Let U, V and W be normed spaces, and suppose that U is dense in V and V is dense in W. Show that U is dense in W.

14.1   Show that the element v_0 in Theorem 14.1 is unique.

14.2   If V is a subset of an inner product space U, show that $V^\perp$ is a closed subspace of U [Hint: let $\{u_n\}$ be a convergent sequence in $V^\perp$ with limit u_0].

14.3   Where in Lemma 14.1 is the completeness of H used?

14.4   If U and V are subsets of an inner product space W and $U \subset V$, show that $V^\perp \subset U^\perp$.

5

Linear operators

In the preceding chapters we have acquainted ourselves with some of the basic structures of arbitrary spaces. We come now to another fundamental concept in functional analysis, namely that of a mapping or operator from one space to another. At the most primitive level one requires only two sets in order to define an operator from one of them to the other; these sets need not have any algebraic or topological structure for the definition to make sense. Obviously, though, the really interesting and useful properties of operators come to the fore only when the two sets are given additional structure: when the two sets are linear spaces, we can introduce the concept of a linear operator, and if the sets are normed spaces as well, then it is possible to construct a rich theory of linear operators on such spaces. After a general introduction to operators in Section 15, we discuss the theory of linear operators on normed spaces in Section 16.

Projection operators are a class of operators that will feature strongly in later chapters when we discuss approximations of boundary-value problems. Apart from this, much of the geometrical structure of Hilbert spaces is laid bare with the aid of projection operators acting on these spaces. For these reasons we devote Section 17 to a discussion of projection operators on Hilbert spaces.

Another special class of operators is that which maps into the real line: these are called functionals, and are discussed in Section 18. Finally, we discuss in Section 19 operators that map pairs of elements into the real line in a linear fashion; these are known as bilinear forms. Linear functionals and bilinear forms play a central role in the study of boundary-value problems, as we will see in Chapter 9 and subsequently.

§15. Operators

The subject of this chapter is not entirely unfamiliar; we have all come across both linear and nonlinear operators in earlier courses on linear

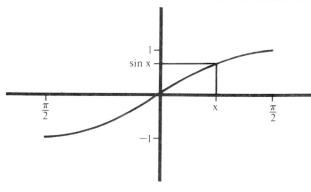

Figure 5.1

algebra, differential equations, and so on. Here we continue the process of generalizing from the familiar.

Consider the function $f(x)$, defined on the interval $I = [-\pi/2,\ \pi/2]$, as shown in Figure 5.1. This familiar situation is really just an example of the action of an operator: specifically, we have defined f to be something that acts on any member x in I, and produces a real number $\sin x$. Furthermore, the *image* $\sin x$ lies in the set $J = [-1, 1]$. More formally, we write all of this as follows:

$$f : I \rightarrow R, \qquad f(x) = \sin x.$$

Here the first expression reads "f maps elements of I to elements of R" while the second expression tells how f does this: f acts on x to produce $\sin x$. The set I is called the *domain* of the operator f, written $D(f)$, and the set $J \subset R$ is called the *range* of f, written $R(f)$. We now generalize.

Let U and V be two sets, and suppose that a rule is given whereby an element u of U is mapped or transformed to an element v of V. This rule is called an *operator* or *transformation* or *mapping* and we write, for an operator T,

$$T : U \rightarrow V, \qquad Tu = v \ (\text{or } T(u) = v), \qquad u \in U, \quad v \in V.$$

The first expression reads "T maps elements of U to elements of V" while the second reads "T acts on u to produce v". We refer to v as the *image* of u, under the mapping T. U is called the *domain* $D(T)$ of T, and we write $R(T)$ for the *range* of T: $R(T)$ consists of all those elements of V that are images of members of U. In other words,

$$R(T) = \{v : v \in V,\ Tu = v \text{ for some } u \in U\}.$$

These concepts are illustrated in Figure 5.2.

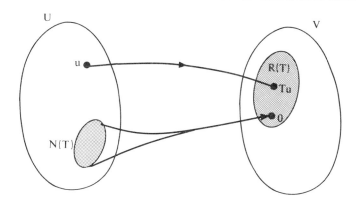

Figure 5.2

If the range of T happens to be all of V, then T is called a *surjective* operator, and we say that T maps U *onto* V. Otherwise T maps U *into* V. The *null space* $N(T)$ of T is the set of all elements of the domain of T whose image is zero:

$$N(T) = \{u \in U : Tu = 0\}.$$

The *inverse image* of $v \in V$ is the set of all $u \in U$ such that $Tu = v$. Likewise, the inverse image of a subset W of V is the set of all $u \in U$ such that $Tu \in W$ (Figure 5.3).

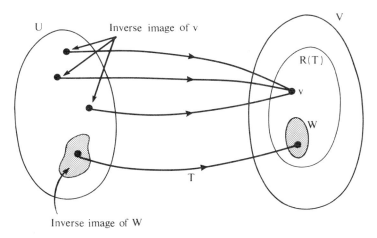

Figure 5.3

Examples

(a) All functions of a real variable are operators from a subset of R to R, for example, the operator or function $f(x) = \sin x$ discussed at the beginning of this section. In the same way, the function

$$f : R^2 \to R, \qquad f(\mathbf{x}) = f(x, y) = x^2 + y^2$$

is an operator which maps the point $\mathbf{x}$ in R^2 to the real number $x^2 + y^2$. Since this number is never negative, we see that $R(f) =$ the set of all non-negative real numbers. Hence f is not surjective. The inverse image of the set $(-1, 0)$ is the empty set, while the inverse image of the point $+1$ is the set of all points on the unit circle.

(b) An $n \times m$ matrix is an operator from R^m to R^n. For example, the operator

$$\mathbf{T} : R^3 \to R^2, \qquad \mathbf{Tx} = \begin{bmatrix} T_{11} & T_{12} & T_{13} \\ T_{21} & T_{22} & T_{23} \end{bmatrix} \begin{bmatrix} x_1 \\ x_2 \\ x_3 \end{bmatrix},$$

is a 2×3 matrix which transforms a member of R^3 to a member of R^2. Whether or not $\mathbf{T}$ is surjective depends on the entries T_{ij} in $\mathbf{T}$; for example, if $T_{11} = T_{21} = 1$ and all other $T_{ij} = 0$, then $\mathbf{Tx} = (x_1, x_1)$ and the range of $\mathbf{T}$ is the subset of R^2 described by the straight line running through the origin at $45°$. The null space of $\mathbf{T}$ consists of all points $\mathbf{x}$ for which $\mathbf{Tx} = (x_1, x_1) = (0, 0)$: thus $N(T) = \{\mathbf{x} \in R^3 : x_1 = 0\}$, which is the $x_2 x_3$ plane.

(c) We shall be seeing a lot of *differential* operators later; these are operators that consist of combinations of ordinary or partial derivative operators. For example, if $u \in C^2(\Omega)$ then the *Laplacian* operator ∇^2 is defined by

$$\nabla^2 : C^2(\Omega) \to C(\Omega), \qquad \nabla^2 u = \partial^2 u / \partial x^2 + \partial^2 u / \partial y^2 \equiv v. \qquad (1)$$

Thus v, the image of u, is a continuous function. To be specific, if $\Omega \subset R^2$ and $u(\mathbf{x}) = x^2 y^3$ then the image of u is the function v defined by

$$v(x, y) = 2y^3 + 6x^2 y.$$

The question of whether or not ∇^2 is surjective amounts to enquiring whether it is possible to find a function u in $C^2(\Omega)$ satisfying (1) for any given v in $C(\Omega)$. That is, the question is one of the *existence* of a solution to the partial differential equation

$$\nabla^2 u = v.$$

We shall consider this problem in greater detail in Section 32. $\qquad \Box$

Two operators $S: U \rightarrow V$ and $T: U \rightarrow V$ are said to be *equal* if for every $u \in U$ we have

$$Su = Tu.$$

When this is the case, we write $S = T$.

The *sum* $S + T$ of two operators $S: U \rightarrow V$ and $T: U \rightarrow V$ is defined to be the operator satisfying

$$(S + T)u = Su + Tu, \qquad u \in U.$$

That is, $T + S$ has the same effect on any member of u as would be obtained by applying T and S separately, and then adding together the result.

The *composition* or *product* TS of two operators $S: U \rightarrow V$ and $T: V \rightarrow W$ is defined to be the operator satisfying

$$TS: U \rightarrow W, \qquad (TS)u = T(Su) \qquad \text{for all } u \in U. \tag{2}$$

That is, the element $(TS)u \in W$ is found by first obtaining the element $Su \in V$, and then by the action of T on Su. Note that the composition TS is meaningless if the element Su does not belong to the domain of T. Furthermore, in general $TS \neq ST$; in fact, ST may be quite meaningless.

Example

Let $U = R^3$, $V = R^2$, $W = R$ and let $\mathbf{T}: U \rightarrow V$, $\mathbf{S}: V \rightarrow W$ be the matrices

$$\mathbf{T} = \begin{bmatrix} 1 & 2 & 1 \\ 2 & 3 & 2 \end{bmatrix}, \qquad \mathbf{S} = [1 \quad 2].$$

Then, for any $\mathbf{x} = (x, y, z)$ in R^3,

$$(\mathbf{ST})\mathbf{x} = \mathbf{S} \begin{bmatrix} 1 & 2 & 1 \\ 2 & 3 & 2 \end{bmatrix} \begin{bmatrix} x \\ y \\ z \end{bmatrix} = [1 \quad 2] \begin{bmatrix} 1 & 2 & 1 \\ 2 & 3 & 2 \end{bmatrix} \begin{bmatrix} x \\ y \\ z \end{bmatrix}$$

$$= 5x + 8y + 5z.$$

It follows that the operator $\mathbf{TS}$ is meaningless since for any $\mathbf{x}$ in R^2 we have $\mathbf{Sx} \in R$ and $\mathbf{T(Sx)}$ makes no sense. $\qquad \square$

The *identity* operator is an operator from a set U into itself, which maps each element of U to the same element. That is,

$$I: U \rightarrow U, \qquad Iu = u \qquad \text{for all } u \in U.$$

The *zero* operator 0 is an operator $0: U \rightarrow V$ which maps every element of U to the zero element in V (we assume of course in this definition that

V has a zero element):

$$0:U\to V, \qquad 0u = 0 \qquad \text{for all } u \in U.$$

Example

Let $U = V = R^3$: then the identity operator $\mathbf{I}:R^3\to R^3$ is simply the 3×3 identity matrix. The zero operator from R^3 to R^2 is the 2×3 matrix containing all zeros. $\qquad\qquad\qquad\qquad\qquad\qquad\qquad\qquad\qquad\square$

One-to-one operators. An operator $T:U\to V$ is *one-to-one* if no two distinct elements of U are mapped to the same element in V. That is, T is *one-to-one* if

$$u_1 \neq u_2 \text{ implies } Tu_1 \neq Tu_2$$

for all $u_1, u_2 \in U$ (Figure 5.4). From this definition it is evident that each v in the range of T is *the image of exactly one element u in U.* We may accordingly define an operator T^{-1}, called the *inverse* of T, which maps v back to u. The inverse is then defined by

$$T^{-1}:R(T)\to U, \qquad T^{-1}(Tu) = u. \tag{3}$$

In view of the definition of the composition of two operators, (3) indicates that

$$T^{-1}T = I.$$

In the same way, by starting with T^{-1} and setting up its inverse we find that

$$TT^{-1} = I.$$

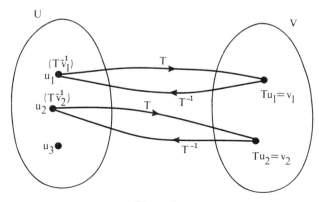

Figure 5.4

If the range of T is all of V (i.e., T is surjective) and T is also one-to-one, then T^{-1} is a one-to-one operator from V onto U, and we say that T *is invertible*.

Examples

(a) The function $f:[-\pi/2, \pi/2] \rightarrow [-1, 1]$, $f(x) = \sin x$, is one-to-one since to each value of $f(x) = \sin x$ there corresponds only one point x. However, if the domain of f is the whole real line, then we see from Figure 5.5 that f is *not* one-to-one since, for any x_1,

$$f(x_1 + 2n\pi) = f(x_1), \qquad n = 1, 2, \ldots .$$

Returning to the case where $D(f) = [-\pi/2, \pi/2]$, the inverse function $f^{-1}:[-1, 1] \rightarrow [-\pi/2, \pi/2]$ is defined by $f^{-1}(y) = \arcsin y$.

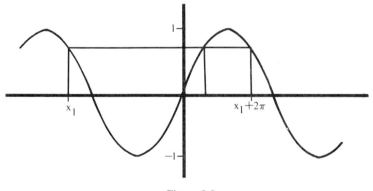

Figure 5.5

(b) Any non-singular $n \times n$ matrix $\mathbf{T}:R^n \rightarrow R^n$ is one-to-one, with inverse $\mathbf{T}^{-1}$ being the usual matrix inverse.

(c) The operator $T = d/dx : C^1[a, b] \rightarrow C[a, b]$ is *not* one-to-one since there are infinitely many functions, all differing from each other by a constant, which have the same image or derivative. However, if we choose the domain of T to be $U = \{u \in C^1[a, b] : u(a) = 0\}$ then T is invertible with inverse T^{-1} defined by

$$T^{-1}(v) = \int_a^x v(y)\, \mathrm{d}y. \qquad \qquad \square$$

Restriction and extension. Suppose that we are given an operator T from U to V, so that $D(T) = U$. Let X be a subset of U: then the *restriction* of T to X is the operator $T|_X$ defined by

$$T|_X : X \to V, \qquad T|_X u = Tu \qquad \text{for all } u \in X.$$

Thus $T|_X$ is an operator with domain X which has the same action on members of X as does T.

Suppose next that U is a subset of a bigger set Y: then the *extension* of T to Y is an operator $\tilde{T}$ with the property

$$\tilde{T} : Y \to V, \qquad \tilde{T}|_U = T.$$

That is, $\tilde{T}u = Tu$ for $u \in U$, so that T is the restriction of $\tilde{T}$ to U.

§16. Linear operators, continuous and bounded operators

Linear operators. A linear operator T is an operator whose domain U is a *linear space*, and which is

(i) additive: $T(u+v) = T(u) + T(v)$ for all $u, v \in U$;

(ii) homogeneous: $T(\alpha u) = \alpha T(u)$ for all $u \in U$, $\alpha \in R$.

We may summarize (i) and (ii) in one statement by defining a linear operator to be an operator that satisfies

$$T(\alpha u + \beta v) = \alpha T(u) + \beta T(v) \qquad \text{for all } u, v \in U, \quad \alpha, \beta \in R.$$

Examples

(a) The differential operator $\mathrm{d}^n / \mathrm{d}x^n : C^n[a, b] \to C[a, b]$ is linear since

$$\frac{\mathrm{d}^n}{\mathrm{d}x^n}(\alpha u + \beta v) = \alpha \frac{\mathrm{d}^n u}{\mathrm{d}x^n} + \beta \frac{\mathrm{d}^n v}{\mathrm{d}x^n}.$$

Similar considerations apply to partial differential operators of all orders, which are also linear operators.

(b) The operator $f : R \to R$, $f(x) = \sin x$, is not a linear operator since $f(x+y) = \sin(x+y) \neq f(x) + f(y) = \sin x + \sin y$.   □

Suppose that $T : U \to V$ is a linear one-to-one operator, and that $Tu_0 = 0$ for some $u_0 \in U$. Since T is linear we have, for any element u, $T(u) = T(u+0) = T(u) + T(0)$ which implies that $T(0) = 0$. But since T is one-to-one, the inverse image of $0 \in V$ must be a unique element in U, from which it follows that $u_0 = 0$.

Conversely, suppose that $T : U \to V$ is a linear operator with the property that $Tu_0 = 0$ only for $u_0 = 0$. Then for any two distinct elements $u, v \in U$, $T(u) - T(v) = T(u-v) \neq 0$ since $u - v \neq 0$ by hypothesis, and so two distinct elements do not have the same image. Hence T is one-to-one. We summarize all of this in the following important theorem.

Theorem 1. A linear operator T is one-to-one if and only if the null space of T is $N(T) = \{0\}$.

Example

Let $\mathbf{T}:R^n \to R^n$ be the operator defined by an $n \times n$ matrix. It is easily shown that $\mathbf{T}$ is a linear operator; the question of whether $\mathbf{T}$ is one-to-one is equivalent to asking whether the equation

$\mathbf{Tx} = \mathbf{y}$

has a unique solution $\mathbf{x}$ for a given $\mathbf{y}$. According to Theorem 1 this question may be answered by considering the equation

$\mathbf{Tx}_0 = \mathbf{0};$

if the only element $\mathbf{x}_0$ satisfying this equation is $\mathbf{x}_0 = \mathbf{0}$, then $\mathbf{T}$ is one-to-one. For example, if $\mathbf{T}:R^2 \times R^2$ is given by

$$\mathbf{T} = \begin{bmatrix} 1 & 2 \\ 2 & 4 \end{bmatrix}$$

then

$$\mathbf{Tx}_0 = \mathbf{0} \Rightarrow \begin{bmatrix} 1 & 2 \\ 2 & 4 \end{bmatrix}\begin{bmatrix} \alpha \\ \beta \end{bmatrix} = \begin{bmatrix} 0 \\ 0 \end{bmatrix}$$

which has the solution $\mathbf{x}_0 = \gamma(-2, 1)$ for any $\gamma \in R$:

$N(\mathbf{T}) = \{\mathbf{x}_0 \in R^2 : \mathbf{x}_0 = \gamma(-2, 1) \quad \text{for all } \gamma \in R\}.$

It follows that the equation $\mathbf{Tx} = \mathbf{y}$ will not have a unique solution; in fact, if $\mathbf{x}_1$ is any solution, then $\mathbf{x}_1 + \gamma(-2, 1)$ is also a solution. $\qquad \square$

All of the above considerations depend only on the algebraic structure of U and V; in other words, we have required no more of U and V than that they be linear spaces. When an operator maps elements from one *normed* space to another, though, many further interesting properties emerge. We take a look first at *continuous operators*.

Before giving a general definition of a continuous operator, it may be helpful to recall the discussion of continuous functions in Section 4. We defined a function $f:R \to R$ to be continuous at x_0 if, given any $\varepsilon > 0$, it is always possible to find a $\delta > 0$ such that $|f(x_0) - f(x)| < \varepsilon$ whenever $|x_0 - x| < \delta$. Now suppose we rephrase this in the language of open sets (Section 13): f is continuous at x_0 if, given any ε, it is always possible to find $\delta > 0$ such that the image of any point in the neighbourhood $N(x_0, \delta)$ lies in the neighbourhood $N(f(x_0), \varepsilon)$. This is precisely how we define a continuous operator in a normed space.

Continuous operator.   Let $T : U \rightarrow V$ be an operator from a normed space U to a normed space V: then T is continuous at $u_0 \in U$ if for every $\varepsilon > 0$ there is a positive number δ, depending possibly on u_0 and ε, such that

$$\| Tu_0 - Tu \| < \varepsilon \qquad \text{whenever } \| u_0 - u \| < \delta. \tag{1}$$

If (1) holds for every $u_0 \in U$ then we simply say that T is continuous. Furthermore, if δ does not depend on u_0 then T is *uniformly continuous*.

The situation is shown schematically in Figure 5.6: choose some point u_0 and a number $\varepsilon > 0$; then T is continuous if a number δ can be found such that the image of the points lying inside the neighbourhood $N(u_0, \delta)$ is contained in the open ball of radius ε and centre Tu_0.

At this point we draw attention to the norms used in (1); since u_0 and u are in U, the norm used when evaluating $\| u_0 - u \|$ is the norm defined on U; on the other hand, the norm used in the evaluation of $\| Tu_0 - Tu \|$ must be the norm defined on V. When wishing to emphasize the distinction we will write, for example, $\| u_0 - u \|_U$ and $\| Tu_0 - Tu \|_V$. Generally, though, it is expected that there will be no confusion about which norm should be used.

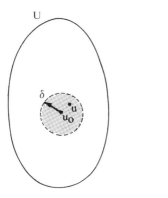 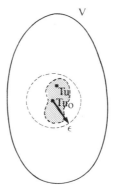

Figure 5.6

Examples

(a)   Let $U = V = R$ and let $f : R \rightarrow R$. Then the definition of continuity given above coincides with that given in Section 4 if we use the norm $\| \cdot \| = | \cdot |$ on R.

(b)   Let $U = V = C[0, 1]$ with the sup-norm, and define $T : C[0, 1] \to C[0, 1]$ by

$$Tu = \int_0^x u(y)\,dy, \qquad x \in [0, 1]$$

(for example, if $u(x) = \cos x$ then Tu is the continuous function $\sin x$). Now

$$\|Tu_0 - Tu\|_\infty = \sup_{x \in [0, 1)} \left| \int_0^x u_0(y)\,dy - \int_0^x u(y)\,dy \right|$$

and so we have to estimate the term inside $|\ldots|$. We find that

$$\left| \int_0^x (u_0(y) - u(y))\,dx \right| \leq \int_0^x |u_0(y) - u(y)|\,dx$$

$$\leq (x - 0) \sup_{y \in [0, 1]} |u_0(y) - u(y)|$$

and so

$$\|Tu_0 - Tu\|_\infty \leq \sup_{x \in [0, 1]} x \sup_{y \in [0, 1]} |u_0(y) - u(y)|$$

$$\leq \sup_{y \in [0, 1]} |u_0(y) - u(y)| = \|u_0 - u\|_\infty .$$

Hence, if $\|u_0 - u\|_\infty < \delta$ then $\|Tu_0 - Tu\|_\infty < \delta$ and so, given any $\varepsilon > 0$ we simply choose $\delta = \varepsilon$ to show that T is continuous. Since δ does not depend on u_0 , we see that T is in fact uniformly continuous.

(c)   Let $U = R^n$ and $V = R^m$ and consider the linear operator T from U to V represented by an $m \times n$ matrix. We endow both R^n and R^m with the Euclidean norms

$$\|\mathbf{x}\|_m = [x_1^2 + \cdots + x_m^2]^{1/2}, \qquad \|\mathbf{x}\|_n = [x_1^2 + \cdots + x_n^2]^{1/2}.$$

Now we consider the image $\mathbf{a}$ of $\mathbf{x}$ under the mapping $\mathbf{T}$: we have

$$\mathbf{a} = \begin{bmatrix} a_1 \\ \vdots \\ a_m \end{bmatrix} = \mathbf{Tx} = \begin{bmatrix} T_{11}x_1 + \cdots + T_{1n}x_n \\ \vdots \\ T_{m1}x_1 + \cdots + T_{mn}x_n \end{bmatrix}$$

or

$$a_i = \sum_{j=1}^n T_{ij}x_j, \qquad 1 \leq i \leq m.$$

If $\mathbf{y}$ is another point in R^n with image $\mathbf{b} \in R^m$, then

$$\|\mathbf{b} - \mathbf{a}\|^2 = \sum_{i=1}^{m} \left(\sum_{j=1}^{n} T_{ij}(y_j - x_j) \right)^2$$

$$\leqslant \sum_{i=1}^{m} \left(\sum_{j=1}^{n} T_{ij}^2 \right) \left(\sum_{j=1}^{n} (y_j - x_j)^2 \right)$$

(using the Schwarz inequality in R^n: $(\sum u_i v_i)^2 \leqslant (\sum u_i^2)(\sum v_i^2)$)

$$\leqslant k^2 \|\mathbf{y} - \mathbf{x}\|^2 \quad \text{where} \quad k^2 = \sum_{i=1}^{m} \sum_{j=1}^{n} T_{ij}^2.$$

Hence

$$\|\mathbf{Ty} - \mathbf{Tx}\| \leqslant k \|\mathbf{y} - \mathbf{x}\|, \tag{3}$$

so for given $\varepsilon > 0$ we simply choose $\delta = \varepsilon/k$: then $\|\mathbf{Ty} - \mathbf{Tx}\| < \varepsilon$ whenever $\|\mathbf{y} - \mathbf{x}\| < \delta$ and $\mathbf{T}$ is thus continuous.   □

Continuous operators can also be characterized in terms of open sets, as the following theorem shows.

Theorem 2. *An operator T (not necessarily linear) from a normed space U into a normed space V is continuous if and only if the inverse image S_0 of any open subset S of V is an open subset of U.*

Proof (see Figure 5.7). Suppose that T is continuous, and for any $u_0 \in S_0$ let $v_0 = Tu_0$. Since S is open, there is a neighbourhood $N(v_0, \varepsilon)$ of v_0 contained entirely in S (Figure 5.7). By the continuity of T, u_0 has a

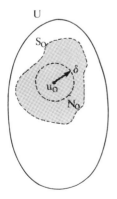
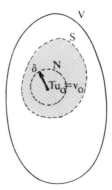

Figure 5.7

neighbourhood $N_0(u_0, \delta)$ which is mapped into $N(v_0, \varepsilon)$. Thus $N_0 \subset S_0$ since N_0 is part of the inverse image of S, so S_0 is open.

Conversely, assume that the inverse image of every open set in V is an open set in U. Then in particular for every $u_0 \in U$ and any neighbourhood $N(Tu_0, \varepsilon)$ of Tu_0, the inverse image N_0, say, of N is open. Hence N_0 also contains a neighbourhood of centre u_0 and radius δ which is mapped into N. Consequently, T is continuous at u_0. Since u_0 was arbitrary, T is continuous. ∎

Bounded operators. The concept of a bounded operator is closely connected with that of a continuous operator. Let $T : U \rightarrow V$ be a *linear* operator: we say that T is *bounded* if it is possible to find a number $K > 0$ such that

$$\|Tu\| \leq K \|u\| \qquad \text{for all } u \text{ in } U. \tag{4}$$

For $u \neq 0$ we see that $K \geq \|Tu\|/\|u\|$. The least upper bound of $\|Tu\|/\|u\|$ for which (4) holds is called the *norm of T,* and is written $\|T\|$. That is,

$$\|T\| = \sup \{\|Tu\|/\|u\|, \qquad u \neq 0\} \tag{5}$$

and we have

$$\|Tu\| \leq \|T\| \|u\| \tag{6}$$

for a bounded linear operator.

We now show that what we have called the "norm" of T does indeed satisfy the norm axioms. First, we note that the *set of all linear operators from U to V forms a linear space,* which we denote by $L(U, V)$. To see this we note that if T and S belong to $L(U, V)$ then $\alpha T + \beta S$, defined by

$$(\alpha T + \beta S)u = \alpha Tu + \beta Su$$

for all $u \in U$, $\alpha, \beta \in R$, also belongs to $L(U,V)$. Also, I is the identity element and the zero operator the zero element.

Secondly, the quantity defined in (5) really does obey all of the norm axioms: $\|T\| \geq 0$ and $\|T\|$ is zero if and only if $T = 0$ (the zero operator). The triangle inequality follows from

$$\|T + S\| = \sup_{u \neq 0} \frac{\|(T + S)u\|}{\|u\|} = \sup_{u \neq 0} \frac{\|Tu + Su\|}{\|u\|}$$

$$\leq \sup_{u \neq 0} \left[\frac{\|Tu\|}{\|u\|} + \frac{\|Su\|}{\|u\|} \right] = \|T\| + \|S\| .$$

Thus $L(U, V)$ is a normed space with norm defined by (5).

Examples

(a) Let $\mathbf{T}:R^n \to R^m$ denote the $m \times n$ matrix operator on R^n. From (3) we know that $\mathbf{T}$ is bounded, hence we may define $L(R^n, R^m)$ to be the linear space of all bounded linear operators from R^n to R^m, that is, the space of all $m \times n$ matrices. If R^n and R^m are provided with the norms defined by

$$\|\mathbf{x}\|_{n,1} = \sum_{i=1}^{n} |x_i| \quad \text{and} \quad \|\mathbf{x}\|_{m,1} = \sum_{i=1}^{m} |x_i| ,$$

respectively, then we have

$$\|T\| = \sup \frac{\|\mathbf{Tx}\|_{m,1}}{\|\mathbf{x}\|_{n,1}}, \quad \mathbf{x} \neq 0,$$

$$= \sup \|\mathbf{Tx}\|_{m,1}, \quad \|\mathbf{x}\|_{n,1} = 1. \quad \text{(See Exercise 16.3.)}$$

Now

$$\|\mathbf{Tx}\|_{m,1} = \sum_{i=1}^{m} \left| \sum_{j=1}^{n} T_{ij} x_j \right| \leqslant \sum_{i=1}^{m} \sum_{j=1}^{n} |T_{ij}| \, |x_j|$$

$$= \sum_{j=1}^{n} |x_j| \sum_{i=1}^{m} |T_{ij}|$$

$$\leqslant \left(\max_{j} \sum_{i=1}^{m} |T_{ij}| \right) \left(\sum_{j=1}^{n} |x_j| \right)$$

$$\leqslant \max_{j} \sum_{i=1}^{m} |T_{ij}|$$

if

$$\sum_{j=1}^{n} |x_j| = \|\mathbf{x}\|_{n,1} = 1. \tag{7}$$

Hence

$$\|T\|_1 = \max \|\mathbf{Tx}\|_{m,1} \leqslant \max_{j} \sum_{i=1}^{m} |T_{ij}| .$$

Suppose that the maximum is attained for $j = p$. Choose $\mathbf{x}$ so that $\mathbf{x} = (0, \dots, 0, 1, 0, \dots, 0)$ (1 is in the pth slot); for this $\mathbf{x}$ we have

$$\|\mathbf{Tx}\|_{m,1} = \sum_{i=1}^{m} |T_{ip}| ,$$

so that we have equality in (7). Hence

$$\|T\|_1 = \max_{j} \sum_{i=1}^{m} |T_{ij}| = \text{maximum absolute column sum.}$$

(b)   Let $T = \mathrm{d}/\mathrm{d}x : C^1[0, 1] \rightarrow C[0, 1]$ with the sup-norm defined on C^1 and C. T is *not* a bounded operator; to show this, we need only consider $u(x) = \sin nx$. Then $\|u\| = 1$ and $\|\mathrm{d}u/\mathrm{d}x\| = \|n \cos nx\| = n$. It follows that there is no constant K such that $\|Tu\| \leqslant K \|u\|$, since $\|u\| = 1$ and $\|Tu\|$ can take on arbitrarily large values (for any chosen constant K, we simply choose n big enough to invalidate the statement $\|Tu\| = n < K$). This result may be extended in an obvious way to show that all ordinary and partial differential operators are unbounded in the sup-norm.   □

The connection between bounded and continuous linear operators is one that is exploited very often. Suppose that $T : U \rightarrow V$ is a bounded linear operator: then

$$\|Tu - Tv\| = \|T(u - v)\| \leqslant K \|u - v\|, \qquad u, v \in U.$$

Given any $\varepsilon > 0$, we set $\delta = \varepsilon/K$ to obtain

$$\|Tu - Tv\| < \varepsilon \quad \text{whenever} \quad \|u - v\| < \delta.$$

Thus T is continuous if it is bounded. Now suppose instead that T is continuous: then, with $u = 0$ and $\varepsilon = 1$ in (1) we can always find a $\delta > 0$ such that

$$\|Tu_0\| \leqslant 1 \quad \text{whenever} \quad \|u_0\| \leqslant \delta.$$

In particular, set $z = \delta u_0/\|u_0\|$: then $\|z\| = \delta$, hence $\|Tz\| \leqslant 1$ and so

$$1 \geqslant \|Tz\| = \|T(\delta u_0/\|u_0\|)\| .$$

That is,

$$\|Tu_0\| \leqslant \delta^{-1} \|u_0\|$$

and T is bounded. We thus have

Theorem 3.   *A linear operator T from a normed space U to a normed space V is continuous if and only if it is bounded.*   ■

Theorem 3 is very useful when it needs to be shown that an operator is continuous; it is frequently more convenient to show boundedness, which in turn implies continuity.

Example

Let $T : C[a, b] \rightarrow C[a, b]$ be defined by $Tu = \int_a^b (x + \xi)u(\xi) \, \mathrm{d}\xi$. Using the sup-norm we have, assuming $b > a > 0$,

$$|Tu| = \left| x \int_a^b u(\xi) \, \mathrm{d}\xi + \int_a^b \xi u(\xi) \, \mathrm{d}\xi \right| \leqslant \left| x \int_a^b u(\xi) \, \mathrm{d}\xi \right| + \left| \int_a^b \xi u(\xi) \, \mathrm{d}\xi \right|$$

$$\leqslant |b| \sup |u(\xi)| \, |b - a| + |b| \sup |u(\xi)| \, |b - a|$$

(see Theorem 6.1). Hence $\|Tu\|_\infty \leqslant 2b(b-a)\|u\|_\infty$ and so T is bounded and consequently continuous. □

It is a remarkably simple but nonetheless extremely important fact that if a linear operator T is continuous, then for any convergent sequence $\{u_n\}$ in its domain, $T(\lim\limits_{n\to\infty} u_n)$ and $\lim\limits_{n\to\infty} T(u_n)$ yield the same result. Note that this says two things: first, if $u_n \to u$ then the sequence $T(u_n)$ in the range of T converges, and secondly the limit is the same as $T(u)$. We now prove this result.

Theorem 4.   *Let $A : U \to V$ be a bounded linear operator and let $\{u_n\}$ be a convergent sequence in U with limit u. Then $u_n \to u$ in U implies that $A(u_n) \to A(u)$ in V.*

Proof.   We are given the fact that $u_n \to u$. Now

$$\|Au_n - Au\|_V = \|A(u_n - u)\|_V \leqslant \|A\| \|u_n - u\|_U .$$

Hence

$$\lim_{n\to\infty} \|Au_n - Au\|_V \leqslant \|A\| \lim_{n\to\infty} \|u_n - u\|_U = 0,$$

so that $Au_n \to Au$. ∎

Open mappings.   In Theorem 2 we saw that a continuous operator T may be characterized by the fact that the inverse image of an open set in the range of T is itself an open set. This of course does not imply that T maps open sets to open sets: for example, the operator T defined by

$$T : (0, 2\pi) \to R, \qquad T(x) = \sin x$$

is continuous, but it maps the *open set* $(0, 2\pi)$ onto the *closed set* $[-1, 1]$.

The operator $T : U \to V$, where U and V are normed spaces, is called an *open mapping* if the image of every open set in U is an open set in V.

The question remains: under what conditions is an operator an open mapping? The answer is provided by one of the important theorems in funtional analysis, the open mapping theorem. We now state this theorem, omitting the rather technical and lengthy proof.

Theorem 5 (The open mapping theorem).   *Let U and V be Banach spaces, and let $T : U \to V$ be a bounded linear operator from U onto V. Then T is an open mapping.*

The open mapping theorem has as a consequence a result which will prove very useful when we study the existence of solutions to boundary-value problems in Chapter 8. This is the so-called *Banach theorem*.

Theorem 6 (The Banach theorem). *A bounded linear one-to-one operator T from a Banach space U onto a Banach space V has a continuous inverse* T^{-1}.

Proof. Since T is one-to-one with range all of V, it remains to show that $T^{-1}: V \to U$ is continuous. From the open mapping theorem we know that T maps any open set $S_0 \subset U$ to an open set $S \subset V$ (Figure 5.8). But the inverse image of S_0 under the mapping T^{-1} is the open set S (T is one-to-one), so that by Theorem 2, T^{-1} is continuous. ■

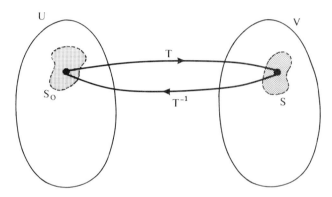

Figure 5.8

§17. Projections

Consider the following situation in R^3, shown in Figure 5.9: given any vector **u**, suppose we define an operator P which has the property that

$$P\mathbf{u} = (u_1, u_2, 0).$$

That is, P projects any vector onto the $x_1 - x_2$ plane. It follows that if **v** is any vector of the form $(v_1, v_2, 0)$ then $P\mathbf{v} = \mathbf{v}$, so that $R(P) = \{\mathbf{v} : \mathbf{v} = (v_1, v_2, 0)\}$ and $N(P) = \{\mathbf{v} : \mathbf{v} = (0, 0, v_3)\}$ and the only vector common to $R(P)$ and $N(P)$ is the zero vector. More generally, P has the property that $P(P\mathbf{u}) \equiv P^2\mathbf{u} = P\mathbf{u}$ for all vectors u in R^3. This is a simple example of a projection operator on R^3; we now generalize these ideas to arbitrary linear spaces.

A linear operator $P : U \to U$ where U is a linear space is called a *projection operator* if

$$P^2 = P, \quad \text{i.e.} \quad P(Pu) = Pu \quad \text{for all} \quad u \in U. \tag{1}$$

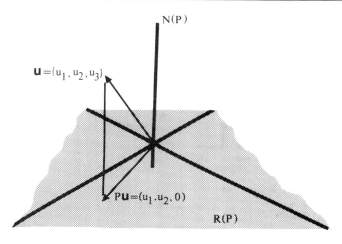

Figure 5.9

Example

Let $U = C[0, 1]$ and define the operator $P: C[0, 1] \to C[0, 1]$ by

$$Pu = u(0)(1 - x) + u(1)x.$$

That is, P maps a continuous function to its *linear interpolate,* shown in Figure 5.10. Then P is a projection operator: to see this, note that

$$P(Pu) = P(v) \quad \text{where} \quad v = Pu = u(0)(1 - x) + u(1)x$$
$$= v(0)(1 - x) + v(1)x = u(0)(1 - x) + u(1)x = Pu. \qquad \square$$

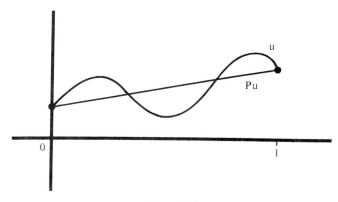

Figure 5.10

The characterization of the range and null space of a projection operator carries over in an obvious way from projections on R^3 to those on linear spaces in general and the main ideas are embodied in the following result.

Theorem 1.   *Let* $P: U \to U$ *be a linear projection operator on a linear space* U. *Then* $R(P) \cap N(P) = \{0\}$ *and every member of* U *has the unique representation* $u = v + w$ *for some* $v \in R(P)$ *and* $w \in N(P)$. *That is,* $U = R(P) \oplus N(P)$.

Proof. We recall from Section 7 the definition of the direct sum $V \oplus W$ of two subspaces V, W of a linear space $U: U = V + W$ and $V \cap W = \{0\}$. Suppose then that $u \in R(P) \cap N(P)$. Then since $u \in R(P)$ there is a $v \in U$ such that $Pv = u$. Hence $P^2v = Pv = Pu = 0$, since $u \in N(P)$ as well. Thus $u = 0$.

To show that $R(P) + N(P) = U$, let u be any member of U, and let $Pu = v$. If we set $w = u - v$, then $Pw = Pu - Pv = P(Pu - Pv) = P(v - Pv) = Pv - Pv = 0$. Hence $u = v + w$ with $v \in R(P)$ and $w \in N(P)$. The uniqueness of the representation follows from Theorem 7.1. ∎

Example

Let $U = C[-1, 1]$ and define P to be the projection that maps any $u \in C[-1, 1]$ to its *even part*:

$$Pu = v \quad \text{where} \quad v(x) = \tfrac{1}{2}(u(x) + u(-x)).$$

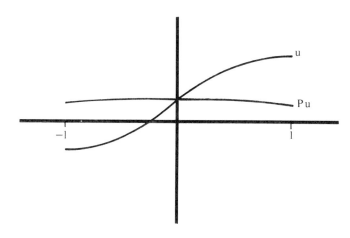

Figure 5.11

The range of P is then

$$R(P) = \{v \in C[-1, 1] : v(x) = v(-x)\},$$

that is, the space of all *even* functions, while the null space of P is the space of all *odd* functions:

$$N(P) = \{v \in C[-1, 1] : v(x) = -v(-x)\}.$$

Clearly $U = R(P) \oplus N(P)$, since every continuous function can be represented as the sum of an odd and an even function, and furthermore the only function that is in $R(P) \cap N(P)$ is the zero function. □

Orthogonal projections. There is a further property of projection operators on R^n that is easily generalized if the linear space is also an *inner product space*: this is the concept of an orthogonal projection. We define an *orthogonal projection operator* to be a projection operator on an inner product space U with the property

$$R(P) \perp N(P), \tag{2}$$

that is, $(u, v) = 0$ for $u \in R(P)$ and $v \in N(P)$.

The situation in R^3 is obvious, as Figure 5.12 shows: let P be the projection operator which maps vectors onto the xy plane: then $R(P)$ is the xy plane, $N(P)$ is the z axis and $R(P) \perp N(P)$. This is in fact the prototype of an orthogonal projection.

Since we now have at our disposal a normed space (the norm being generated by the inner product), it is natural to enquire into the continuity of projection operators. We have the following result.

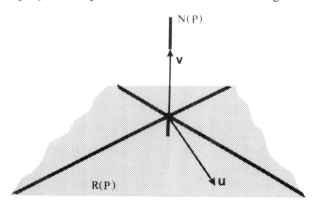

Figure 5.12

Theorem 2. *An orthogonal projection $P : U \to U$ on an inner product space U is continuous.*

Proof. Any $u \in U$ can be expressed in the form $u = v + w$ where $v \in R(P)$ and $w \in N(P)$. Furthermore, $(v, w) = 0$, so it follows (why?) that $\|u\|^2 = \|v\|^2 + \|w\|^2$, hence $\|Pu\|^2 = \|v\|^2 \leq \|u\|^2$. Thus P is bounded, hence continuous. ∎

Up to now we have discussed the situation that obtains when we are given a projection operator. What of the converse situation: suppose we are given a subspace V of an inner product space H; is it possible to define an orthogonal projection P with the property that $R(P) = V$? The answer lies in a logical extension to Theorem 14.2, the projection theorem, as we now show.

Theorem 3. *Let V be a closed linear subspace of a Hilbert space H. Then there is exactly one orthogonal projection $P : H \rightarrow H$ with $R(P) = V$.*

Proof. We know that $H = V + V^{\perp}$ and that every $u \in H$ can be represented uniquely as

$u = v + w$ where $v \in V$, $w \in V^{\perp}$.

Now define $P : H \rightarrow H$ by

$P(v + w) = v,$

that is, P is a projection with range $R(P) = V$. It is not difficult to show that P is an orthogonal projection, so all that remains is to show that P is unique. Let Q be another orthogonal projection with $R(Q) = V$. Now $N(Q) = V^{\perp}$, and

$Pv = v = Qv$ for $v \in V,$

$Pw = 0 = Qw$ for $w \in V^{\perp}.$

Hence $P(v + w) = Q(v + w)$ for all $u = v + w \in H$, i.e. $P = Q$. ∎

We see then that provided H is a Hilbert space, for any given closed subspace V we can set up a unique orthogonal projection P onto V. Going back to Theorem 1, and using Exercise 17.4, we then have

$$H = R(P) \oplus N(P) = V \oplus V^{\perp}. \tag{3}$$

Note the close relationship between Theorem 3 and Theorem 14.2. We conclude this section with a similar extension of Theorem 14.1.

Theorem 4. *Let V be a closed subspace of a Hilbert space H, and let P be the orthogonal projection from H onto V. Then for any $u_0 \in H$*

$\|u_0 - Pu_0\| \leq \|u_0 - v\|$ *for all $v \in V$,*

that is,

$$\|u_0 - Pu_0\| = \inf \|u_0 - v\| \quad \textit{for all } v \in V.$$

Proof. Since $H = R(P) \oplus N(P)$ from Theorem 1, clearly $u_0 - Pu_0 \perp N(P) = V^\perp$. Hence

$$(u_0 - Pu_0, Pu_0 - v) = 0 \quad \text{for all } v \in V$$

since $Pu_0 - v \in V$ also, and so

$$\begin{aligned}
\|u_0 - v\|^2 &= \|u_0 - Pu_0 + Pu_0 - v\|^2 \\
&= \|u_0 - Pu_0\|^2 + \|Pu_0 - v\|^2 \\
&\geq \|u_0 - Pu_0\|^2,
\end{aligned}$$

which proves the theorem. ∎

Example

Consider the space $L_2(-1, 1)$. Let V be the subspace of L_2 consisting of all *even* functions, that is,

$$V = \{v \in L_2(-1, 1) : v(-x) = v(x) \quad \text{a.e.} \quad \text{on } (-1, 1)\}.$$

Then V is a subspace of $L_2(-1, 1)$, and in fact V is closed (show this). According to Theorem 3 there is an orthogonal projection from P onto V, i.e. $R(P) = V$. This projection is defined by

$$P: L_2 \to L_2, \qquad Pv = \tfrac{1}{2}[v(x) + v(-x)].$$

P is clearly a projection (see the example after Theorem 1): it is linear and $P^2 = P$. The null space of P is the set of functions satisfying

$$Pv = 0 \quad \text{or} \quad v(x) = -v(-x),$$

that is, $N(P)$ is the space of *odd* functions:

$$N(P) = \{v \in L_2(-1, 1) : v(x) = -v(-x)\}.$$

We easily verify that P is an orthogonal projection: if $u \in R(P)$ and $v \in N(P)$ then

$$(u, v) = \int_{-1}^{1} u(x)v(x)\, dx = 0$$

since the product uv is odd. Hence $R(P) \perp N(P)$. Finally, we have by Theorem 1,

$$L_2(-1, 1) = R(P) \oplus N(P),$$

that is, every function in $L_2(-1, 1)$ can be represented as the unique sum of an even and an odd function. □

§18. Linear functionals

Dual space. We have seen that the set $L(U, V)$ of all linear operators from a normed space U to a normed space V is itself a normed space, with norm defined by (16.5). When $V = R$ we refer to $L(U, R)$ as the space of *linear functionals* on U; that is, a linear functional f on U is any linear operator that maps elements of a normed space to R, $f : U \rightarrow R$. The space of all continuous linear functionals on a normed space U is called the *dual space* of U, and is denoted by U'. That is,

$$U' = L(U, R),$$

and for any $f \in U'$ we have

$$\|f(u)\| = |f(u)| \leqslant K \|u\| \qquad \text{for all } u \in U. \tag{1}$$

The second expression states that f is bounded and hence continuous. Using the definition (22.4) of an operator norm we see that the norm $\|f\|_{U'}$ of a member of U' is given by

$$\|f\|_{U'} = \sup \frac{|f(v)|}{\|v\|}, \qquad v \neq 0. \tag{2}$$

Examples

(a)   Let $f : L_2(a, b) \rightarrow R$ be defined by

$$f(u) = \int_a^b u(x)\,\mathrm{d}x.$$

Then f is clearly a linear functional: $f(\alpha u + \beta v) = \alpha f(u) + \beta f(v)$. Furthermore, since $u \in L_2$ we have

$$|f(u)| = \left| \int_a^b 1 \cdot u(x)\,\mathrm{d}x \right| \qquad \leqslant \|1\|_{L_2} \|u\|_{L_2} = |b - a| \|u\|_{L_2}$$

and so f is bounded. Therefore f is a member of the dual space $L_2(a, b)'$ of $L_2(a, b)$.

(b)   One reason for the importance of functionals is exemplified by the Dirac delta "function" $\delta(x)$. This is commonly treated as a function with the property

$$\int_{-\infty}^{\infty} \delta(x)u(x)\,\mathrm{d}x = u(0),$$

and δ is described as a function which is zero everywhere with a "spike" at the origin: $\delta(x) = 0$ for $x \neq 0$, $\delta(x) \rightarrow \infty$ at $x = 0$ but $\int_{-\infty}^{\infty} \delta(x)\,\mathrm{d}x = 1$ (Figure 5.13).

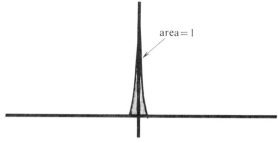

Figure 5.13

However, we are unable to construct a function in the ordinary sense having the properties just described. In fact, δ is a *continuous linear functional* on the space of continuous functions $C[a, b]$ (here $b > 0 > a$):

$$\delta : C(a, b) \to R, \qquad \delta(u) = u(0).$$

The continuity of δ follows from

$$|\delta(u)| = |u(0)| \leqslant \sup |u(x)| = \|u\|_\infty.$$

That is, δ is more correctly viewed as an object that acts on a continuous function to produce its value at the origin. □

A result of crucial importance in functional analysis states that if f is a continuous linear functional on a *Hilbert* space H, then it is always possible to find an element u in H such that

$$f(v) = (u, v) \qquad \text{for all } v \in H$$

where $(.\,,.)$ is the inner product on H. Before proving this, note that if we are given an element u of an inner product space U then we can always define a linear functional f by

$$f(v) = (u, v) \qquad \text{for all } v \in U.$$

Linearity of f is obvious; we can show that f is *continuous* and that $\|f\| = \|u\|_U$ by observing firstly that

$$|f(v)| = |(u, v)| \leqslant K \|v\| \qquad \text{(using the Schwarz inequality)}$$

where $K = \|u\|$, so that $\|f\| \leqslant \|u\|$ (cf. equation (1)), and secondly that $|f(u)| = |(u, u)| = \|u\|^2$ so that $\|f\| \geqslant |(u, u)|/\|u\| = \|u\|$. Thus $\|f\| = \|u\|$. What we will show is that if the space concerned is a *Hilbert* space, then the converse result also holds. This is the famous *Riesz representation theorem*:

Theorem 1 (Riesz representation theorem). Let H be a Hilbert *space and let f be a continuous linear functional on H. Then there exists a unique element u in H such that*

$$f(v) = (u, v) \qquad \text{for all } v \in H. \tag{3}$$

Furthermore, $\|f\| = \|u\|$. $\qquad\qquad\qquad\qquad\qquad\qquad\qquad\qquad\qquad$ (4)

Proof. We assume that $f \neq 0$ since, if $f = 0$, then (3) and (4) hold if we set $u = 0$. Also, observe that if a representation (3) exists, then the element u must be non-zero. Secondly, for any v in H for which $f(v) = 0$ we must have $(u, v) = 0$. This implies that u must be orthogonal to any member of the null space of f, that is, $u \in N(f)^{\perp}$. We are thus led to show the existence of u by considering $N(f)$ and $N(f)^{\perp}$.

Now $N(f)$ is a *closed* linear space. Furthermore, since $f \neq 0$ by hypothesis, $N(f) \neq H$ (if $N(f)$ were equal to H this would imply $f(v) = 0$ for all $v \in H$).

Thus $N(f)$ is a proper subset of H and so, by the projection theorem, $N(f) \neq \{0\}$. Hence there must be at least one non-zero element u_0, say, in $N(f)^{\perp}$. Set

$$z = f(v)u_0 - f(u_0)v \qquad \text{for any } v \in H;$$

then

$$f(z) = f(v)f(u_0) - f(u_0)f(v) = 0$$

and so $z \in N(f)$. Also, since $u_0 \in N(f)^{\perp}$ we have

$$0 = (u_0, z) = (u_0, f(v)u_0 - f(u_0)v) = f(v)(u_0, u_0) - f(u_0)(u_0, v)$$

which implies that

$$f(v) = f(u_0)(u_0, v)/\|u_0\|^2. \tag{5}$$

Finally, we set

$$u = \frac{f(u_0)}{\|u_0\|^2} u_0, \tag{6}$$

and from (5) we see that the element u defined by (6) satisfies (3). The existence of u has been proved.

The proof that u is unique and the derivation of (4) are more straightforward than the existence proof, and are left as exercises (see Exercise 18.6). ■

In many instances it is a difficult task to construct in practice the element u related to a linear functional f by (3). In certain cases, though, this can be done; we give some examples.

Examples

(a)   Let f be a linear functional on on R^n: then $f(\mathbf{x})$ is a real number and according to Theorem 1 we can always find a point $\mathbf{y} \in R^n$ such that

$$f(\mathbf{x}) = \mathbf{x} \cdot \mathbf{y}.$$

For example, if f is defined by

$$f(\mathbf{x}) = x_1 + \cdots + x_n \quad \text{for} \quad \mathbf{x} = (x, \ldots, x_n) \in R^n,$$

then we may find $\mathbf{y} = (y, \ldots, y_n)$ from

$$y_i = \mathbf{y} \cdot \mathbf{e}_i = f(\mathbf{e}_i), \qquad i = 1, \ldots, n$$

where $\mathbf{e}_i = (0 \ldots \ldots 1 \ldots \ldots 0)$ are the basis elements of R^n. Hence

$$\uparrow$$

$$i\text{th position}$$

$$y_i = 0 + \cdots + 1 + \cdots + 0 = 1 \text{ and so } \mathbf{y} = (1, 1, \ldots, 1).$$

(b)   Let f be a linear functional on $L_2(0, 1)$ defined by

$$f : L_2 \to R, \qquad f(v) = \int_0^{1/2} v(x)\,\mathrm{d}x.$$

Then according to Theorem 1 there exists a unique $u \in L_2(\Omega)$ with the property that

$$f(v) = (u, v) \quad \text{or} \quad \int_0^1 u(x)v(x)\,\mathrm{d}x = \int_0^{1/2} v(x)\,\mathrm{d}x.$$

Clearly $u(x)$ is the function

$$u(x) = \begin{cases} 1, & 0 < x \le \tfrac{1}{2} \\ 0, & \tfrac{1}{2} < x < 1 \end{cases}.$$ $\qquad\qquad\square$

We have seen from the Riesz representation theorem that there is a *unique one-to-one correspondence* between functionals on H and members of H. That is, we can set up a *bijective map* (i.e., one-to-one with range all of H) $F : H \to H'$, called the *Riesz map* of H. The Riesz map has the following properties:

(i) F is an *isomorphism* (an isomorphism is a map which is one-to-one and onto);
(ii) F is an *isometry*, that is, $\|Fu\| = \|u\|$, so that F preserves the lengths of elements.

These properties follow from Theorem 1. Of importance to us is the fact that, since members of H and of its dual are in one-to-one correspondence and, furthermore, since any two members $u \in H$ and $f \in H'$ thus related have the same "length", $\|u\| = \|f\|$, we may identify a Hilbert space with its dual. That is, though H and H' are different spaces, their members possess identical properties: if we know all about H, then we automatically know everything about H' via the Riesz map. We summarize by saying that H' *is isometrically isomorphic to* H, and write

$$H = H', \tag{7}$$

meaning that we identify H with its dual.

Example

(a)   In view of the above remarks we may write

$$L_2(\Omega)' = L_2(\Omega). \tag{8}$$

For example, if $f : L_2(0, 1) \to R$ with

$$f(v) = \int_0^{1/2} v(x)\, dx$$

then according to the previous Example (b) we have

$$u(x) = \begin{cases} 1, & 0 < x \leq \tfrac{1}{2} \\ 0, & \tfrac{1}{2} < x < 1 \end{cases}$$

as the unique member of $L_2(0, 1)$ corresponding to f, that is, we identify f with u. Of course,

$$\|f\| = \sup \frac{|f(v)|}{\|v\|} = \|u\|_{L_2}.$$

(b)   The equivalence (8) has its counterpart in the case of the spaces $L_p(\Omega)$, which are not Hilbert spaces for $p \neq 2$. Indeed, it can be shown that if $f \in L_p(\Omega)'$ where $1 < p < \infty$, then it is possible to find a $u \in L_q(\Omega)$, where $\dfrac{1}{p} + \dfrac{1}{q} = 1$, such that

$$f(v) = \int_\Omega uv\, dx,$$

and that furthermore

$$\|f\|_{L_p'} = \|u\|_{L_q}.$$

Thus elements of L_p' and L_q are in one-to-one correspondence and may be connected by an isometric isomorphism. We thus write

$$L_p(\Omega)' = L_q(\Omega), \qquad \frac{1}{p} + \frac{1}{q} = 1, \qquad 1 < p < \infty.$$ □

§19. Bilinear forms

Another special type of operator that will occur very frequently in our study of boundary-value problems is one that maps a *pair* of elements to the real numbers, and which is linear in each of its slots. This is called a *bilinear form*: B is a bilinear form if

$$B : U \times V \to R \qquad (U, V \text{ are linear spaces}),$$
$$B(\alpha u + \beta w, v) = \alpha B(u, v) + \beta B(w, v), \qquad u, w \in U, v \in V,$$
$$B(u, \alpha v + \beta w) = \alpha B(u, v) + B(u, w), \qquad u \in U, v, w \in V.$$

Examples

(a)   Let $U = V = R^3$; then the operator defined by $B(\mathbf{x}, \mathbf{y}) = \mathbf{x} \cdot \mathbf{y}$, is a bilinear form. Indeed, for any inner product space U the inner product $(., .) : U \times U \to R$ is a bilinear form. This follows from the axioms of the inner product. An example of a nonlinear form is

$$B : R^3 \times R^3 \to R, \qquad B(\mathbf{x}, \mathbf{y}) = |\mathbf{x}| + |\mathbf{y}| \; ;$$

here

$$B(\alpha \mathbf{x} + \beta \mathbf{z}, \mathbf{y}) = |\alpha \mathbf{x} + \beta \mathbf{z}| + |\mathbf{y}| \neq \alpha B(\mathbf{x}, \mathbf{y}) + \beta B(\mathbf{z}, \mathbf{y})$$

(this last expression being equal to $\alpha(|\mathbf{x}| + |\mathbf{y}|) + \beta(|\mathbf{z}| + |\mathbf{y}|)$).

(b)   Let $U = V = C^1[a, b]$: then the operator defined by

$$B : C^1[a, b] \times C^1[a, b] = \int_a^b (uv + u'v') \, dx$$

is a bilinear form.   □

Continuous bilinear forms. Suppose that we are given a bilinear form $B : U \times V \to R$ where now U and V are *normed linear spaces*. Consider the expression $B(u, v)$; if there is a positive number K such that

$$|B(u, v)| \leqslant K \|u\| \|v\| \qquad \text{for all } u \in U, v \in V, \tag{1}$$

then B is called a *continuous* bilinear form (cf. continuous or bounded functionals). Later on we will see that all differential equations have associated bilinear forms, and in order for problems to be well posed in a certain sense it is essential for these bilinear forms to be continuous.

Example

We discuss here an example that is typical of a class of problems that will appear later. Denote by $H^1(a, b)$ the linear space of functions which, together with their first derivatives, are square-integrable. We will show later in Chapter 7 that $H^1(a, b)$ is in fact a complete normed space if we give it the norm $\|u\|_{H^1}$ defined by $\|u\|_{H^1}^2 = \int_a^b [u^2 + (u')^2]\, dx$. Now define the bilinear form B by

$$B : H^1(a, b) \times H^1(a, b) \to R, \qquad B(u, v) = \int_a^b [u'v' + cuv]\, dx$$

where $c(x)$ is a bounded positive function with $c_1 \geq c(x) \geq c_2 > 0$ for $x \in (a, b)$. To show that B is continuous, consider

$$|B(u, v)| = \left| \int_a^b [u'v' + cuv]\, dx \right| \leq \left| \int_a^b [u'v' + c_1 uv]\, dx \right|$$

$$= |(u', v')_{L_2} + c_1 (u, v)_{L_2}|$$

$$\leq |(u', v')_{L_2}| + c_1 |(u, v)_{L_2}|$$

$$\leq \|u'\|_{L_2} \|v'\|_{L_2} + c_1 \|u\|_{L_2} \|v\|_{L_2}$$

(using the Schwarz inequality).

Hence

$$|B(u, v)|^2 \leq [\|u'\|_{L_2} \|v'\|_{L_2} + c_1 \|u\|_{L_2} \|v\|_{L_2}]^2. \tag{2}$$

Next, we use the Schwarz inequality for R^n:

$$\left(\sum a_i b_i \right)^2 \leq \left(\sum a_i^2 \right)\left(\sum b_i^2 \right) \quad \text{for} \quad \mathbf{a}, \mathbf{b} \in R^n.$$

If we set $n = 2$ with $\mathbf{a} = (c_1 \|u\|_{L_2}, \|u'\|_{L_2})$ and $\mathbf{b} = (\|v\|_{L_2}, \|v'\|_{L_2})$ then (2) becomes

$$|B(u, v)|^2 \leq [c_1^2 \|u\|_{L_2}^2 + \|u'\|_{L_2}^2][\|v\|_{L_2}^2 + \|v'\|_{L_2}^2]$$

$$= [c_1^2 \|u\|_{L_2}^2 + \|u'\|_{L_2}^2] \|v\|_{H^1}^2$$

$$\leq K^2 (\|u\|_{L_2}^2 + \|u'\|_{L_2}^2) \|v\|_{H^1}^2 \quad (K = \max(1, c_1))$$

$$= K^2 \|u\|_{H^1}^2 \|v\|_{H^1}^2.$$

By taking the square root of both sides we see that B is continuous. $\square$

H-elliptic bilinear forms. Given a bilinear form $B : H \times H \to R$, where H is an inner product space, we say that B is *H-elliptic* if there exists a constant $\alpha > 0$ such that

$$B(v, v) \geq \alpha \|v\|_H^2 \qquad \text{for all } v \in H. \tag{3}$$

Thus an H-elliptic form is one which is bounded below, in the sense of (3).

Example

Consider the previous example again; we have

$$
\begin{aligned}
|B(v, v)| &= \left| \int_a^b [(v')^2 + cv^2] \, dx \right| \\
&\geq \left| \int_a^b [(v')^2 + c_2 v^2] \, dx \right| \\
&\geq \alpha \left| \int_a^b [(v')^2 + v^2] \, dx \right| = \alpha \|v\|_{H^1}^2
\end{aligned}
$$

where $\alpha = \min(1, c_2)$, and so B is H^1-elliptic. $\qquad\square$

The Riesz representation theorem for linear functionals has a counterpart for bilinear forms which will prove useful later on. Suppose that we are given an inner product space H, a linear functional f on H and a member u of H; then we can define a continuous, H-elliptic bilinear form B on H according to the rule

$$B : H \times H \to R, \qquad B(u, v) = f(v) \qquad \text{for all } v \in H. \tag{4}$$

From the H-ellipticity of B and continuity of f we have

$$\alpha \|u\|^2 \leq B(u, u) = f(u) \leq \|f\| \, \|u\|$$

and so

$$\|u\|_H \leq \frac{1}{\alpha} \|f\|_{H'}. \tag{5}$$

In the above sense, then, u and f generate the bilinear form B. We now prove the converse assertion: namely, given a bilinear form B and a linear functional with suitable properties, there is a unique $u \in H$ satisfying (4) and (5). This is the *Lax–Milgram theorem*.

Theorem 1 (the Lax–Milgram theorem). *Let H be a Hilbert space and let $B : H \times H \to R$ be a continuous, H-elliptic bilinear form defined on H. Then, given any continuous linear functional f on H, there exists a* *unique element u in H such that*

$$B(u, v) = f(v) \qquad \text{for all } v \in H. \tag{6}$$

Furthermore,

$$\|u\|_H \leq \alpha^{-1} \|f\|_{H'} \tag{7}$$

The proof of this theorem is rather lengthy, so in order to make it more digestible we break it up into a series of five lemmas.

Lemma 1. *Given any u in H there is a unique element w in H such that*

$$B(u, v) = (w, v) \qquad \text{for all } v \in H. \tag{8}$$

Proof. Given any $u \in H$, $B(u, .)$ is a bounded linear functional on H since

$$B(u, .) : H \to R, \qquad B(u, v) \leq K' \|v\|$$

where $K' = K \|u\|$. Hence, according to the Riesz representation theorem there is a unique element w in H such that $B(u, v) = (w, v)$. ∎

Lemma 2. *Let A be the operator that associates u with w:*

$$A : H \to H, \qquad Au = w. \tag{9}$$

Then A is a bounded linear operator.

Proof. Let $w_1, w_2 \in H$; then according to Lemma 1 we may associate with these elements u_1 and u_2 in H defined by

$$B(u_1, v) = (w_1, v), \qquad B(u_2, v) = (w_2, v) \qquad \text{for all } v \text{ in } H.$$

Since B is bilinear,

$$B(\alpha u_1 + \beta u_2, v) = \alpha B(u_1, v) + \beta B(u_2, v)$$
$$= (\alpha w_1 + \beta w_2, v). \tag{10}$$

Furthermore, from the definition of A we have

$$Au_1 = w_1, \qquad Au_2 = w_2 \quad \text{so} \quad \alpha Au_1 + \beta Au_2 = \alpha w_1 + \beta w_2 \tag{11}$$

But from (8)–(10) we see that A maps $\alpha u_1 + \beta u_2$ to $\alpha w_1 + \beta w_2$:

$$A(\alpha u_1 + \beta u_2) = \alpha w_1 + \beta w_2. \tag{12}$$

The linearity of A thus follows from (11) and (12). A is bounded since, setting $v = Au$ in (8) and using the continuity of B and (9),

$$K \|u\| \|Au\| \geq B(u, Au) = (w, Au) = \|Au\|^2 \Rightarrow \|Au\| \leq K \|u\|. \qquad ■$$

Lemma 3. *A is one-to-one with bounded inverse* A^{-1}.

Proof. Let $R(A)$ denote the range of A (of course, $R(A) \subset H$). We will show that $Az = 0$ only for $z = 0$ and use Theorem 16.1 to show that A is one-to-one. Let z be such that $Az = 0$. Then, since A by definition maps z to a member Az of H such that $B(z, v) = (Az, v)$ we have

$$B(z, v) = (0, v) = 0 \qquad \text{for all } v \in H.$$

In particular, for $v = z$,

$$0 = B(z, z) \geqslant \alpha \, \|z\|^2$$

so that $\|z\| = 0$ or $z = 0$. Hence A is one-to-one, and its inverse $A^{-1} : R(A) \to H$ exists. Furthermore, A^{-1} is linear since A is linear (see Exercise 16.2), and A^{-1} is bounded since

$$\alpha \, \|u\|^2 \leqslant B(u, u) = (w, u) \leqslant \|w\| \, \|u\|, \qquad \text{(using the Schwarz inequality)}$$

whence

$$\|u\| = \|A^{-1}w\| \leqslant \alpha^{-1} \|w\|. \qquad \blacksquare$$

Lemma 4. $R(A)$ *is a complete space.*

Proof. Let $\{w_k\}$ be a Cauchy sequence in $R(A)$. Since $R(A)$ is a subset of H, $\{w_k\}$ is a Cauchy sequence in H too, and so it converges in H:

$$\lim_{k \to \infty} \|w_k - w\| = 0 \quad \text{in} \quad H.$$

We have to show that w is in $R(A)$. To do this, let u_k be defined by $Au_k = w_k$. Then

$$\|u_k - u_l\| = \|A^{-1}w_k - A^{-1}w_l\| = \|A^{-1}(w_k - w_l)\|$$
$$\leqslant \|A^{-1}\| \, \|w_k - w_l\|$$

so that

$$\lim_{k,l \to \infty} \|u_k - u_l\| \leqslant \|A^{-1}\| \lim_{k,l \to \infty} \|w_k - w_l\| = 0$$

($\{w_k\}$ is a Cauchy sequence in H). Hence $\{u_k\}$ is also a Cauchy sequence in H, with limit u in H. Furthermore, since $Au_k = w_k$ we have

$$\lim_{k \to \infty} Au_k = \lim_{k \to \infty} w_k = w \quad \text{or} \quad w = A(\lim u_k) = Au$$

(using Theorem 16.3). Hence w is in the range of A, and since w is the limit of an arbitrary Cauchy sequence, $R(A)$ is complete. $\blacksquare$

Lemma 5. $R(A) = H$, *that is,* A *is* bijective.

Proof. Suppose that $R(A)$ is a proper subspace of H, so that there is a non-zero element u_0 which lies in $R(A)^\perp$ (recall the projection theorem, Theorem 14.2). Then

$$(u_0, t) = 0 \qquad \text{for all } t \in R(A).$$

Using Lemmas 1 and 3 we set $Au_0 = w_0$ so that $w_0 \in R(A)$, and furthermore from (8) we have $B(u_0, v) = (w_0, v)$ for all v in H. In particular, if we set $v = u_0$ then

$$\alpha \|u_0\|^2 \leqslant B(u_0, u_0) = (w_0, u_0) = 0 \quad \text{since} \quad w \in R(A),\ u_0 \in R(A)^\perp.$$

Hence $u_0 = 0$, which is a contradiction, so that $R(A)^\perp = \{0\}$ and $R(A) = H$. ∎

Finally, we gather together all the pieces of information to give the

Proof of the Lax–Milgram theorem. Lemma 1 shows that for any given $u \in H$ there is a unique $w \in H$ defined by (8). This lemma does *not* prove the converse: indeed, we define the operator A by (9), and in order to prove that the converse is true it is necessary to show that A is *bijective*. This is done in Lemmas 3, 4 and 5. Hence we conclude that given any $w \in H$ there exists a unique $u \in H$ such that

$$B(u, v) = (w, v). \tag{13}$$

By the Riesz theorem, every bounded linear functional f can be expressed in the form

$$f(v) = (w, v) \qquad \text{for all } v \in H, \tag{14}$$

with $\|f\| = \|w\|$. Thus (13) and (14) imply (6), and (7) follows from the H-ellipticity of B and the continuity of f, as in (4) and (5). This proves the theorem. ∎

Bibliographical remarks

Operators from one set to another are usually known as functions when the sets have no structure. Much of the material covered in Section 15 falls into this category, and can usually be found in texts which discuss set theory. Good accounts of the material in Section 15 may be found in Naylor and Sell [29], Oden [32], Kreyszig [25] and Roman [40].

    Section 16 makes use of both algebraic and topological properties of sets. The definition of a linear operator requires only the algebraic notion of a linear space, while the definition of a continuous operator obviously needs a normed space. The above mentioned texts by Naylor and Sell [29], Oden [32] and Kreyszig [25] are good references for further reading, as is Volume 2 of the pair of texts by Roman [41]. The same applies to the material of Sections 17 and 18; these texts are all good references. A detailed account, including proofs, of the equivalence of $L_p(\Omega)'$ and $L_q(\Omega)$ may be found in the book by Goffman and Pedrick [17].

A good source for discussions of bilinear forms, including the Lax–Milgram theorem, is Rektorys [39]. This theorem has been generalized to the case of a bilinear form $B : H \times V \to R$ where H and V are distinct Hilbert spaces, by Babuška (see Babuška and Aziz [3] for this result); an account of this generalization is also given by Oden [32].

Exercises

15.1 Describe the range and null space of the following operators:

(i) $M : (-1, 1) \to R^2$, $M(x) = (x, \sqrt{(1 - x^2)})$

(ii) $K : L_2(0, 1) \to R$, $Ku = \int_0^1 [u(x)]^2 \, dx$

(iii) $f : (0, \pi/2) \to R$, $f(x) = \tan x$

15.2 Which of the following operators is one-to-one? surjective?

(i) $K : C[0, 1] \to C[0, 1]$, $Ku = \int_0^x u(y) \, dy$

(ii) $T : R^2 \to R^2$, $T(\mathbf{x}) = (y, x)$

15.3 Describe the compositions ST and TS for the operators

(i) $T : R^2 \to R^2$, $S : R^2 \to R^2$

$T(\mathbf{x}) = (x, -y)$, $S(\mathbf{x}) = (2y, x)$;

(ii) $T : R \to R$, $T(x) = \sin x$ and $S : R \to R$, $S(x) = x^2 - 1$.

15.4 Suppose that $S : U \to V$ and $T : V \to W$ are invertible operators. Show that TS is invertible and $(TS)^{-1} = S^{-1}T^{-1}$.

16.1 Which of the following are linear operators?

(i) $T : L_2(-1, 1) \to L_2(-1, 1)$, $Tu = \int_{-1}^1 K(x, y)u(y) \, dy$;

(ii) $T : C^1[a, b] \to C[a, b]$, $Tu = x^2\, \partial u / \partial x + 2u$;

(iii) $M : R^2 \to R$, $M(\mathbf{x}) = xy$.

16.2   If $T : U \to V$ is an invertible linear operator from U to V, where U and V are linear spaces, show that T^{-1} is linear.

16.3   Show that the norm of an operator can equivalently be defined by

$$\|T\| = \sup \|Tu\|, \qquad \|u\| = 1$$

or

$$\|T\| = \sup \|Tu\|, \qquad \|u\| \leqslant 1.$$

16.4   Let U be the space R^n with the norm $\|\mathbf{x}\|_\infty = \max\limits_{1 \leqslant j \leqslant n} |x_j|$. If $\mathbf{A} : U \to U$ is a linear operator represented by an $n \times n$ matrix, show that $\|\mathbf{A}\|_\infty = \max\limits_{i} \sum_{j=1}^{n} |A_{ij}|$. Determine $\|\mathbf{A}\|_\infty$ if $[\mathbf{A}] = \begin{bmatrix} -2 & 4 \\ 1 & 3 \end{bmatrix}$.

16.5   Show that the identity operator $I : U \to U$ is continuous, where U is *any* normed space. If V is the normed space $C^1[a, b]$ with the norm

$$\|u\|_V = \|u\|_\infty + \|u'\|_\infty$$

and W is the space $C^1[a, b]$ with the sup-norm, produce an example to show that $I : V \to W$ is not continuous.

16.6   If $T : U \to V$ and $S : V \to W$ are bounded linear operators, show that $ST : U \to W$ is bounded with $\|ST\| \leqslant \|S\|\,\|T\|$.

16.7   Show that the null space $N(T)$ of an operator $T : U \to V$ is closed if T is a bounded linear operator.

16.8   An operator $T : U \to V$ is *bounded below* if there exists a constant K such that

$$\|Tu\|_V \geqslant K \|u\|_U, \qquad u \in U.$$

If T is a bounded below linear operator, show that T is one-to-one, and that $T^{-1} : R(T) \to U$ is a bounded operator.
   Show that the operator $D : C_0^1[0, 1] \to C[0, 1]$, $Du = du/dx$ is bounded below, where C_0^1 is the space of functions in C^1 that are zero at $x = 0$ and $x = 1$. Use the sup-norm. [Hint: Consider $u(x) = \int_0^x u'(y)\, dy$.]

17.1   If $P : U \to U$ is a projection operator, show that $(I - P)$ is also a projection. How are $R(I - P)$ and $N(I - P)$ related to $R(P)$ and $N(P)$?

17.2 Show that $\|P\| = 1$ if P is an orthogonal projection.

17.3 Give an example of a nonlinear operator P that satisfies $P^2 = P$.

17.4 Show that $N(P) = R(P)^\perp$ and $R(P) = N(P)^\perp$ if P is an orthogonal projection on an inner product space.

17.5 Let T be the transformation defined by

$$T : L_2(R) \rightarrow L_2(R), \qquad Tu = \begin{cases} u(x) & \text{if } |x| < 1, \\ 0 & \text{otherwise.} \end{cases}$$

Show that T is an orthogonal projection. What are the range and null space of T?

18.1 Let $\mathbf{A}$ be a positive-definite symmetric $n \times n$ matrix, i.e. $\mathbf{x}'\mathbf{A}\mathbf{x} > 0$ for all non-zero vectors $\mathbf{x}$. Then the space R^n is a Hilbert space when endowed with the inner product

$$(\mathbf{x}, \mathbf{y}) = \sum_{i,j=1}^{n} A_{ij} x_i y_j.$$

Given a functional $f : R^n \rightarrow R$, find the element $\mathbf{x}$ such that

$$f(\mathbf{y}) = (\mathbf{x}, \mathbf{y})$$

when f is defined by:

(i) $f(\mathbf{y}) = y_1 + y_2 + \cdots + y_n$;

(ii) $f(\mathbf{y}) = y_1$.

18.2 For each $f \in L_2(0, 1)$ let $u(x)$ be the solution of $u'' + u' - 2u = f$ with $u(0) = u(1) = 0$. Define the functional l by

$$l : L_2(0, 1) \rightarrow R, \qquad l(f) = \int_0^1 u(x) \, dx.$$

Show that l is a bounded linear functional, and find the function u, the value of $l(f)$, and the element g such that $l(f) = (g, f)$, when $f(x) = 2x$.

18.3 Repeat Exercise 18.2 for the differential equation $u'' - 2u' + u = f$.

18.4 If U is a normed space (not necessarily complete), prove that U' is a Banach space.

18.5 Where in the proof of the Riesz representation theorem is the completeness of the Hilbert space H first used, and where is it used subsequently?

18.6 Complete the proof of Theorem 18.1 by showing that u is unique and that $\|f\| = \|u\|$.

19.1 If $B : U \times U \to R$ is a continuous bilinear form on an inner product space U, show that

$$\lim_{n \to \infty} B(u_n, v_n) = B(u, v)$$

if $u_n \to u$ and $v_n \to v$.

19.2 Let $f : H_0^1(0, 1) \to R$ and $B : H_0^1(0, 1) \times H_0^1(0, 1) \to R$ be defined by

$$f(v) = \int_0^1 (-1 - 4x)v \, dx, \qquad B(u, v) = \int_0^1 (x + 1)u'v' \, dx$$

where

$$H_0^1(0, 1) = \{v \in L_2(0, 1) : v' \in L_2(0, 1), \qquad v(0) = v(1) = 0\};$$

this is a Hilbert space with the inner product

$$(u, v)_{H_0^1} = \int_0^1 (uv + u'v') \, dx = (u, v)_{L_2} + (u', v')_{L_2}.$$

Show that f is continuous, that B is continuous and H_0^1-elliptic, and verify that the unique element u satisfying

$$B(u, v) = f(v)$$

is $u(x) = x^2 - x$. [Hint: it may be necessary to use integration by parts. You may assume that a constant $C > 0$ exists such that $\|v\|_{L_2} \leqslant C \|v'\|_{L_2}$.]

19.3 Let $B : U \times U \to R$ be a continuous, U-elliptic bilinear form, and define the bilinear form $\bar{B} : U \times U \to R$ by

$$\bar{B}(u, v) = B(u, v) + (u, cv)_{L_2}.$$

If $U = H_0^1(0, 1)$ and $c(x)$ satisfies $0 < c_1 \leqslant c(x) \leqslant c_2$, show that $\bar{B}$ is continuous and U-elliptic.

6

Orthonormal bases and Fourier series

In vector algebra we are accustomed to carrying out computations using the components of vectors. A set of three mutually orthogonal unit vectors, **i**, **j**, **k** is selected as a basis, and every vector **u** can then be written as $\mathbf{u} = \alpha\mathbf{i} + \beta\mathbf{j} + \gamma\mathbf{k}$, the coefficients α, β, γ being the components of **u** relative to the chosen basis. In Section 20 we start the process of extending this notion to linear spaces in general, by introducing finite-dimensional linear spaces. In Section 21 we endow the linear space with an inner product or a norm, and describe various properties which such inner product or normed spaces have by virtue of their being finite-dimensional. In Section 22 we examine linear operators which act on finite-dimensional spaces. We find here that such operators are always continuous, and that in general they inherit the simple nature of their domains. Finally, in Section 23 all of these concepts are extended to infinite-dimensional spaces. If the space concerned is a Hilbert space, then the idea of an orthonormal basis carries over in a natural way from the finite-dimensional situation.

§20. Finite-dimensional spaces

In this section we discuss linear spaces that have the property that every member can be expressed as a *finite* sum of multiples (that is, a linear combination) of a selected subset of members of that space. The motivation for endowing linear spaces with this property once again comes from the space of vectors: every vector in three dimensions can be represented as a sum of multiples of three non-coplanar vectors.

We start our discussion by defining a linear combination of elements.

Linear   combination. Let U be   a   linear   space,   and   let

125

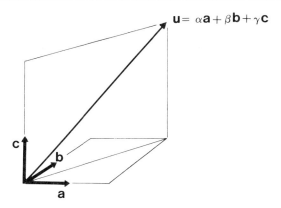

Figure 6.1

$\{u_1, u_2, \ldots, u_n\}$ be a set of elements in U. The expression

$$\alpha_1 u_1 + \alpha_2 u_2 + \cdots + \alpha_n u_n, \tag{1}$$

where $\alpha_1, \ldots, \alpha_n$ are real numbers, is said to be a *linear combination* of the elements $u_1, \ldots, u_n$. Note that $\alpha_1 u_1 + \cdots + \alpha_n u_n \in U$.

In this section we restrict attention to *finite* linear combinations; as long as we do this, the theory that arises is purely algebraic. Infinite linear combinations of the form $\sum_{i=1}^{\infty} \alpha_i u_i$ require topological tools for their treatment and we postpone our discussion of this to Section 23.

Linear dependence, independence. Let U be a linear space, and let $\{u_1, \ldots, u_n\}$ be a finite set of elements of U. Then this set is *linearly dependent* if there exist real numbers $\alpha_1, \alpha_2, \ldots, \alpha_n$, not all zero, such that

$$\alpha_1 u_1 + \cdots + \alpha_n u_n = 0. \tag{2}$$

The set $\{u_1, \ldots, u_n\}$ is *linearly independent* if (2) holds only when all of the α_i are zero.

Examples

(a) Let $U = R^2$, and consider the vectors $\mathbf{u}_1 = (2, 1)$ and $\mathbf{u}_2 = (1, 2)$. To test for linear dependence, we consider

$$\alpha_1 \mathbf{u}_1 + \alpha_2 \mathbf{u}_2 = \mathbf{0} \quad \text{or} \quad (2\alpha_1 + \alpha_2)\mathbf{e}_1 + (\alpha_1 + 2\alpha_2)\mathbf{e}_2 = \mathbf{0}$$

where $\mathbf{e}_1 = (1, 0)$ and $\mathbf{e}_2 = (0, 1)$. We must accordingly have

$$2\alpha_1 + \alpha_2 = 0 \quad \text{and} \quad \alpha_1 + 2\alpha_2 = 0.$$

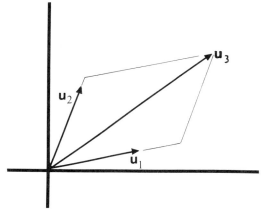

Figure 6.2

The only possible solution to these two equations is $\alpha_1 = \alpha_2 = 0$, and so $\mathbf{u}_1$ and $\mathbf{u}_2$ are *linearly independent*. Graphically this is easy to see: it is not possible to express $\mathbf{u}_2$ as a multiple of $\mathbf{u}_1$. Now suppose that we also have the vector $\mathbf{u}_3$ as shown in Figure 6.2. The set is linearly dependent since, whatever the length and direction of $\mathbf{u}_3$, it is always possible to express it in the form $\mathbf{u}_3 = \beta_1 \mathbf{u}_1 + \beta_2 \mathbf{u}_2$ for some β_1, β_2. Hence there exist scalars β_1, β_2 and $\beta_3 = -1$ such that $\beta_1 \mathbf{u}_1 + \beta_2 \mathbf{u}_2 + \beta_3 \mathbf{u}_3 = 0$.

(b)   Let $U = L_2(0, 1)$ and consider $u_1 = \cosh x$, $u_2 = \sinh x$, $u_3 = e^x$. Then the equation

$$\sum_{i=1}^{3} \alpha_i u_i = 0 \Rightarrow \alpha_1 \cosh x + \alpha_2 \sinh x + \alpha_3 e^x = 0$$

is satisfied for any non-zero α_i that are related to each other by $\alpha_1 = \alpha_2$, $\alpha_3 = -2\alpha_1$, and so the set is linearly *dependent*.   □

Basis, dimension.   A finite set $\{u_1, \ldots, u_k\}$ of elements of a linear space U is said to *span* U if *every* $u \in U$ can be written as $u = \alpha_1 u_1 + \cdots + \alpha_k u_k$ for some scalars α_i, $i = 1, \ldots, k$. A set $\{u_1, \ldots, u_k\}$ of elements of U is said to be a *basis* of U if and only if (i) the set spans U and (ii) the elements $\{u_1, \ldots, u_k\}$ are linearly independent. The number of elements that form a basis is called the *dimension* of U. We write dim U for the dimension of U. If $\{u_i\}_{i=1}^{k}$ is a basis for a linear space U and

$$u = \sum_{i=1}^{k} \alpha_i u_i,$$

(3)

then α_i $(i = 1, \ldots, k)$ are called the *components* of u relative to the basis u_i. Note that the components change with a change of basis.

We stress the fact that the above definition applies only when $\{u_1 \cdots u_k\}$ is a finite set; exactly what is meant by an infinite-dimensional space will become clear later. We also note that, while the dimension of a space is fixed, it is possible to construct many different bases. These points should become clearer in the following examples.

Examples

(a)   Consider the space R^3: the set $\{e_i\} = \{(1, 0, 0), (0, 1, 0), (0, 0, 1)\}$ is linearly independent and also spans R^3, hence $\{e_i\}_{i=1}^3$ is a basis for R^3 and dim $R^3 = 3$. The point $\mathbf{x} = (2, 0, 3)$ has components $\{x_i\} = \{2, 0, 3\}$ relative to e_i. But if we choose the basis $\{f_i\} = \{(1, 1, 2), (1, -1, 1), (2, 1, 0)\}$, then the components of $\mathbf{x}$ are $\{x_i\} = \{1, 1, 0\}$, so (Figure 6.3)

$$\mathbf{x} = 1(1, 1, 2) + 1(1, -1, 1).$$

(b)   Consider the space $P_3[0, 1]$ of polynomials of degree $\leqslant 3$ defined on $[0, 1]$. The set $E = \{p_i\}_{i=0}^3$, where $p_i(x) = x^i$, forms a basis: E spans $P_3[0, 1]$ since every polynomial can be written as

$$p(x) = \sum_{i=0}^3 \alpha_i p_i = \alpha_0 + \alpha_1 x + \alpha_2 x^2 + \alpha_3 x^3.$$

Furthermore, E is linearly independent since

$$\sum_{i=0}^3 \alpha_i p_i = 0 \Rightarrow \alpha_0 + \alpha_1 x + \alpha_2 x^2 + \alpha_3 x^3 = 0$$

holds only if all α_i are zero. Thus dim $P_3[0, 1] = 4$. The components of the

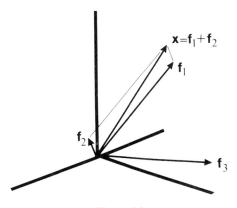

Figure 6.3

polynomial $p(x) = 2x - x^2 + x^3$ relative to the basis E are $\{\alpha_i\} = \{0, 2, -1, 1\}$. But relative to the basis $F = \{(1 - x), (1 + x), x^2, x^3\}$ the components of p are easily shown to be $\{-1, 1, -1, 1\}$ so that

$$p(x) = -1 \cdot (1 - x) + 1 \cdot (1 + x) - x^2 + x^3.$$

$\square$

The following theorem describes an obvious but important property of finite-dimensional spaces.

Theorem 1.   *Let U be a finite-dimensional linear space with* $\dim U = n$. *Then any subset of U containing more than n members is linearly dependent.*

Proof.   Let $B = \{v_1, v_2, \ldots, v_n\}$ be a basis for U, and let $S = \{u_1, \ldots, u_n, u_{n+1}, \ldots, u_{n+k}\}$ be any set of $(n + k)$ elements in U. Then by definition there are scalars A_{ij} such that

$$u_i = \sum_{j=1}^{n} A_{ij} v_j, \qquad i = 1, \ldots, n + k.$$

For any set of scalars $\beta_1, \ldots, \beta_{n+k}$ we have

$$\beta_1 u_1 + \cdots + \beta_{n+k} u_{n+k} = \sum_{i=1}^{n+k} \beta_i u_i$$

$$= \sum_{j=1}^{n+k} \beta_i \sum_{i=1}^{n} A_{ij} v_j$$

$$= \sum_{j=1}^{n} \left(\sum_{i=1}^{n+k} A_{ij} \beta_i \right) v_j.$$

Thus if $\beta_1 u_1 + \cdots + \beta_{n+k} u_{n+k} = 0$, then we must have

$$\sum_{j=1}^{n} \gamma_j v_j = 0 \quad \text{where} \quad \gamma_j = \sum_{i=1}^{n+k} A_{ij} \beta_i;$$

but since $\{v_j\}$ is linearly independent, this implies that $\gamma_j = 0$, or

$$\sum_{i=1}^{n+k} A_{ij} \beta_i = 0, \quad \text{i.e.} \quad \mathbf{A}^T \boldsymbol{\beta} = \mathbf{0}$$

where $\mathbf{A}$ is an $(n + k) \times n$ matrix and $\boldsymbol{\beta}$ an $(n + k) \times 1$ column vector.

From a standard result for sets of linear algebraic equations, every set of n homogeneous (i.e. right-hand side equal to 0) equations in $(n + k)$ unknowns has a non-trivial solution: hence there are scalars $\beta_1, \ldots, \beta_m$, not all zero, such that $\sum \beta_i u_i = 0$, so that $\{u_i\}$ is linearly dependent.   ∎

§21. Finite-dimensional inner product and normed spaces

Concepts such as linear dependence of a set and (finite) dimension of a space are algebraic: they require for their definition only the concept of a linear space. But if the linear space happens also to be an *inner product space*, it is possible to deduce a number of useful properties of finite-dimensional inner product spaces.

First, it is simple to check whether or not a set $\{u_i\}_{i=1}^k$ is linearly dependent. To see this, suppose that

$$\alpha_1 u_1 + \cdots + \alpha_k u_k = 0, \qquad u_i \in U, \, \alpha_i \in R, \qquad i = 1, \ldots, k. \qquad (1)$$

Take the inner product of both sides of this equation with u_1 to obtain

$$A_{11}\alpha_1 + A_{12}\alpha_2 + \cdots + A_{1k}\alpha_k = 0$$

where $A_{ij} = (u_i, u_j) = (u_j, u_i) = A_{ji}$. By successively taking the inner product of (1) with each of the members u_i, we eventually find that

$$\sum_{j=1}^k A_{ij}\alpha_j = 0 \quad \text{or} \quad \mathbf{A}\boldsymbol{\alpha} = \mathbf{0} \qquad (2)$$

where $\mathbf{A}$ is the symmetric matrix with entries A_{ij} and $\boldsymbol{\alpha}$ is the column vector $(\alpha_1, \ldots, \alpha_k)$. Now a necessary and sufficient condition for a non-trivial solution of (2) is $\det \mathbf{A} = 0$: hence, *the set $\{u_1, \ldots, u_k\}$ is linearly dependent if and only if $\det \mathbf{A} = 0$.*

Examples

(a) Let $U = R^2$ with $\mathbf{u}_1$, $\mathbf{u}_2$ and $\mathbf{u}_3$ as in Example (a) on page 126. Then, with $(\mathbf{u}, \mathbf{v}) \equiv \mathbf{u} \cdot \mathbf{v}$ we have

$$\det \mathbf{A} = \det \begin{bmatrix} 5 & 4 & 2\beta_1 + \beta_2 \\ 4 & 5 & \beta_1 + 2\beta_2 \\ 2\beta_1 + \beta_2 & \beta_1 + 2\beta_2 & \beta_1^2 + \beta_2^2 \end{bmatrix}$$

which is easily shown to be identically zero for any values of β_1 and β_2. Hence the set $\{\mathbf{u}_1, \mathbf{u}_2, \mathbf{u}_3\}$ is linearly dependent.

(b) The functions $u_1 = 1$, $u_2 = x$, $u_3 = x^2$ are linearly independent in $L_2(-1, 1)$ since

$$\det \mathbf{A} = \det \left(\int_{-1}^{1} u_i(x) u_j(x) \, dx \right) = \det \begin{bmatrix} 2 & 0 & 2/3 \\ 0 & 2/3 & 0 \\ 2/3 & 0 & 2/5 \end{bmatrix} \neq 0. \qquad \square$$

Orthonormal sets and bases.   If U is an inner product space, a set

$\{\phi_1, \ldots, \phi_k, \ldots\}$ of elements in U is said to be an *orthonormal set* if the elements are mutually orthogonal and have unit length:

$$(\phi_i, \phi_j) = \begin{cases} 0 & \text{if } i \neq j, \\ 1 & \text{if } i = j. \end{cases} \tag{3}$$

Any orthonormal set is linearly independent: if we use (2), we find that $\mathbf{A} = \mathbf{I}$. Now suppose that U is a finite-dimensional inner product space with dim $U = n$. Then a basis $\{\phi_1, \ldots, \phi_n\}$ of U whose elements satisfy (3) is said to be an *orthonormal basis*.

Examples

(a)   The set $\{(1, 0, 0), (0, 1, 0), (0, 0, 1)\}$ forms an orthonormal basis for R^3.

(b)   Consider the space $L_2(-1, 1)$: the infinite set

$$\{\phi_k : \phi_k(x) = \sin k\pi x, \quad k = 1, 2, \ldots\}$$

is an orthonormal set since

$$(\phi_k, \phi_l)_{L_2} = \int_{-1}^{1} \sin k\pi x \sin l\pi x \, dx = \begin{cases} 0 & \text{if } k \neq l, \\ 1 & \text{if } k = l. \end{cases}$$

But $L_2(-1, 1)$ is not finite-dimensional, so talk of an orthonormal basis is premature at this stage.   □

One of the main advantages of orthonormal bases over other bases is that computations involving the former are much simpler. For example, if $\{\phi_i\}_{i=1}^n$ is an orthonormal basis for U and $u, v \in U$, then

$$(u, v) = \left(\sum_{i=1}^n u_i\phi_i, \sum_{j=1}^n v_j\phi_j \right)$$

$$= \sum_{i=1}^n \sum_{j=1}^n u_i v_j (\phi_i, \phi_j) = \sum_{i=1}^n u_i v_i.$$

Given any non-orthonormal basis $S = \{\psi_i\}_{i=1}^n$, it is possible to construct from S an orthonormal basis $\{\phi_i\}_{i=1}^n$ using the *Gram–Schmidt orthonormalization* procedure. We illustrate the procedure for vectors in three dimensions and then generalize from that.

Let $\{\psi_1, \psi_2, \psi_3\}$ be any basis: we construct ϕ_1 from

$$\phi_1 = \psi_1 / |\psi_1|.$$

Next, project ψ_2 on to the plane orthogonal to ϕ_1 (Figure 6.4). We use

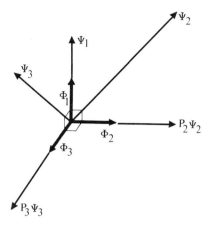

Figure 6.4

for this purpose the projection operator P_2 defined by

$$P_2\mathbf{u} = \mathbf{u} - (\mathbf{u} \cdot \boldsymbol{\phi}_1)\boldsymbol{\phi}_1.$$

Then set

$$\boldsymbol{\phi}_2 = P_2\boldsymbol{\psi}_2/|P_2\boldsymbol{\psi}_2|.$$

Finally, project $\boldsymbol{\psi}_3$ on to the line orthogonal to both $\boldsymbol{\phi}_1$ and $\boldsymbol{\phi}_2$, using the projection operator P_3 defined by

$$P_3\mathbf{u} = \mathbf{u} - (\mathbf{u} \cdot \boldsymbol{\phi}_1)\boldsymbol{\phi}_1 - (\mathbf{u} \cdot \boldsymbol{\phi}_2)\boldsymbol{\phi}_2.$$

Then set

$$\boldsymbol{\phi}_3 = P_3\boldsymbol{\psi}_3/|P_3\boldsymbol{\psi}_3|.$$

Generally, the procedure may be summarized as follows: given a basis $\{\psi_i\}_{i=1}^n$, we form an orthonormal basis $\{\phi_i\}_{i=1}^n$ from

$$\phi_i = P_i\psi_i/\|P_i\psi_i\| \tag{4}$$

where

$$P_1 u = u,$$

$$P_i u = u - \sum_{k=1}^{i-1} (u, \phi_k)\phi_k \quad \text{for} \quad i = 2, 3, \ldots, n.$$

We conclude with an important result, which tells us that all finite-dimensional normed spaces are complete. To prove this result we need a lemma, which is recorded without proof (the proof may be found in references given at the end of the chapter).

Lemma 1. *Let $\{u_1, \ldots, u_n\}$ be a linearly independent set of members of a normed space U. Then there is a constant $c > 0$ such that for every choice of scalars $\alpha_1, \ldots, \alpha_n$, we have*

$$\|\alpha_1 u_1 + \cdots + \alpha_n u_n\| \geq c[|\alpha_1| + \cdots + |\alpha_n|]. \tag{5}$$

Theorem 1. *Every finite-dimensional normed space U is complete.*

Proof. Let $\{u_k\}_{k=1}^{\infty}$ be a Cauchy sequence in U: we will show that $u_k \to u$ in U. Let $\dim U = n$, and let $\{e_1, \ldots, e_n\}$ be any basis for U. Then each u_k can be written as

$$u_k = \sum_{i=1}^{n} \alpha_{ki} e_i$$

where α_{ki} are the components of u_k. Using (5) and the fact that $\{u_k\}$ is Cauchy, we have, for any given $\varepsilon > 0$,

$$\varepsilon > \|u_k - u_l\| = \left\|\sum_{i=1}^{n} (\alpha_{ki} - \alpha_{li})e_i\right\| \geq c \sum_{i=1}^{n} |\alpha_{ki} - \alpha_{li}|$$

for $k, l > N$. Since

$$|\alpha_{ki} - \alpha_{li}| \leq \sum_{i=1}^{n} |\alpha_{ki} - \alpha_{li}| < \frac{\varepsilon}{c}$$

we see that $\{\alpha_{ki}\}$ is a Cauchy sequence in R for each i. Hence $\alpha_{ki} \to \alpha_k$, say. Now define

$$u = \alpha_1 e_1 + \cdots + \alpha_n e_n.$$

We have

$$\|u_k - u\| = \left\|\sum_{i=1}^{n} (\alpha_{ki} - \alpha_i)e_i\right\| \leq \sum_{i=1}^{n} |\alpha_{ki} - \alpha_i| \, \|e_i\|;$$

we know that $\alpha_{ki} \to \alpha_i$, hence $u_k \to u$ in U, and so U is complete.   ■

§22. Linear operators on finite-dimensional spaces

We turn now to the consideration of linear operators whose domains are finite-dimensional spaces. We will show in this section that the nature of such operators is heavily influenced by the fact that their domains have finite dimension.

We show first that if T is a linear operator on a finite-dimensional normed space, then T is always continuous.

Theorem 1.   *Let* $T : U \to V$ *be a linear operator where* U *and* V *are normed spaces, and* U *has finite dimension. Then* T *is continuous.*

Proof.   Let $\{e_1, \ldots, e_n\}$ be a basis for U: then any $u \in U$ has the representation $u = \alpha_1 e_1 + \cdots + \alpha_n e_n$ for certain scalars $\alpha_1, \ldots, \alpha_n$, and so

$$\|Tu\| = \|T(\alpha_1 e_1 + \cdots + \alpha_n e_n)\| = \|\alpha_1 Te_1 + \cdots + \alpha_n Te_n\|$$
$$\leqslant |\alpha_1| \, \|Te_1\| + \cdots + |\alpha_n| \, \|Te_n\|$$
$$\leqslant M(|\alpha_1| + \cdots + |\alpha_n|)$$

where $M = \max \{\|Te_1\|, \ldots, \|Te_n\|\}$. From Lemma 21.1 there is a constant $C > 0$ such that

$$C(|\alpha_1| + \cdots + |\alpha_n|) \leqslant \|u\|,$$

so that

$$\|Tu\| \leqslant \frac{M}{C} \|u\|.$$

Thus T is bounded, hence continuous.   ■

There is a very simple relationship between the dimensions of the domain, null space and range of a linear operator when the operator acts on a finite dimensional space, as we now show.

Theorem 2.   *Let* $T : U \to V$ *be a linear operator with* $\dim U = n$ *and* $\dim N(T) = k \leqslant n$. *Then*
(a)   *if* $\{e_1, \ldots, e_k\}$ *is a basis for* $N(T)$ *and* $\{e_1, \ldots, e_k, e_{k+1}, \ldots e_n\}$ *is a basis for* U, $\{Te_{k+1}, \ldots, Te_n\}$ *is a basis for* $R(T)$.
(b) $\dim N(T) + \dim R(T) = \dim U$.

Proof.   (a) The elements $Tu_1, Tu_2, \ldots, Tu_n$ certainly span $R(T)$ since any $v \in R(T)$ satisfies, for some $u \in U$,

$$v = Tu = T \left(\sum_{i=1}^{n} \alpha_i e_i \right) = \sum_{i=1}^{n} \alpha_i Te_i,$$

where α_i are the components of u relative to the basis e_i. Since $e_1, \ldots, e_k$ are in $N(T)$ we have $Te_1 = \cdots = Te_k = 0$, so $\{Te_{k+1}, \ldots, Te_n\}$ spans $R(T)$. We show next that this set is linearly independent. Suppose that there are scalars $\beta_{k+1}, \ldots, \beta_n$ such that

$$\sum_{i=k+1}^{n} \beta_i (Te_i) = 0;$$

then we have

$$T\left(\sum_{i=k+1}^{n} \beta_i e_i \right) = 0$$

so that $\sum_{i=k+1}^{n} \beta_i e_i \in N(T)$. This element is thus expressible as

$$\sum_{i=k+1}^{n} \beta_i e_i = \sum_{j=1}^{k} \gamma_j e_j$$

for some scalars $\gamma_1, \ldots, \gamma_k$, so that

$$\sum_{i=1}^{n} \beta_i e_i = 0,$$

where we have set $\beta_1 = -\gamma_1, \ldots, \beta_k = -\gamma_k$. But $\{e_1, \ldots, e_n\}$ is linearly independent, hence $\beta_1 = \cdots = \beta_n = 0$. So $\{Te_{k+1}, \ldots, Te_n\}$ is linearly independent and, since it spans $R(T)$, it forms a basis for $R(T)$. Part (b) is a trivial consequence of (a). ■

For the special case in which $T : U \rightarrow V$ with dim $U =$ dim $V = n$, we can deduce from Theorem 2 the following.

Corollary to Theorem 2. Let $T : U \rightarrow V$ be a linear operator with dim $U =$ dim $V = n$. Then $N(T) = \{0\}$ if and only if $R(T) = V$, and when this is so, T is one-to-one and onto, with a unique inverse $T^{-1} : V \rightarrow U$ which is also one-to-one and onto.

Example

Let $U = R^3$, $V = R^2$ and let $\mathbf{T} : U \rightarrow V$ be the matrix $\mathbf{T} = \begin{bmatrix} 1 & 0 & 2 \\ 3 & 4 & 2 \end{bmatrix}$. The null space of $\mathbf{T}$ consists of all vectors $\mathbf{x}$ satisfying

$$\mathbf{Tx} = \mathbf{0} \quad \text{or} \quad \begin{aligned} x_1 + 2x_3 &= 0, \\ 3x_1 + 4x_2 + 2x_3 &= 0. \end{aligned}$$

It is not difficult to see that $N(\mathbf{T})$ consists of all vectors of the form $\mathbf{x} = \alpha(-2, 1, 1)$ so that dim $N(T) = 1$. We should then have dim $R(T) = 3 - 1 = 2$; this is borne out by the fact, first of all, that $\{\mathbf{x}, \mathbf{y}, \mathbf{z}\} = \{(-2, 1, 1), (1, 2, 0), (1, 1, 1)\}$ forms a basis for U. Now $\mathbf{x}$ spans $N(T)$ so $\{\mathbf{Ty}, \mathbf{Tz}\}$ forms a basis for V, as is readily verified. □

Isomorphisms. We have seen that finite-dimensional spaces all "look" the same, in that their elements are uniquely described by the specification of the components relative to a given basis. The situation is even simpler, as it turns out: it is possible to set up a one-to-one correspondence between the elements of any n-dimensional inner product space U and the elements of R^n, in such a way that the elements thus related have the same lengths. More precisely, let $T: U \rightarrow R^n$ be a linear operator from U to R^n; we assert that it is possible to define T in such a way that T is one-to-one with range all of R^n and furthermore that T is *isometric*: if $Tu = v$, then $\|u\| = \|Tu\| = \|v\|$.

Any operator T satisfying these conditions is called an *isometric isomorphism*, and the spaces U and R^n are said to be *isometrically isomorphic* (see also Section 18). The important point is that U and R^n are, to all intents and purposes, one and the same thing and so we write $U = R^n$. We summarize all of this in the following theorem.

Theorem 3. *Let U be any finite-dimensional inner product space with* $\dim U = n$. *Then* $U = R^n$, *that is, there exists an isometric isomorphism from U to R^n.*

The proof of this theorem is outlined in Exercise 22.4.

Representation of linear operators by matrices. We have seen in various places that an $m \times n$ matrix is a linear operator from R^n to R^m. It is natural to ask, then, whether there exists any way in which a linear operator from one arbitrary finite-dimensional space to another can be represented by a matrix. This is easily done, as we now show.

Let $T: U \rightarrow V$ where $\dim U = n$ and $\dim V = m$. Let $\{e_1, \ldots, e_n\}$ and $\{f_1, \ldots, f_m\}$ be bases for U and V, respectively. Then if u is any member of U with image $v \in V$ under the mapping T, there are scalars $\alpha_1, \ldots, \alpha_n$ and $\beta_1, \ldots, \beta_m$ such that u and v can be represented as

$$u = \alpha_1 e_1 + \cdots + \alpha_n e_n, \qquad v = \beta_1 f_1 + \cdots + \beta_m f_m.$$

Since $Tu = v$ and T is linear, we have

$$\alpha_1 Te_1 + \cdots + \alpha_n Te_n = \beta_1 f_1 + \cdots + \beta_m f_m. \tag{1}$$

Now Te_j is in V for $j = 1, \ldots, n$, and so it is possible to express Te_j in the form

$$Te_j = \sum_{i=1}^{m} T_{ij} f_i \tag{2}$$

where T_{ij} are scalars. We form a matrix $\mathbf{T}$ with components T_{ij}: then $\mathbf{T}$ is the *matrix of T relative to the bases* $\{e_i\}$ *and* $\{f_j\}$, and (1) becomes

$$\sum_{i=1}^{m} \sum_{j=1}^{n} T_{ij} \alpha_j f_i = \sum_{i=1}^{m} \beta_i f_i$$

or

$$\sum_{i=1}^{m} \left(\sum_{j=1}^{n} T_{ij}\alpha_j - \beta_i \right) f_i = 0.$$

Since $\{f_i\}$ is linearly independent we have

$$\sum_{j=1}^{n} T_{ij}\alpha_j = \beta_i \quad \text{or} \quad \mathbf{T}\alpha = \beta \tag{3}$$

where $\alpha = (\alpha_1, \ldots, \alpha_n)$ and $\beta = (\beta_1, \ldots, \beta_m)$. It follows that, if we know the matrix corresponding to a linear operator, then (3) can be used to find the components of the image of any member of the domain of the operator.

Example

Let $U = P_2[0, 1]$ and $V = P_1[0, 1]$, where $P_k[0, 1]$ is the set of polynomials of degree $\leq k$ on $[0, 1]$. We know that $\dim P_k[0, 1] = k + 1$. Suppose that we choose as bases for U and V

$$\{e_1, e_2, e_3\} \quad \text{and} \quad \{f_1, f_2\}$$

where $e_1 = f_1 = 1$, $e_2 = f_2 = x$, $e_3 = x^2$, and let T be the derivative operator d/dx: then

$$Te_1 = 0, \qquad Te_2 = 1, \qquad Te_3 = 2x.$$

The matrix $\mathbf{T}$ corresponding to d/dx is found from (2):

$$0 = T_{11} + T_{21}x,$$
$$1 = T_{12} + T_{22}x,$$
$$2x = T_{13} + T_{23}x.$$

Equating coefficients of x^0 and x^1 we see that

$$\begin{array}{l} T_{11} = T_{21} = T_{22} = T_{13} = 0, \\ T_{12} = 1, \qquad T_{23} = 2, \end{array} \quad \text{or} \quad \mathbf{T} = \begin{bmatrix} 0 & 1 & 0 \\ 0 & 0 & 2 \end{bmatrix}. \qquad \square$$

Linear functionals. Linear functionals on finite-dimensional spaces have a particularly simple structure; in fact, they inherit the finite-dimensionality of their domain, so that $\dim U' = \dim U$ if $\dim U$ is finite. To see this, let $\{e_1, \ldots, e_n\}$ be a basis for the n-dimensional normed space U, and define a total of n linear functionals $l_1, \ldots, l_n$ on U by

$$l_i : U \to R, \qquad l_i(e_j) = \begin{cases} 1 & \text{if } i = j, \\ 0 & \text{if } i \neq j. \end{cases} \tag{4}$$

We claim that $L = \{l_1, \ldots, l_n\}$ thus defined is a basis for U': indeed, L is linearly independent since, if

$$\alpha_1 l_1 + \alpha_2 l_2 + \cdots + \alpha_n l_n = 0, \tag{5}$$

then this implies that

$$\sum_{i=1}^{n} \alpha_i l_i(u) = 0 \qquad \text{for all } u \in U,$$

which in turn gives

$$0 = \sum_{i=1}^{n} \alpha_i l_i(e_j) = \alpha_j,$$

using (4). Thus (5) holds only for all $\alpha_j = 0$. Secondly, every $l \in U'$ has the unique representation

$$l = \beta_1 l_1 + \cdots + \beta_n l_n$$

where $\beta_i = l(e_i)$. To see this, let $u = \alpha_1 e_1 + \cdots + \alpha_n e_n \in U$: then

$$l(u) = l(\alpha_1 e_1 + \cdots + \alpha_n e_n) = \sum_{i=1}^{n} \alpha_i \beta_i.$$

On the other hand,

$$l_i(u) = l_i(\alpha_1 e_1 + \cdots + \alpha_n e_n) = \alpha_i.$$

Hence

$$l(u) = \sum_{i=1}^{n} \beta_i l_i(u),$$

or

$$l = \beta_1 l_1 + \beta_2 l_2 + \cdots + \beta_n l_n,$$

as asserted. We see then that $\{l_1, \ldots, l_n\}$ spans U', so that dim $U' = n$. With the aid of Theorem 3 we conclude that U' is isometrically isomorphic to U, and we say that U is *self-dual*, with $U' = U$. Of course, if U is an inner product space (finite-dimensional or not) then by the Riesz representation theorem it is self-dual.

§23. Fourier series in Hilbert spaces

Our main aim in this section is to extend the idea of a basis to arbitrary Hilbert spaces, including spaces of functions. Now, generally speaking these spaces are *not finite-dimensional*. For example, it is not possible to find a finite set of functions in $L_2(\Omega)$ which spans $L_2(\Omega)$. The best we can do is to construct an infinite sequence of functions with the property that any member of the space can be approximated arbitrarily closely by a finite linear combination of these functions provided that a sufficiently large number of functions is used. This leads us to the idea of a basis consisting of a countably infinite set. We will be working with such sets in inner product spaces and, though not necessary, the resulting theory is rendered more concise if we confine attention to *orthonormal sets*, which are sets of the form $\{\phi_1, \phi_2, \ldots, \phi_k, \ldots\}$ for which

$$(\phi_i, \phi_j) = \begin{cases} 1 & \text{if } i = j, \\ 0 & \text{if } i \neq j. \end{cases}$$

Maximal orthonormal set, basis.   Let U be any inner product space and let $\Phi = \{\phi_i\}_{i=1}^{\infty}$ be an orthonormal set in U. Then we say that Φ is a *maximal orthonormal set* in U if there is no other non-zero member ϕ in U that is orthogonal to all ϕ_i. That is, Φ is maximal if $(\phi, \phi_i) = 0$ for all i implies that $\phi = 0$. A maximal orthonormal set in a *Hilbert space H* is called an *orthonormal basis* for H.

It is clear that we are generalizing from the finite-dimensional case; indeed, since every finite-dimensional inner product space is complete and therefore a Hilbert space, the above definition of an orthonormal basis holds true trivially.

Example

Let $H = L_2(-1, 1)$ and consider the set $\Phi_1 = \{\sin \pi x, \sin 2\pi x, \ldots\}$. Φ_1 is an orthonormal set since

$$(\phi_k, \phi_l) = \int_{-1}^{1} \sin k\pi x \sin l\pi x \, dx = \begin{cases} 1 & \text{if } k = l, \\ 0 & \text{if } k \neq l. \end{cases}$$

But Φ_1 is *not maximal*: in particular, any even function $u_e(x)$ is orthogonal to all ϕ_k since $\int_{-1}^{1} u_e(x) \sin k\pi x \, dx = 0$, $\sin k\pi x$ being an odd function. But it can be shown that the set

$$\Phi_2 = \{1/\sqrt{2}\} \cup \{\sin k\pi x, \cos k\pi x\}_{k=1}^{\infty}$$

is a maximal orthonormal set. Since $L_2(-1, 1)$ is complete, Φ_2 is an orthonormal basis.   □

Let $\{\phi_k\}$ be an orthonormal set in an inner product space U. Then, for any $u \in U$ the numbers $u_k = (u, \phi_k)$ are called the *Fourier coefficients* of u. These are the infinite-dimensional counterparts of the components u_k of an element u of a finite-dimensional space U: if $\{\phi_1, \ldots, \phi_n\}$ is an orthonormal basis for U then

$$u = \sum_{k=1}^{n} u_k \phi_k \quad \text{where} \quad u_k = (u, \phi_k).$$

Precisely under what conditions the expression

$$u = \sum_{k=1}^{\infty} (u, \phi_k)\phi_k \tag{1}$$

is valid for an element u of an infinite-dimensional inner product space U is essentially the subject of this section. We shall state these conditions in a moment, but first we must digress and make clear exactly what is meant by an infinite sum of the form $\sum_{k=1}^{\infty} \alpha_k u_k$. Suppose that $\{u_k\}$ is a sequence in an inner product space, and define the nth *partial* sum s_n by

$$s_n = \sum_{k=1}^{n} \alpha_k u_k, \tag{2}$$

where $\alpha_1, \alpha_2, \dots$ is a set of real numbers. Now suppose we obtain $s_1, s_2, \dots$ using (2); the series $\sum_{k=1}^{\infty} \alpha_k u_k$ is then said to *converge* to an element u if the sequence $\{s_n\}$ of partial sums converges to u. That is, we write $u = \sum_{k=1}^{\infty} \alpha_k u_k$ if, given any $\varepsilon > 0$, it is possible to find a number N such that

$$\|s_n - u\| < \varepsilon \quad \text{whenever } n > N \tag{3}$$

or, more briefly,

$$\lim_{n \to \infty} \|s_n - u\| = 0.$$

It is in this sense that we must interpret the expression (1).

The first thing we look at is the issue of projections in Hilbert spaces. Suppose that $\{\phi_k\}_{k=1}^{\infty}$ is an orthonormal set in a Hilbert space H, and let V be the subspace of H spanned by the first n members $\phi_1, \dots, \phi_n$:

$$V = \left\{ v \in H : v = \sum_{k=1}^{n} \alpha_k \phi_k, \ \alpha_k \in R \right\}.$$

According to Theorem 17.3 there is a unique orthogonal projection P from H onto V (V is finite-dimensional, hence complete). We now characterize this projection.

Theorem 1. *Let $\{\phi_1, \phi_2, \dots, \phi_n\}$ be an orthonormal set in a Hilbert space H, and let $V = \text{span}\{\phi_1, \dots, \phi_n\}$. Then the orthogonal projection P of H onto V is given by*

$$Pu = \sum_{k=1}^{n} (u, \phi_k)\phi_k.$$

Proof. P is clearly linear; to show that P is a projection consider

$$P^2 u = P(Pu) = P\left(\sum_{k=1}^{n} (u, \phi_k)\phi_k \right)$$

$$= \sum_{k=1}^{n} (u, \phi_k)P\phi_k = \sum_{k=1}^{n} (u, \phi_k)\sum_{l=1}^{n} (\phi_k, \phi_l)\phi_l$$

$$= \sum_{k=1}^{n} (u, \phi_k)\phi_k = Pu.$$

Hence P is a projection. Obviously $R(P) \subset V$ and all that is left is to show that $R(P) = V$. To do this, take any $v \in V$; then v can be written as $v = \alpha_1\phi_1 + \cdots + \alpha_n\phi_n$, and $Pv = v$, as is easily confirmed. Hence $v \in R(P)$, so $R(P) = V$.

Finally, to show that P is orthogonal consider $u \in N(P)$, $v \in R(P)$. Then $Pv = v$ and

$$(u, v) = (u, Pv) = \left(u, \sum_{k=1}^{n} (v, \phi_k)\phi_k \right)$$

$$= \sum_{k=1}^{n} (v, \phi_k)(u, \phi_k) = \left(\sum_{k=1}^{n} (u, \phi_k)\phi_k, v \right) = (Pu, v) = 0$$

so P is orthogonal. ∎

Clearly we may regard Pv as the *approximation* of v in the subspace V. If $v \in V$ then $Pv = v$ and the approximation coincides with v itself. Also, from Theorem 17.4 we know that the *error* $\|v - Pv\|$ in the approximation is the smallest distance from v to the subspace V.

Example

Let $H = L_2(-1, 1)$ and let V be the subspace of H spanned by the orthonormal set $\{1/\sqrt{2}, \cos \pi x, \sin \pi x\}$. Consider the function $v(x) = x^2$: then the Fourier coefficients of v are

$$(v, 1/\sqrt{2}) = \int_{-1}^{1} \frac{1}{\sqrt{2}} x^2 \, dx = \frac{\sqrt{2}}{3},$$

$$(v, \cos \pi x) = \int_{-1}^{1} x^2 \cos \pi x \, dx = -4/\pi^2,$$

$$(v, \sin \pi x) = \int_{-1}^{1} x^2 \sin \pi x \, dx = 0,$$

and so

$$Pv = \frac{\sqrt{2}}{3} - \frac{4}{\pi^2} \cos \pi x.$$

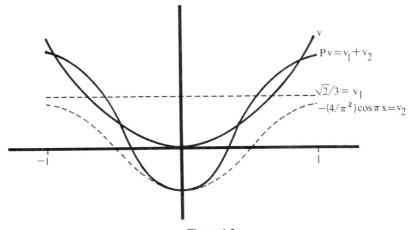

Figure 6.5

The error in the approximation is found from

$$\|v - Pv\|_{L_2}^2 = \int_{-1}^{1} \left[x^2 - \left(\frac{\sqrt{2}}{3} - \frac{4}{\pi^2} \cos \pi x \right) \right]^2 dx = 0.0516$$

so that the relative error is

$$\|v - Pv\|_{L_2} / \|v\|_{L_2} = 0.36. \qquad\qquad \square$$

The next thing we wish to do is to extend Theorem 1 to the case in which the orthonormal set is infinite. Let $\{\phi_k\}_{k=1}^{\infty}$ be an orthonormal set in a Hilbert space H and set

$$U = \left\{ v \in H : v = \sum_{k=1}^{\infty} \alpha_k \phi_k, \ \alpha_k \in R \right\} \qquad\qquad (4)$$

(the infinite sum here is interpreted as in (3)). Next, let V be the *closure* of U, that is $V = \bar{U}$. Then we have the following theorem.

Theorem 2. *Let $\{\phi_k\}_{k=1}^{\infty}$ be a countable orthonormal set in a Hilbert space H and let V be the closed subspace of H defined by (4). Then every $v \in V$ can be written as*

$$v = \sum_{k=1}^{\infty} (v, \phi_k) \phi_k. \qquad\qquad (5)$$

Furthermore, the orthogonal projection of H onto V is given by

$$Pu = \sum_{k=1}^{\infty} (u, \phi_k) \phi_k, \qquad u \in H.$$

Proof.   Let $v \in V$. Then by definition of V we can set

$$v = \lim_{n \to \infty} v_n$$

where each v_n is a finite linear combination of members of V, that is, $v_n = \sum_{k=1}^{m} \alpha_k \phi_k$, where the α_k's and m depend on n (the upper limit of the sum is $m \geqslant n$ because v_n is not necessarily a linear combination of the first n ϕ's). From Theorem 1 and Theorem 17.4 we have

$$\left\| v - \sum_{k=1}^{m} (v, \phi_k)\phi_k \right\| \leqslant \left\| v - \sum_{k=1}^{m} \alpha_k \phi_k \right\| = \| v - v_n \|.$$

But $\| v - v_n \| \to 0$ as $n \to \infty$, so that $v = \sum_{k=1}^{\infty} (v, \phi_k)\phi_k$. The proof that P is an orthogonal projection is similar to the proof of Theorem 1, with one exception: we need to justify the interchange with summations, that is we need to show that

$$\sum_{k=1}^{\infty} (u, \alpha_k \phi_k) = \left(u, \sum_{k=1}^{\infty} \alpha_k \phi_k \right)$$

holds. This part of the proof is outlined in Exercise 23.4(b).   ∎

We have shown in Theorem 2 that it is possible to express a member v of a Hilbert space H in the form (5) if v belongs to the closed subspace V generated by $\{\phi_k\}$. This brings us to the following question: when can we do this for any v in H? The answer is provided by the *Fourier series theorem*:

Theorem 3.   *Let H be a* Hilbert *space, and let* $\Phi = \{\phi_i\}_{i=1}^{\infty}$ *be a countable orthonormal set in H. Then any $u \in H$ can be expressed in the form*

$$u = \sum_{k=1}^{\infty} (u, \phi_k)\phi_k \tag{6}$$

if and only if Φ is an orthonormal basis, that is, a maximal orthonormal set.

Proof.   Assume first that Φ is an orthonormal basis, and let V be the closed subspace of H generated by Φ. We must show that $V = H$. Let $u \in V^{\perp}$, so that $u \perp \phi_i$ for all i. But Φ is maximal, so $u = 0$ and hence $V^{\perp} = \{0\}$. From Theorem 14.2 it follows that $V = H$, so that the orthogonal projection onto V is the identity. From Theorem 2,

$$Iu = u = \sum_{i} (u, \phi_i)\phi_i \qquad \text{for all } u \in H.$$

Conversely, suppose that every $u \in H$ has the form (6). Then we have, for $u \in H$,

$$\|u\|^2 = (u, u) = \left(\sum_i u_i \phi_i, \sum_j u_j \phi_j \right)$$

$$= \sum_i \sum_j u_i u_j (\phi_i, \phi_j) = \sum_i u_i u_i = \sum_i (u, \phi_i)^2. \tag{7}$$

Now if Φ is not maximal then there is a vector ϕ_0, with $\|\phi_0\| = 1$, such that $(\phi_0, \phi_i) = 0$ for all i. But from (7) we have

$$1 = \|\phi_0\|^2 = \sum_i (\phi_0, \phi_i)^2 = 0,$$

a contradiction. Hence Φ is an orthonormal basis. ∎

Examples

(a) Let $H = L_2(-1, 1)$ and consider the orthonormal set $\Phi_1 = \{\sin n\pi x, n = 1, 2, \ldots\}$. We have seen earlier that Φ_1 is not an orthonormal basis. To show that (6) does not hold, consider $u(x) = x^2$: then

$$(u, \phi_k) = \int_{-1}^{1} x^2 \sin k\pi x \, dx = 0 \quad \text{for} \quad k = 1, 2, \ldots .$$

Hence $\sum_{k=1}^{\infty} (u, \phi_k)\phi_k = \sum_{k=1}^{\infty} 0 \cdot \phi_k = 0 \neq x^2$.

On the other hand, $\Phi = \{1/\sqrt{2}\} \cup \{\sin n\pi x, \cos n\pi x, n = 1, 2, \ldots\}$ is an orthonormal basis; using Φ in conjunction with (6), we obtain the Fourier series representation

$$u(x) = x^2 = \sum_{k=0}^{\infty} \alpha_k \phi_k$$

where $\phi_0 = 1/\sqrt{2}$, $\phi_{2k-1} = \sin k\pi x$, $\phi_{2k} = \cos k\pi x$, $k = 1, 2, \ldots$, and

$$\alpha_k = \int_{-1}^{1} x^2 \phi_k \, dx.$$

The basis discussed in this example is of course the traditional trigonometric Fourier series basis which may be familiar from earlier courses. Theorem 2 tells us that a Fourier series representation is possible using any orthonormal basis in a Hilbert space; we give examples below of other orthonormal bases. The proofs that these orthonormal sets are bases are rather lengthy, and are omitted.

(b) The *Legendre polynomials* are defined by

$$P_n(x) = \frac{1}{2^n n!} \frac{d^n}{dx^n} (x^2 - 1)^n, \qquad n = 0, 1, 2, \ldots$$

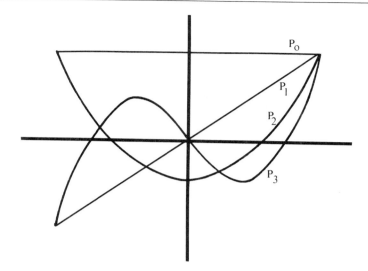

Figure 6.6

and form an *orthogonal* set on $L_2(-1, 1)$. Since

$$(P_n, P_n) = \int_{-1}^{1} P_n^2(x)\, \mathrm{d}x = 2/(2n+1)$$

the corresponding *orthonormal* set is $\{\phi_n(x),\ n = 0, 1, \ldots\}$ where $\phi_n = [(2n+1)/2]^{1/2} P_n$. This orthonormal set can in fact be shown to be maximal, so that $\{\phi_n(x)\}$ forms an orthonormal basis. It is also worth noting that the Legendre polynomials can be obtained by applying the Gram–Schmidt procedure to the set $\{1, x, x^2, \ldots\}$ of functions (cf. Exercise 21.1).

(c)   The *Laguerre polynomials* $L_n(x)$ are defined by

$$L_n(x) = e^{(x/2)}\frac{\mathrm{d}^n}{\mathrm{d}x^n}(x^n e^{-x})$$

and form an orthogonal set on $L_2(0, \infty)$. Since $\|L_n\| = n!$ the set

$$\{\phi_n = (1/n!)L_n\}$$

is orthonormal, and is also a basis for $L_2(0, \infty)$.   □

Bibliographical remarks

Finite-dimensional spaces are given a detailed treatment in Halmos [19] and in Hoffman and Kunze [21] and Roman [40], [41]. The texts by

Noble [31] and Strang [47] also give good expositions of linear algebra, and are more applications-oriented. The subject of orthonormal bases in Hilbert spaces is developed in a comprehensive fashion in Naylor and Sell [29] and Oden [32], as are the other topics in this chapter. Another very good treatment of the topic of orthonormal bases is to be found in Kufner and Kadlec [26].

Exercises

20.1 Which of the following subsets of R^3 is linearly independent?

(a) $(1, 4, 9)$, $(1, 0, 9)$, $(-1, 4, -9)$;
(b) $(1, 4, 9)$, $(1, 0, 9)$, $(2, 4, 8)$.

20.2 Let $p_k(x) = e^{ikx}$. Is $\{p_1, p_2, \ldots, p_n\}$ linearly independent in $L_2(0, 2\pi)$?

20.3 Let U be the set of solutions to the ODE

$$u'' - 2u' + u = 0, \qquad u \in C^2[0, 1].$$

Show that U is a linear space, find dim U and display a basis for U.

20.4 Let V and W be proper subspaces of a linear space U such that $U = V \oplus W$. Show that

$$\dim (V \oplus W) = \dim V + \dim W.$$

20.5 Let V be a subspace of a linear space U, with dim $U = n$. Show (a) that every linearly independent subset of V is part of a basis for U; (b) that if V is a proper subspace of U, then dim $V < $ dim U.

21.1 Let V be the four-dimensional subspace of $L_2(-1, 1)$ spanned by $\{1, x, x^2, x^3\}$. Use the Gram–Schmidt procedure to construct an orthonormal basis for V.

21.2 Apply the Gram–Schmidt procedure to the vectors

$$\{(1, 0, 1), (1, 0, -1), (0, 3, 4)\}$$

to obtain an orthonormal basis for R^3.

21.3 Test the set $\{u_1, u_2\} = \{e^x, e^{-3x}\}$ for linear dependence in $L_2(0, 1)$ by evaluating det $\mathbf{A}$ where $A_{ij} = (u_i, u_j)_{L_2}$.

21.4 Let U be a normed space. A norm $\|\cdot\|_1$ on U is said to be equivalent to another norm $\|\cdot\|_2$ if there are numbers k, $K > 0$ such that

$$k \|u\|_2 \leqslant \|u\|_1 \leqslant K \|u\|_2.$$

Show that any norm $\|\cdot\|_1$ on a finite-dimensional space U is equivalent to any other norm $\|\cdot\|_2$ on U. [Lemma 21.1 may be useful.]

22.1 Let U and V be the spaces of polynomials of degree 3 and 1, respectively, and let $\{1 + t, \, t(1 + t), \, 1 + t^3\}$ and $\{1, t\}$ be bases for U and V, respectively. Let $T: U \to V$ be the linear operator defined by

$$Tp = \mathrm{d}^2 p / \mathrm{d}x^2.$$

Find the matrix corresponding to T.

22.2 Let U be the linear space of all functions of the form $u(x) = \alpha + \beta \cos x + \gamma \sin x$, $0 \leqslant x \leqslant 2\pi$, and define $T: U \to U$ by

$$Tu = \int_0^{2\pi} [1 + \cos (x - \zeta)] u(\zeta) \, \mathrm{d}\zeta.$$

Find the matrix corresponding to T.

22.3 Let $\mathbf{T}$ be an $n \times m$ matrix with transpose $\mathbf{T}'$, and consider the equation

$$\mathbf{T a} = \mathbf{b}$$

where $\mathbf{a} \in R^m$ and $\mathbf{b} \in R^n$. Suppose that we wish to solve for $\mathbf{a}$: show that a necessary condition for such a solution to exist is

$$(\mathbf{c}, \mathbf{b}) = 0 \qquad \text{for all } \mathbf{c} \in N(\mathbf{T}'),$$

that is, $\mathbf{b} \in N(\mathbf{T}')^\perp$. (Note that $(\mathbf{x}, \mathbf{Ty}) = (\mathbf{T}'\mathbf{x}, \mathbf{y})$.)
This shows that $R(\mathbf{T}) \subset N(\mathbf{T}')^\perp$. Show that $R(\mathbf{T}) = N(\mathbf{T}')^\perp$. Determine $N(\mathbf{T}')$ for the matrix

$$\mathbf{T} = \begin{bmatrix} 3 & -1 & 1 \\ 1 & 2 & 1 \\ 4 & 1 & 2 \end{bmatrix}$$

and hence find the general form of the vector $\mathbf{b}$ such that $\mathbf{Ta} = \mathbf{b}$.

22.4 Prove Theorem 22.3.

23.1   The set $\{1/\sqrt{2\pi},\ \cos kx,\ \sin kx,\ k = 1, 2, \ldots\}$ is an orthonormal basis for $L_2(-\pi, \pi)$. Find the Fourier coefficients u_i if

(i) $u(x) = 1$;   (ii) $u(x) = \begin{cases} -1, & -\pi \leqslant x \leqslant 0 \\ 1, & 0 < x \leqslant \pi \end{cases}$

23.2   If $\Phi = \{\phi_1, \phi_2, \ldots\}$ is an orthonormal basis of a Hilbert space H, prove the *Parseval equality*:

$$(u, v) = \sum_{i=1}^{\infty} (u, \phi_i)(v, \phi_i).$$

Deduce that $\|u\|^2 = \sum_{i=1}^{\infty} (u, \phi_i)^2$.

23.3   If Φ is an orthonormal *set* in an inner product space U, prove the *Bessel inequality*:

$$\sum_{i=1}^{\infty} (u, \phi_i)^2 \leqslant \|u\|^2 \qquad \text{for any } u \in U. \qquad\qquad (*)$$

[Note that Exercise 23.3 implies that we have equality if U is a *Hilbert* space and if Φ is an orthonormal *basis*.]

[To prove (*), consider $0 \leqslant \|u - \sum_{i=1}^{N} (u, \phi_i)\phi_i\|^2$; expand and show that (*) holds for finite sums. Then deduce that (*) holds for infinite sums.]

23.4   The aim of this exercise is to show that the projection P defined in Theorem 23.2 is continuous, and is an orthogonal projection.

(a)   Show that if $\sum_{k=1}^{\infty} (u, \phi_k)^2$ is convergent, then so is $\sum_{k=1}^{\infty} (u, \phi_k)\phi_k$, and hence show that $\|Pu\| \leqslant \|u\|$, so that P is continuous.

(b)   Show that $(v, \sum_{k=1}^{\infty} \alpha_k \phi_k) = \sum_{k=1}^{\infty} (v, \alpha_k \phi_k)$ for any convergent series $\sum_{k=1}^{\infty} \alpha_k \phi_k$ (consider (v, s_n) where $s_n = \sum_{k=1}^{n} \alpha_k \phi_k$ and use Exercise 10.5. Hence show that P is an orthogonal projection.

23.5   Let V be the subspace of $L_2(-1, 1)$ spanned by the first four orthonormal Legendre polynomials $\phi_n = (n + \frac{1}{2})^{1/2} P_n(x)$, $n = 0, \ldots, 3$. Write out ϕ_n explicitly, find the orthogonal projection of $u(x) = x^4$ onto V, and verify the inequality in Theorem 17.4 for any suitable choice of v.

7

Distributions and the Sobolev spaces $H^m(\Omega)$

An important aspect of the theory of boundary-value problems is that concerned with the nature of the solution to the problem. Given the problem of finding a function u that satisfies a partial differential equation and one or more boundary conditions, it is clearly of great value to know beforehand whether such a solution exists and, if so, whether it is unique, and finally how smooth this function is. When approaching such questions one is essentially dealing with the properties of an operator (in this case a partial differential operator) A from one space of functions U to another space V, so to start with it is necessary to choose suitable spaces for U and V. The spaces $C^m(\Omega)$ appear to be appropriate since they are spaces of m-times continuously differentiable functions and we are dealing with differential equations. However, they suffer from the drawback that, with the exception of $C(\Omega)$ with the sup-norm, they are not complete. This deficiency would place unnecessarily severe restrictions on our future developments.

The Sobolev spaces $H^m(\Omega)$ provide, as we shall see, a very natural setting for boundary-value problems because, first of all, they are complete; secondly, it is possible to obtain quite general results regarding existence and uniqueness of solutions to differential equations, using these spaces. A third advantage is the fact that Sobolev spaces provide a means of characterizing the degree of smoothness of functions. Finally, and perhaps of most importance is the fact that approximate solution methods such as the Galerkin and finite element methods are most conveniently and correctly formulated in finite-dimensional subspaces of Sobolev spaces.

In order to discuss Sobolev spaces we need to know a little about distributions. This necessary background is provided in Sections 24 and 25. Then in Section 26 we introduce Sobolev spaces, and discuss some of their more important properties. The apparently innocuous question of how one obtains the value of a function on the boundary of a domain, given the function on the domain, is shown in Section 27 to be a non-trivial problem. We show that a function has to satisfy certain requirements in order for its values on the boundary to be defined unambiguously. Finally, in Section 28 we discuss the space $H_0^m(\Omega)$ of functions in $H^m(\Omega)$ which, together with their derivatives of order less than m, vanish on the boundary. We also introduce in this section the space $H^{-m}(\Omega)$ (for $m > 0$) which is defined to be the dual space of $H_0^m(\Omega)$.

§24. Distributions

In this section and in those that follow we shall often need to deal with partial derivatives of all orders, and when discussing general ideas the notation can sometimes become very clumsy. As a prelude to the main topic of this chapter we therefore introduce the very useful multi-index notation for partial derivatives.

Multi-index notation.   Let Z_+^n denote the set of all ordered n-tuples of non-negative integers: a member of Z_+^n will usually be denoted by α or β where, for example,

$$\alpha = (\alpha_1, \alpha_2, \ldots, \alpha_n),$$

each component α_i being a non-negative integer.

We denote by $|\alpha|$ the sum $|\alpha| = \alpha_1 + \alpha_2 + \cdots + \alpha_n$ and by $D^\alpha u$ the partial derivative

$$D^\alpha u = \frac{\partial^{|\alpha|} u}{\partial x_1^{\alpha_1} \partial x_2^{\alpha_2} \cdots \partial x_n^{\alpha_n}}.$$

Thus if $|\alpha| = m$, then $D^\alpha u$ will denote one of the mth partial derivatives of u.

Examples

(a)   If $n = 3$ then a multi-index $\alpha \in Z_+^3$ is an ordered triple of non-negative integers. For example, $\alpha = (1, 0, 3)$ belongs to Z_+^3, with $|\alpha| = \alpha_1 + \alpha_2 + \alpha_3 = 1 + 0 + 3 = 4$. Furthermore, in this case the partial derivative $D^\alpha u$ is the fourth derivative defined by

$$D^\alpha u = \frac{\partial^4 u}{\partial x^{\alpha_1} \partial y^{\alpha_2} \partial z^{\alpha_3}} = \frac{\partial^4 u}{\partial x^1 \partial y^0 \partial z^3} = \frac{\partial^4 u}{\partial x \, \partial z^3}.$$

(b)   Let $n = 2$, and consider the expression

$$I = \sum_{|\alpha| \leqslant 2} a_\alpha D^\alpha u \tag{1}$$

where a_α are given functions of x and y. Then (1) is equivalent to

$$I = \sum_{|\alpha|=0} a_\alpha \, D^\alpha u + \sum_{|\alpha|=1} a_\alpha \, D^\alpha u + \sum_{|\alpha|=2} a_\alpha \, D^\alpha u. \tag{2}$$

When $|\alpha| = 0$ the only possibility is $\alpha = (0, 0)$ (remember $n = 2$ here, so we are dealing with ordered pairs); the other values are

$$|\alpha| = 1: \qquad \alpha = (0, 1) \quad \text{and} \quad (1, 0),$$
$$|\alpha| = 2: \qquad \alpha = (2, 0) \quad \text{and} \quad (1, 1) \text{ and } (0, 2).$$

Suppose that the functions a_α are given as

$$a_{00} = a_{20} = a_{02} = 1, \qquad a_{10} = a_{01} = 2x, \qquad a_{11} = x^2,$$

where we have written, for example, a_{10} for $a_{(1,0)}$. Then

$$\sum_{|\alpha|=0} a_\alpha \, D^\alpha u = a_{00} \frac{\partial^0 u}{\partial x^0 \, \partial y^0} = 1 \cdot u,$$

$$\sum_{|\alpha|=1} a_\alpha \, D^\alpha u = a_{10} \frac{\partial^1 u}{\partial x^1 \, \partial y^0} + a_{01} \frac{\partial^1 u}{\partial x^0 \, \partial y^1} = 2x \left(\frac{\partial u}{\partial x} + \frac{\partial u}{\partial y} \right),$$

$$\sum_{|\alpha|=2} a_\alpha \, D^\alpha u = a_{20} \frac{\partial^2 u}{\partial x^2 \, \partial y^0} + a_{11} \frac{\partial^2 u}{\partial x^1 \, \partial y^1} + a_{02} \frac{\partial^2 u}{\partial x^0 \, \partial y^2}$$

$$= 1 \cdot \frac{\partial^2 u}{\partial x^2} + x^2 \frac{\partial^2 u}{\partial x \, \partial y} + 1 \cdot \frac{\partial^2 u}{\partial y^2}.$$

Collecting all terms, we find that

$$\sum_{|\alpha| \le 2} a_\alpha \, D^\alpha u = \frac{\partial^2 u}{\partial x^2} + x^2 \frac{\partial^2 u}{\partial x \, \partial y} + \frac{\partial^2 u}{\partial y^2} + 2x \left(\frac{\partial u}{\partial x} + \frac{\partial u}{\partial y} \right) + u.$$

Hence $\sum_{|\alpha| \le k} a_\alpha \, D^\alpha u$ is, in general, shorthand for a linear combination of partial derivatives of u, up to and including those of order k.

The advantages of using multi-index notation should be evident from this simple example. □

In Section 21 we discussed an example that showed that the Dirac delta δ is not a function at all, but is instead a continuous linear *functional*, in that it operates on a continuous function u to produce a real number, namely $u(0)$:

$$\delta : C[-1, 1] \to R, \qquad \delta(u) = u(0).$$

Here we are going to show that the Dirac delta belongs to a linear space of functionals called the space of *distributions*, which will in turn play a central role in the definition of Sobolev spaces. In order to introduce distributions formally, we first set up a space of very smooth functions on which these distributions can operate.

Functions with compact support.   For reasons that will be evident later, it is desirable to consider the action of distributions not on all of $C(\Omega)$, but on only a small subset of $C(\Omega)$ consisting of very smooth functions which, roughly speaking, are zero in a neighbourhood of the boundary.

To start with, suppose that a function u defined on a domain Ω is non-zero only for points belonging to a subset K of Ω. Let $\bar{K}$ be the *closure* of K. Then $\bar{K}$ is called the *support* of u. We say that u has *compact support* on Ω if its support $\bar{K}$ is a closed, bounded subset of Ω.

Example

(a)   Let $\Omega = (-1, 1)$ and define u to be the function

$$u(x) = \begin{cases} \sin 2\pi x, & 0 \le |x| \le 1/2, \\ 0, & 1/2 \le |x| < 1. \end{cases}$$

Then K will be the open set given by $K = (-1/2, 0) \cup (0, 1/2)$ and $\bar{K}$ is given by $\bar{K} = [-1/2, 1/2]$ (note that $x = 0$ and $x = \pm 1/2$ are limit points of K). Since $\bar{K}$ is a closed bounded subset of Ω, it follows that $u(x)$ has compact support on Ω (see Figure 7.1).

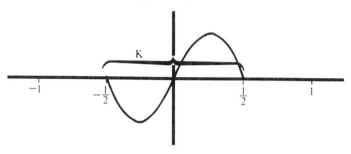

Figure 7.1

(b)   Now consider the function $u(x) = \sin \pi x$ on $\Omega = (-1, 1)$. Here $K = (-1, 0) \cup (0, 1)$ and $\bar{K} = [-1, 1]$ which is *not* a subset of Ω. Hence $u(x)$ does not have compact support on Ω. Remember that Ω is a domain, that is, an open connected set. $\square$

The space $C_0^\infty(\Omega)$.   We define $C_0^\infty(\Omega)$ to be the space of all infinitely differentiable functions which, together with all of their derivatives, have compact support on Ω, where Ω is an open bounded subset of R^n.

Example

A common example of a member of C_0^∞ is the function

$$\phi(x) = \begin{cases} 0, & |x| \geq a, \\ \exp \dfrac{1}{x^2 - a^2}, & |x| < a, \end{cases}$$

defined on $\Omega = (-b, b)$ where $b > a > 0$, as shown in Figure 7.2. It is not difficult to show that u is infinitely differentiable, and that the support of u and all of its derivatives is the set $[-a, a]$. □

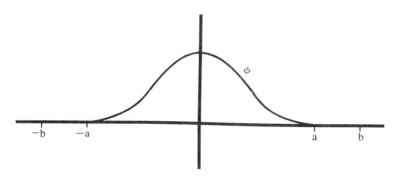

ϕ

$-b$ $-a$ a b

Figure 7.2

Distributions. We define a distribution on a domain Ω in R^n to be a *continuous linear functional* on $C_0^\infty(\Omega)$. That is, a distribution is a continuous linear map from $C_0^\infty(\Omega)$ to R. Thus the space of distributions is the *dual space* (see Section 18) of C_0^∞ and, in keeping with the notation introduced in Section 18 for dual spaces, we denote the space of distributions by $C_0^\infty(\Omega)'$.

Example

By now we are familiar with the idea of the Dirac delta being a distribution. In fact, δ belongs to $C_0^\infty(-a, a)'$ for any $a > 0$ since it is defined by

$$\delta : C_0^\infty(-a, a) \to R, \qquad \delta(\phi) = \phi(0), \qquad \phi \in C_0^\infty(-a, a),$$

and is therefore a continuous linear functional on $C_0^\infty(-a, a)$. □

Regular distributions. It is not only highly irregular objects like the Dirac delta that are distributions. In fact, there are many ordinary functions that can be identified with distributions. All we require of a function f is that the integral $\int_K |f(\mathbf{x})|\, d\mathbf{x}$ be finite on every closed subset K of Ω. When this is so, f is said to be *locally integrable* on Ω, and we can then define in a very natural way a distribution F associated with f by

$$F: C_0^\infty \to R, \qquad F(\phi) = \int_\Omega f\phi\, d\mathbf{x}, \qquad \phi \in C_0^\infty(\Omega). \tag{3}$$

If the support of ϕ is $K \subset \Omega$, then

$$|F(\phi)| = \left| \int_\Omega f\phi\, d\mathbf{x} \right| = \left| \int_K f\phi\, d\mathbf{x} \right| \leqslant \sup_{\mathbf{x} \in K} |\phi(\mathbf{x})| \int_K |f(\mathbf{x})|\, d\mathbf{x}$$

which is bounded, and so we are assured that $F(\phi)$ has meaning. We say that F is the distribution *generated* by f. In future the different notation for a function (f) and its associated distribution (F) will be dispensed with, and we simply write f for both quantities. Whether f is the function or the distribution that it generates will be clear from the context; for example, f in the expression $\int_\Omega f\phi\, d\mathbf{x}$ is clearly a function while f in $f(\phi)$ is a distribution.

Examples

(a)   Every continuous function is locally integrable and hence generates a distribution, but there are also many irregular and discontinuous functions which are locally integrable. One such example is the function $f(x) = |x|^{-1/2}$ on $[-1, 1]$. This function has a singularity at the origin, but is locally integrable since

$$\int_a^b |f(x)|\, dx = \int_a^b |x|^{-1/2}\, dx$$

is bounded for every closed interval $[a, b]$ in $[-1, 1]$. Thus f generates the distribution, also denoted f, defined by

$$f(\phi) = \int_{-1}^1 |x|^{-1/2}\phi\, dx.$$

(b)   The step function $H(x)$, defined on $[-1, 1]$ by

$$H(x) = \begin{cases} 0, & -1 \leqslant x < 0, \\ 1 & 0 \leqslant x \leqslant 1, \end{cases}$$

is locally integrable and generates the distribution H which satisfies

$$H(\phi) = \int_{-1}^1 H(x)\phi(x)\, dx \quad \text{or} \quad H(\phi) = \int_0^1 \phi(x)\, dx. \qquad \square$$

A distribution that is generated by a locally integrable function is called a *regular distribution*. If a distribution cannot be generated by a locally integrable function, it is then said to be a *singular* distribution. An important example of a singular distribution is the Dirac delta: it is not difficult to show (see Exercise 24.2) that there does not exist a locally integrable function which generates δ.

It is possible to define in a very natural way the *product* of a continuous function and a distribution. Specifically, if u is continuous and f is a distribution on $\Omega \subset R^n$ then by uf we understand the distribution satisfying

$$uf(\phi) = f(u\phi) \qquad \text{for all } \phi \in C_0^\infty(\Omega). \tag{4}$$

Note that (4) is a generalization of the trivial identity

$$\int_\Omega [u(\mathbf{x})f(\mathbf{x})]\phi(\mathbf{x}) \, dx = \int_\Omega f(\mathbf{x})[u(\mathbf{x})\phi(\mathbf{x})] \, dx$$

that holds when f is locally integrable.

Example

The distribution $u\delta$ on $(-1, 1)$, where $u(x) = x$, satisfies

$$x\delta(\phi) = \delta(x\phi)$$
$$= (x\phi)|_{x=0} = 0 \qquad \text{for all } \phi \in C_0^\infty(-1, 1).$$

Hence $x\delta = 0$, the zero distribution. □

§25. Derivatives of distributions

Quantities such as the Dirac delta and the Heaviside step function do not have derivatives in the ordinary sense. However, if they are treated as distributions it is possible to extend the concept of a derivative in such a way that a derivative can be defined for these quantities and, indeed, for any distribution. Furthermore, we can define the derivative of a distribution in such a way that, should the distribution be (generated by) a continuously differentiable function, then we recover the usual notion of a derivative. To this end we appeal to *Green's theorem*: the classical version of this well-known theorem states that the identity

$$\int_\Omega u \frac{\partial v}{\partial x_i} \, dx = \int_\Gamma uv v_i \, ds - \int_\Omega v \frac{\partial u}{\partial x_i} \, dx \tag{1}$$

holds for all functions u, v in $C^1(\bar{\Omega})$, where v_i is the ith component of the outward unit normal vector $\mathbf{v}$ to Γ and $\Omega \subset R^n$ (Figure 7.3). The

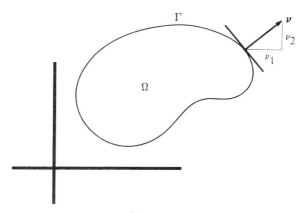

Figure 7.3

one-dimensional version of (1) is the integration-by-parts formula:

$$\int_a^b uv'\,dx = [u\,v]_a^b - \int_a^b vu'\,dx, \qquad u, v \in C^1[a, b].$$ (2)

Indeed, (2) is just a special case of (1) with $\Omega = (a, b)$, $\Gamma = \{a, b\}$ and $\mathbf{v} = \pm 1$ at $x = b$ and a, respectively.

The theorem is easily generalized to a result involving partial derivatives of order m of functions $u, v \in C^m(\bar{\Omega})$ (see Exercise 25.1): by replacing u by $D^\alpha u$ in (1), with $|\alpha| = m$, we can show that

$$\int_\Omega (D^\alpha u)v\,dx = (-1)^{|\alpha|}\int_\Omega u\,D^\alpha v\,dx + \int_\Gamma H(u, v)\,ds$$ (3)

where $H(u, v)$ is an expression involving a sum of products of derivatives of u and v of order less than m.

Now replace v in (3) by a function ϕ belonging to $C_0^\infty(\Omega)$. Then since $\phi = 0$ on the boundary, (3) becomes

$$\int_\Omega (D^\alpha u)\phi\,dx = (-1)^{|\alpha|}\int_\Omega u\,D^\alpha\phi\,dx.$$ (4)

Since u is m-times continuously differentiable it generates a distribution, also denoted by u, so that

$$u(\phi) = \int_\Omega u\phi\,dx$$

or, since $D^\alpha\phi$ also belongs to $C_0^\infty(\Omega)$,

$$u(D^\alpha\phi) = \int_\Omega u\,D^\alpha\phi\,dx.$$

Furthermore, $D^\alpha u$ is continuous so it is able to generate a regular distribution (denoted by $D^\alpha u$) satisfying

$$D^\alpha u(\phi) = \int_\Omega (D^\alpha u)\phi \, dx.$$

Hence (4) can be written as

$$D^\alpha u(\phi) = (-1)^{|\alpha|}u(D^\alpha \phi) \qquad \text{for all } \phi \in C_0^\infty(\Omega). \tag{5}$$

We take (5) as the basis for defining the derivative of any distribution f, as follows: the αth *distributional* or *generalized partial derivative* of a distribution f is *defined* to be a distribution, denoted by $D^\alpha f$, that satisfies

$$D^\alpha f(\phi) = (-1)^{|\alpha|}f(D^\alpha \phi), \qquad \phi \in C_0^\infty(\Omega). \tag{6}$$

Thus we use the same notation for the generalized derivative of a distribution as that used for the conventional derivative of a function. Of course, if the function belongs to $C^m(\bar{\Omega})$ then the generalized derivative coincides with the conventional αth partial derivative for $|\alpha| \le m$, as can be seen immediately from (4) and (5).

For the special case of first derivatives we can dispense with index notation and write (6) as

$$\frac{\partial f}{\partial x_i}(\phi) = -f\left(\frac{\partial \phi}{\partial x_i}\right), \qquad i = 1, \ldots, n. \tag{7}$$

Furthermore, for the case $\Omega = (a, b) \subset R$ all derivatives are with respect to x only, and so (6) becomes

$$\frac{d^k f}{dx^k}(\phi) = (-1)^k f\left(\frac{\partial^k \phi}{\partial x^k}\right). \tag{8}$$

Examples

(a) The first generalized derivative of the Heaviside step function $H(x)$ is the distribution H satisfying

$$H'(\phi) = (-1)^1 H\left(\frac{d\phi}{dx}\right)$$

$$= -\int_{-1}^{1} H(x) \frac{d\phi}{dx} \, dx \qquad (H \text{ is locally integrable})$$

$$= -\int_{0}^{1} \frac{d\phi}{dx} \, dx = -[\phi]_0^1 = \phi(0) = \delta(\phi)$$

so that, symbolically, $H' = \delta$, that is, the derivative H' of the step function is the *Dirac delta*.

(b)   The ramp function $R(\mathbf{x})$ on $\Omega = (-1, 1) \times (-1, 1) \subset R^2$ is defined by

$$R(\mathbf{x}) = \begin{cases} xy & \text{if} \quad x \geqslant 0, \quad y \geqslant 0, \\ 0 & \text{if} \quad x < 0 \text{ or } y < 0. \end{cases}$$

The generalized derivative $D^{(1,0)}R = \partial R/\partial x$ is found from

$$\frac{\partial R}{\partial x}(\phi) = -R\left(\frac{\partial \phi}{\partial x}\right) = -\int_{-1}^{1} \int_{-1}^{1} R(\mathbf{x}) \frac{\partial \phi}{\partial x} \, dx \, dy \quad (R \text{ is locally integrable})$$

$$= -\int_{-1}^{1} \int_{-1}^{1} xy \frac{\partial \phi}{\partial x} \, dx \, dy = \int_{0}^{1} \int_{0}^{1} y\phi \, dx \, dy$$

after using Green's theorem. Furthermore, $D^{(1,1)}R = \partial^2 R/\partial x \, \partial y$ is found from

$$\frac{\partial^2 R}{\partial x \, \partial y}(\phi) = (-1)^2 R\left(\frac{\partial^2 \phi}{\partial x \, \partial y}\right) = \int_{0}^{1} \int_{0}^{1} xy \frac{\partial^2 \phi}{\partial x \, \partial y} \, dx \, dy$$

$$= \int_{0}^{1} \int_{0}^{1} \phi \, dx \, dy \quad \text{(applying Green's theorem twice)}$$

$$= \int_{\Omega} H(\mathbf{x})\phi(x) \, dx \, dy$$

where $H(x)$ is the two-dimensional step function:

$$H(x) = \begin{cases} 1 & \text{if} \quad x \geqslant 0, \quad y \geqslant 0, \\ 0 & \text{if} \quad x < 0 \text{ or } y < 0. \end{cases}$$

Hence

$$\frac{\partial^2 R}{\partial x \, \partial y}(\phi) = H(\phi) \text{ so that } D^{(1,1)}R = H. \qquad \square$$

Weak derivatives.   Suppose that a function u is locally integrable so that it generates a distribution, also denoted by u, that satisfies (24.3). Furthermore, the distribution u possesses distributional derivatives of all orders: in particular, the derivative $D^\alpha u$ is defined by (6). Of course $D^\alpha u$ may or may not be a regular distribution; if it is a regular distribution, then naturally it is generated by a locally integrable function $D^\alpha u(\mathbf{x})$ so that

$$D^\alpha u(\phi) = \int_{\Omega} D^\alpha u(\mathbf{x})\phi(\mathbf{x}) \, dx. \tag{9}$$

It follows in this case from (6) and (9) that the functions u and $D^\alpha u$ are related by

$$\int_\Omega D^\alpha u(\mathbf{x})\phi(\mathbf{x})\,dx = (-1)^m \int_\Omega u(\mathbf{x})D^\alpha\phi(\mathbf{x})\,dx \tag{10}$$

for $|\alpha| = m$. We call the *function* $D^\alpha u$ obtained in this way the αth *weak derivative* of the function u. Of course, if u is sufficiently smooth to belong to $C^m(\bar\Omega)$, then its weak derivatives $D^\alpha u$ coincide with its classical derivatives for $|\alpha| \le m$.

A remark concerning notation is in order here. We have reached the stage where $D^\alpha u$ stands for the classical partial derivative of a function, or for the weak partial derivative of a function, or for the generalized derivative of a distribution (possibly generated by a function). For the most part it should be clear from the context exactly which derivative is being used, but should there be any danger of ambiguity it will be made quite clear exactly what $D^\alpha u$ stands for. The same applies to the notation $\partial u/\partial x$, etc: we allow this to represent any one of the various types of derivatives.

Example

The function $u(x) = |x|$ belongs to $C[-1, 1]$, but the classical derivative u' does not exist at the origin. However, the *weak* derivative of u is the function

$$u' = \begin{cases} -1 & \text{for} \quad -1 \le x < 0, \\ +1 & \text{for} \quad\ \ 0 \le x \le 1. \end{cases}$$

(see Figure 7.4) since the identity $\int u'\phi\,dx = -\int u\phi'\,dx$ is easily shown to hold. Note furthermore that $u' \in L_2(-1, 1)$, which is of course locally integrable.   □

The above example illustrates one fundamental difference between classical and weak derivatives. The classical derivative, if it exists, is a function defined *pointwise* on an interval, so it must be at least continuous. A weak derivative, on the other hand, need only be locally integrable. Thus any function v differing from a weak derivative u' on a set of measure zero (e.g. at a finite number of points on the real line) is itself a weak derivative of u.

Distributional differential equations. Since we now have at our disposal the concept of the derivative of a distribution, it is natural to consider next differential equations involving distributions. For example,

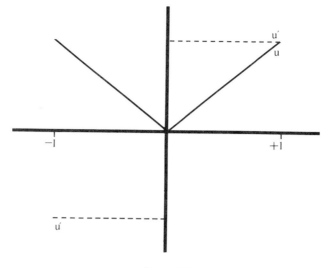

Figure 7.4

suppose that we are required to find the distribution g that satisfies

$$g' = f \tag{11}$$

for a given distribution f, on some interval of the real line. If f and g were ordinary functions (e.g. $f \in C[a, b]$, $g \in C^1[a, b]$), then (11) would be a simple first order differential equation. Since f and g are actually distributions, we go back to the definition (8) of a generalized derivative: then (11) really means that we have to find the distribution g satisfying

$$g'(\phi) = f(\phi) \quad \text{or} \quad -g(\phi') = f(\phi) \qquad \text{for all } \phi \in C_0^\infty(a, b). \tag{12}$$

If by (11) we understand (12), then (11) is said to be a *distributional differential equation*.

The same procedure applies in the case of more general differential equations. For example, suppose that we are required to find the distribution g satisfying

$$Ag = f \tag{13}$$

where A is the (generalized) differential operator given by

$$A = a_0(x) \frac{d^k}{dx^k} + a_1(x) \frac{d^{k-1}}{dx^{k-1}} + \cdots + a_k(x). \tag{14}$$

We interpret (13) as a differential equation involving *generalized* derivatives of g, and seek g such that

$$Ag(\phi) = f(\phi) \qquad \text{for all } \phi \in C_0^\infty(a, b),$$

which is equivalent to

$$g(A^*\phi) = f(\phi), \qquad \phi \in C_0^\infty(a, b),$$

the operator A^* resulting from successive applications of (6) and (24.4): thus

$$A^*\phi = (-1)^k \frac{d^k}{dx^k}(a_0\phi) + \cdots + a_k\phi. \tag{15}$$

Generally for partial differential equations involving distributions the same procedure is adopted: the problem of finding a distribution g satisfying

$$Ag = f \tag{16}$$

where $Ag = \sum_{|\alpha| \le k} a_\alpha D^\alpha g$, is equivalent to the problem of finding g such that

$$Ag(\phi) = f(\phi) \quad \text{or} \quad g(A^*\phi) = f(\phi), \tag{17}$$

where A^* is obtained, as in (15), by repeated application of (6) and (24.4).

Naturally one would expect that if f is continuous (that is, a distribution generated by a continuous function), then the solution g should be a function which is k times continuously differentiable. This is indeed so: in other words, when the distributions involved are generated by sufficiently differentiable functions, we recover the classical concept of a differential equation. In this case g is called a *classical* solution. More generally, though, if f is a regular distribution generated by a function which is locally integrable but not continuous, or indeed if it is a singular distribution, then equation (16) cannot be expected to have any meaning in the classical sense. The solution in this case is called a *weak* or *generalized* solution.

When g satisfies an equation of the form (16) or (17) we say that $Ag = f$ *in the sense of distributions*, or that g satisfies (16) *distributionally*.

Examples

(a) The equation

$$xg' = 0 \quad \text{on} \quad \Omega = (-1, 1) \tag{18}$$

has the classical solution $g = $ constant. But if we regard (18) as a *distributional* differential equation then the weak solution is

$$g(x) = c_1 H(x) + c_2$$

where c_1 and c_2 are constants. We check as follows: $g' = c_1 \delta$ so that

$$(xg')(\phi) = g'(x\phi) = c_1\delta(x\phi) = c_1[(x\phi)(0)] = 0;$$

hence $xg' = 0$ in the sense of distributions.

(b)   The equation $g'' = \delta'$ has no classical solution on $(-1, 1)$ but its weak solution is

$$g = H + c_1 x + c_2. \tag{19}$$

This is verified by considering

$$g''(\phi) = H''(\phi) = H(\phi'') = \int_0^1 \phi'' \, dx = -\phi'(0) = \delta'(\phi).$$

Hence g defined by (19) satisfies $g'' = \delta'$ distributionally.   □

§26.   The Sobolev spaces $H^m(\Omega)$

Before we actually get down to discussing Sobolev spaces, it is appropriate at this stage to elaborate on the degree of smoothness that we expect the boundary Γ of a domain Ω in R^n $(n \geq 2)$ to have, since some results concerning Sobolev spaces hold only when the boundary is sufficiently smooth. Let Ω be a domain in R^n $(n \geq 2)$ with boundary Γ. Let x_0 be an arbitrary point on Γ and construct $B(x_0, \varepsilon)$, the open ball of radius ε, centre x_0, for some $\varepsilon > 0$, i.e. $B(x_0, \varepsilon) = \{x \in R^n : |x - x_0| < \varepsilon\}$.

Next, set up a coordinate system $(\xi_1, \ldots, \xi_n)$ such that the segment $\Gamma \cap B(x_0, \varepsilon)$ can be expressed as

$$\xi_1 = f(\xi_2, \ldots, \xi_n).$$

If the function f is m-times continuously differentiable for every $x_0 \in \Gamma$ we say that Γ *is of class* C^m. We say that Γ is *Lipschitzian* if f is Lipschitz-continuous, that is, if there is a constant k such that

$$|f(\hat{\xi}) - f(\hat{\eta})| < k \, |\hat{\xi} - \hat{\eta}|,$$

where $\hat{\xi} = (\xi_2, \ldots, \xi_n)$ and $\hat{\eta} = (\eta_2, \ldots, \eta_n)$ (note that a Lipschitz-continuous function is continuous). The situation is illustrated in Figure 7.5 for $n = 2$.

Unless otherwise stated we assume that Γ is Lipschitzian; this includes, in R^2, boundaries that are triangular, rectangular and annular, while, in R^3, tetrahedra and cubes are Lipschitzian. Boundaries which are not

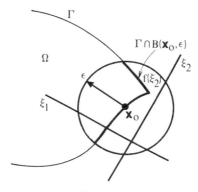

Figure 7.5

Lipschitzian include those with cusps and those which have the domain Ω on both sides (Figure 7.6).

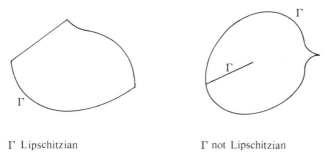

Γ Lipschitzian Γ not Lipschitzian

Figure 7.6

Sobolev spaces.   We define the Sobolev space of order m, denoted by $H^m(\Omega)$, to be the space consisting of those functions in $L_2(\Omega)$ which, together with all their *weak* partial derivatives up to and including those of order m, belong to $L_2(\Omega)$:

$$H^m(\Omega) = \{u : D^\alpha u \in L_2(\Omega) \text{ for all } \alpha \text{ such that } |\alpha| \leq m\}. \tag{1}$$

We make $H^m(\Omega)$ an *inner product space* by introducing the Sobolev inner product $(.\,,.)_{H^m}$ defined by

$$(u, v)_{H^m} = \int_\Omega \sum_{|\alpha| \leq m} D^\alpha u \, D^\alpha v \, dx \quad \text{for} \quad u, v \in H^m(\Omega) \tag{2}$$

This inner product in turn generates the *Sobolev norm* $\|\cdot\|_{H^m}$ defined by

$$\|u\|_{H^m}^2 = (u, u)_{H^m} = \int_\Omega \sum_{|\alpha| \leq m} (D^\alpha u)^2 \, dx. \tag{3}$$

Note that $H^0(\Omega) = L_2(\Omega)$, and furthermore that we may write $(u, v)_{H^m}$ as

$$(u, v)_{H^m} = \sum_{|\alpha| \leq m} (D^\alpha u, D^\alpha v)_{L_2}; \tag{4}$$

in words, the Sobolev inner product $(u, v)_{H^m}$ is equal to the sum of the L_2 inner products of $D^\alpha u$ and $D^\alpha v$ over all α such that $|\alpha| \leq m$. Of course, we could also write

$$\|u\|_{H^m}^2 = \sum_{|\alpha| \leq m} \|D^\alpha u\|_{L_2}^2. \tag{5}$$

Example

(a)   Consider the function $u(x)$ defined on $\Omega = (0, 2)$ by

$$u(x) = \begin{cases} x^2, & 0 < x \leq 1, \\ 2x^2 - 2x + 1, & 1 < x < 2. \end{cases}$$

We have

$$v(x) \equiv u'(x) = \begin{cases} 2x, & 0 < x \leq 1, \\ 4x - 2, & 1 < x < 2 \end{cases}$$

which is a continuous function. The (weak) derivative of this function is

$$w(x) \equiv u''(x) = \begin{cases} 2, & 0 < x \leq 1, \\ 4, & 1 < x < 2. \end{cases}$$

By inspection we see that u, u' and u'' all belong to $L_2(0, 2)$; however, the (generalized) derivative of u'' is $u''' = 2\delta \notin L_2(0, 2)$. Hence u is a member of $H^2(0, 2)$, the function v belongs to $H^1(0, 2)$ and w belongs to $L_2(0, 2) = H^0(0, 2)$ (see Figure 7.7).
The respective norms of these functions are

$$\|u\|_{H^2}^2 = \int_0^2 [u^2 + (u')^2 + (u'')^2] \, dx = 71.37,$$

$$\|v\|_{H^1}^2 = \int_0^2 [v^2 + (v')^2] \, dx = 39,$$

$$\|w\|_{L_2}^2 = \int_0^2 w^2 \, dx = 20.$$

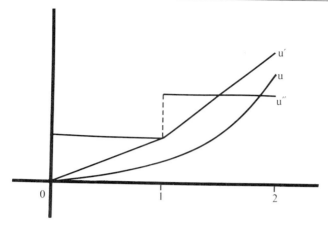

Figure 7.7

(b)   The function u defined on $\Omega = (-1, 1) \times (-1, 1)$ by

$$u(\mathbf{x}) = \begin{cases} y & \text{for} \quad y > 0, \\ 0 & \text{for} \quad y \leq 0, \end{cases}$$

belongs to $H^1(\Omega)$ (see Figure 7.8). To see this, we start by evaluating its

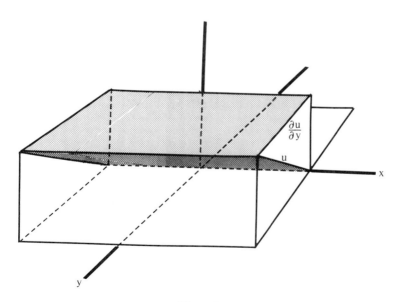

Figure 7.8

first derivatives: for $\phi \in C_0^\infty(\Omega)$,

$$\int_\Omega \frac{\partial u}{\partial x} \phi \, dx \, dy = -\int_\Omega u \frac{\partial \phi}{\partial x} \, dx \, dy = -\int_0^1 y \left[\int_{-1}^1 \frac{\partial \phi}{\partial x} \, dx \right] dy$$

$$= -\int_0^1 y[\phi(1, y) - \phi(-1, y)] \, dy = 0.$$

Hence $\partial u / \partial x = 0$. Secondly,

$$\int_\Omega \frac{\partial u}{\partial y} \phi \, dx \, dy = -\int_\Omega u \frac{\partial \phi}{\partial y} \, dx \, dy = -\int_{-1}^1 \left(\int_0^1 y \frac{\partial \phi}{\partial y} \, dy \right) dx$$

$$= -\int_1^1 \left([y\phi]_0^1 - \int_0^1 \phi \, dy \right) dx$$

$$= \int_{-1}^1 \int_0^1 \phi \, dy \, dx = \int_\Omega H(y)\phi(x, y) \, dx \, dy.$$

Hence $\partial u / \partial y = H(y)$. We can show next that $\partial^2 u / \partial y^2 = \delta_y$, the two-dimensional Dirac delta defined by $\delta_y(\phi) = \phi(x, 0)$, so that $\partial^2 u / \partial y^2 \notin L_2(\Omega)$. Hence $u \in H^1(\Omega)$. □

The picture that emerges is that the spaces $H^m(\Omega)$ provide a very logical means for characterizing the degree of smoothness of a function. When dealing with the spaces $C^m(\bar{\Omega})$ we understand "degree of smoothness" to mean "how many times can the function be differentiated?" In the case of Sobolev spaces "degree of smoothness" is understood to mean "how many times can the function be differentiated (weakly) before it ceases to belong to $L_2(\Omega)$?" The following theorem summarizes the most important properties of the space $H^m(\Omega)$.

Theorem 1. *Let $H^m(\Omega)$ be the Sobolev space of order m. Then*
(i) $H^r(\Omega) \subseteq H^m(\Omega)$ *if $r \geq m$;*
(ii) $H^m(\Omega)$ *is a Hilbert space with respect to the norm $\|\cdot\|_{H^m}$;*
(iii) $H^m(\Omega)$ *is the completion or closure, with respect to the norm $\|\cdot\|_{H^m}$, of the space $\hat{C}^m(\Omega)$ of m-times continuously differentiable functions which have a finite norm $\|\cdot\|_{H^m}$;*
(iv) $H^m(\Omega)$ *is also the completion or closure, with respect to the norm $\|\cdot\|_{H^m}$, of the space $\hat{C}^\infty(\Omega)$ of infinitely differentiable functions which have a finite norm $\|\cdot\|_{H^m}$.*

Proof. We prove only (i) and (ii); the proofs of (iii) and (iv) are rather long and technical, and are omitted.
(i) If $u \in H^r(\Omega)$ then $D^\alpha u$ belongs to $L_2(\Omega)$ for all α such that $|\alpha| \leq r$.

But $m \leqslant r$, hence $D^\alpha u$ also belongs to $L_2(\Omega)$ for all α such that $|\alpha| \leqslant m$. Thus $u \in H^m(\Omega)$ and so $H^r(\Omega) \subseteq H^m(\Omega)$.

(ii) We know that $H^m(\Omega)$ is an inner product space, so what remains to be shown is that $H^m(\Omega)$ is complete. Let $\{u_k\}$ be a Cauchy sequence in $H^m(\Omega)$. We have to show that u_k converges to a function u in $H^m(\Omega)$. First, by definition

$$\lim_{k,l \to \infty} \|u_k - u_l\|_{H^m} = 0$$

or, using the definition of the H^m norm,

$$\lim_{k,l \to \infty} \sum_{|\alpha| \leqslant m} \|D^\alpha u_k - D^\alpha u_l\|_{L_2}^2 = 0.$$

Since each term in this sum is positive, it follows that we must have

$$\lim_{k,l \to \infty} \|D^\alpha u_k - D^\alpha u_l\|_{L_2} = 0 \qquad \text{for all } \alpha, \ |\alpha| \leqslant m.$$

Hence $\{D^\alpha u_k\}$ is a Cauchy sequence in $L_2(\Omega)$ for each α such that $|\alpha| \leqslant m$. Since L_2 is complete, it follows that $D^\alpha u_k$ converges to a function $u^{(\alpha)}$, say, which belongs to L_2. In particular, for $|\alpha| = 0$, u_k converges to a function u, say, in L_2.

We show next that u is in $H^m(\Omega)$. Consider

$$\int_\Omega u^{(\alpha)} \phi \, dx = \int_\Omega \left(\lim_{k \to \infty} D^\alpha u_k \right) \phi \, dx = \left(\lim_{k \to \infty} D^\alpha u_k, \phi \right)_{L_2}$$

$$= \lim_{k \to \infty} (D^\alpha u_k, \phi)_{L_2} = \lim_{k \to \infty} (-1)^{|\alpha|} (u_k, D^\alpha \phi)_{L_2}$$

$$= (-1)^{|\alpha|} \left(\lim_{k \to \infty} u_k, \phi \right)_{L_2} = (-1)^{|\alpha|} \int_\Omega u D^\alpha \phi \, dx$$

where we have used the result in Exercise 10.5. Thus $u^{(\alpha)}$ is the αth weak derivative of u and since u, as well as all of its weak derivatives of order $\leqslant m$, is in $L_2(\Omega)$, u belongs to $H^m(\Omega)$. Hence $H^m(\Omega)$ is complete. ∎

Parts (iii) and (iv) of the theorem have an important interpretation: from the definition of the completion of a space (Section 13) we know that $\hat{C}^m(\Omega)$ and $\hat{C}^\infty(\Omega)$ are *dense* in $H^m(\Omega)$; hence, for any $u \in H^m(\Omega)$ it is always possible to find m-times continuously differentiable and infinitely differentiable functions $f(x)$ and $g(x)$, say, which are arbitrarily close to u in the sense that

$$\|u - f\|_{H^m} < \varepsilon \quad \text{and} \quad \|u - g\|_{H^m} < \varepsilon$$

for any given $\varepsilon > 0$. In other words, *every member of $H^m(\Omega)$ is either a* *member of $C^m(\Omega)$ or $C^\infty(\Omega)$, or it may be approximated arbitrarily closely by functions from either of these two spaces.*

Example

Refer to the previous example (a): from what was said there we conclude that, given any $\varepsilon > 0$, it is possible to find functions f, g, h, in $C^\infty(0, 1)$ that satisfy

$$\|u - f\|_{H^2} = \left[\int_0^1 (u - f)^2 + (u' - f')^2 + (u'' - f'')^2 \, dx \right]^{1/2} < \varepsilon, \tag{6}$$

$$\|v - g\|_{H^1} = \left[\int_0^1 (v - g)^2 + (v' - g')^2 \, dx \right]^{1/2} < \varepsilon, \tag{7}$$

$$\|w - k\|_{L_2} = \left[\int_0^1 (w - h)^2 \, dx \right]^{1/2} < \varepsilon. \tag{8}$$

Furthermore, it is also possible to find functions f, g and h in $C^3(0, 1)$, $C^2(0, 1)$ and $C^1(0, 1)$, respectively, such that (6)–(8) hold.

When $m = 0$ we can deduce properties of $H^0(\Omega) = L_2(\Omega)$ from Theorem 1. These are summarized in the following corollary.

Corollary to Theorem 1. (a) $L_2(\Omega)$ *is the completion of $C(\Omega)$ in the L_2-norm*; (b) $L_2(\Omega)$ *is also the completion, with respect to the L_2-norm, of the space $C^\infty(\Omega)$.*

The Sobolev embedding theorem. A glance at the examples discussed earlier in this section may lead one to wonder whether it is true that members of $H^m(\Omega)$ are simply functions which, together with their derivatives of order $\leq m - 1$, are continuous. After all, it is not easy, for example, to conceive of a function in $H^1(\Omega)$ that is not continuous. A famous theorem due to Sobolev asserts that, as we would expect, all members of $H^1(a, b)$ are indeed continuous functions, *but that this does not hold for higher-dimensional domains.* Before stating the theorem we give a simple example: let Ω be the disc of radius $1/2$ in R^2, and let $u = \ln(\ln(1/r))$ where $r^2 = x^2 + y^2$. Then, using polar coordinates (r, θ),

$$\iint_\Omega u^2 \, dx \, dy = \int_0^{1/2} \int_0^{2\pi} [\ln(\ln(1/r))]^2 r \, dr \, d\theta = \int_0^{2\pi} \int_{\ln 2}^\infty (e^{-t} \ln t)^2 \, dt \, d\theta$$

$(t = -\ln r)$ which is easily shown to be bounded, while

$$\iint_\Omega [(\partial u/\partial x)^2 + (\partial u/\partial y)^2] r \, dr \, d\theta = \iint_\Omega (\ln r)^{-2} \, d(\ln r) \, d\theta = 2\pi/\ln 2.$$

Hence $\|u\|_{H^1}$ is finite and so u belongs to $H^1(\Omega)$. But u is *not* continuous at the origin. We now state the *Sobolev embedding theorem*, the proof of which may be found in the references at the end of this chapter.

Theorem 2. Let Ω be a bounded domain in R^n with a Lipschitz boundary Γ. If $m > n/2$, then every function v in $H^m(\Omega)$ is continuous. Furthermore,

$$\sup |v(\mathbf{x})| \leqslant C \|v\|_{H^m}. \tag{9}$$

According to the theorem, if $n = 1$ so that Ω is a subset of the real line, then functions in $H^1(\Omega)$ are continuous. For domains that are subsets of the plane, though, $n = 2$ and we require that a function be a member of $H^2(\Omega)$ in order to guarantee its continuity. Furthermore, (9) states that the identity operator from $H^m(\Omega)$ with the H^m-norm to $C(\Omega)$ with the sup-norm is a *bounded operator*.

The Poincaré inequality.   We conclude this section with an important inequality which is frequently very useful.

Theorem 3 (The Poincaré inequality).   Let Ω be a domain in R^n with a Lipschitz boundary. Then for any $u \in H^k(\Omega)$ there are constants c_1 and c_2 such that

$$\|u\|_{H^k}^2 \leqslant c_1 \sum_{|\alpha|=k} \int_\Omega (D^\alpha u)^2 \, dx + c_2 \sum_{|\alpha|<k} \left(\int_\Omega D^\alpha u \, dx \right)^2. \tag{10}$$

Proof.   We will prove (10) for the case $\Omega = (a, b) \subset R$; the higher-dimensional results follow in a similar way. Thus for $n = 1$ we have to derive the inequality

$$\|u\|_{H^k}^2 \leqslant c_1 \int_a^b \left(\frac{d^k u}{dx^k} \right)^2 dx + c_2 \sum_{j<k} \left(\int_a^b \frac{d^j u}{dx^j} dx \right)^2. \tag{11}$$

Let ξ and η be two points in (a, b) with $\eta < \xi$; then we have

$$u(\xi) - u(\eta) = \int_\eta^\xi u' \, dx$$

and so, using the Schwarz inequality,

$$[u(\xi) - u(\eta)]^2 = \left[\int_\eta^\xi u' \, dx \right]^2 \leqslant \left[\int_\eta^\xi 1^2 \, dx \right] \left[\int_\eta^\xi (u')^2 \, dx \right]$$

$$\leqslant (b - a) \int_a^b (u')^2 \, dx.$$

Now we integrate with respect to ξ, keeping η fixed, and then with respect to η; we get

$$(b-a)\left[\int_a^b u^2(\xi)\,d\xi + \int_a^b u^2(\eta)\,d\eta\right] - 2\int_a^b u(\xi)\,d\xi\int_a^b u(\eta)\,d\eta$$

$$\leq (b-a)^3\int_a^b (u')^2\,dx$$

or

$$\int_a^b u^2\,dx \leq C_1\int_a^b (u')^2\,dx + C_2\left(\int_a^b u\,dx\right)^2 \tag{12}$$

where $C_1 = \frac{1}{2}(b-a)^2$ and $C_2 = 1/(b-a)$. Since $u \in H^k(a, b)$, (12) is still valid if we replace u by u', or by u'', and so on, up to $d^{k-1}u/dx^{k-1}$. That is, we also have

$$\int_a^b (u')^2\,dx \leq C_1\int_a^b (u'')^2\,dx + C_2\left(\int_a^b u'\,dx\right)^2, \tag{13}$$

$$\vdots$$

$$\int_a^b \left(\frac{d^{k-1}u}{dx^{k-1}}\right)^2\,dx \leq C_1\int_a^b \left(\frac{d^k u}{dx^k}\right)^2\,dx + C_2\left(\int_a^b \frac{d^{k-1}u}{dx^{k-1}}\,dx\right)^2.$$

To obtain (11) for $k = 1$ we add $\int_a^b (u')^2\,dx$ to both sides of (12) to get

$$\|u\|_{H^1}^2 \leq (1 + C_1)\int_a^b (u')^2\,dx + C_2\left(\int_a^b u\,dx\right)^2. \tag{14}$$

Next, to get (11) for $k = 2$ we add $\int_a^b (u'')^2\,dx$ to both sides of (14) and use (13), to obtain

$$\|u\|_{H^2}^2 \leq (1 + C_1(1 + C_1))\int_a^b (u'')^2\,dx + C_2\left(\int_a^b u\,dx\right)^2$$

$$+ C_2(1 + C_1)\left(\int_a^b u'\,dx\right)^2.$$

Continuing in this manner we can derive (11) for any value of k.

§27.   Boundary values of functions in $H^m(\Omega)$

Traces of functions in $H^m(\Omega)$. Later on, when dealing with boundary-value problems, we will be concerned not only with values of functions on an open domain Ω, but also with their values on the

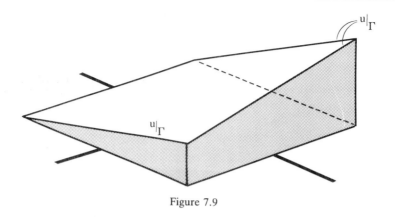

Figure 7.9

boundary Γ. Now, in the case of continuous functions defined on $\bar{\Omega} = \Omega \cup \Gamma$, one simply finds the values of a function u on the boundary by evaluating u on Γ: we then write this as $u\mid_\Gamma$. For example, if $\bar{\Omega} = [0, 1]$ and $u = x + 2$, then $u\mid_\Gamma$ consists of u evaluated at $x = 0$ and $x = 1$: $u\mid_\Gamma = \{u(0), u(1)\} = \{2, 3\}$. Similarly, if Ω is the square $\{(x, y) : |x| < 1, |y| < 1\}$ with boundary $\Gamma = \{(x, y) : |x| = 1, |y| \leqslant 1\} \cup \{(x, y) : |y| = 1, |x| \leqslant 1\}$, and we set $u \in C(\bar{\Omega})$, $u(x, y) = (1 + x)(3 + y)$, then the restriction of u to the boundary is the function $u\mid_\Gamma$ shown in Figure 7.9. This procedure may be formalized by introducing an operator γ, called the *trace operator*: γ is a *linear* operator which acts on a continuous function $u \in C(\bar{\Omega})$ to produce its restriction to the boundary Γ, that is,

$$\gamma : C(\bar{\Omega}) \to C(\Gamma), \qquad \gamma(u) = u\mid_\Gamma. \tag{1}$$

Note that, since u is continuous, we expect its restriction to Γ to be a continuous function on Γ, so that $u\mid_\Gamma \in C(\Gamma)$. Thus if $\Omega \subset R^2$, then the graph of $u\mid_\Gamma$ can be drawn as a *continuous* curve above Γ (as in Figure 7.9). The case of $\Omega \subset R$ is of course a degenerate special case.

We are interested in the problem of how to define $u\mid_\Gamma$ when u belongs to $L_2(\Omega)$ or, more generally, to one of the Sobolev space $H^m(\Omega)$. Note that functions belonging to $H^m(\Omega)$ are only defined uniquely on Ω and not on $\bar{\Omega}$ since Γ is a set of measure zero and functions in $H^m(\Omega)$ differing on a set of measure zero are regarded as being identical. One way of defining $u\mid_\Gamma$ (or $\gamma(u)$) for a function u in $L_2(\Omega)$ would be to set up a Cauchy sequence $\{u_k\}$ of *continuous* functions on $\bar{\Omega}$ which converges to u in the L_2-norm (recall from the corollary to Theorem 26.1 that $C(\Omega)$ is dense in $L_2(\Omega)$). Since u_k is in $C(\bar{\Omega})$ we can unambiguously define, as in (1),

$$\gamma(u_k) = u_k\mid_\Gamma, \tag{2}$$

so we hope to be able to define $\gamma(u)$ by

$$\gamma(u) = \lim_{k\to\infty} \gamma(u_k) \quad \text{or} \quad \gamma\Big(\lim_{k\to\infty} u_k\Big) = \lim_{k\to\infty} \gamma(u_k). \tag{3}$$

However, a glance back to Theorem 16.4 shows that we want γ to be a *continuous* (or bounded) linear operator from $C(\bar{\Omega})$ to $C(\Gamma)$, and we hope to extend γ to a bounded operator from $L_2(\Omega)$ to $L_2(\Gamma)$. But this is not possible: it can be shown that there is no continuous mapping of the kind we are looking for. This means that it is impossible to define *unambiguously* $u\,|_\Gamma$ when u is in $L_2(\Omega)$, and this is further illustrated by the following example.

Example

Let $u(x) = 1$ on $\Omega = (0, 1)$; u is actually continuous and it would seem logical to define $u\,|_\Gamma$ by $u(0) = u(1) = 1$. But if we set up the Cauchy sequence $\{u_k\}$ defined by (Figure 7.10)

$$u_k(x) = \begin{cases} (-1)^{k+1}(1-kx), & 0 \leqslant x < 1/k, \\ 0, & 1/k \leqslant x < 1 - 1/k, \\ (-1)^{k+1}(1 + k(x-1)), & 1 - 1/k \leqslant x \leqslant 1, \end{cases}$$

we find that $\gamma(u_k) = u_k\,|_\Gamma = \{u_k(0), u_k(1)\} = \{(-1)^{k+1}, (-1)^{k+1}\}$; thus $u_k\,|_\Gamma$ oscillates between $+1$ and -1 and $\lim_{k\to\infty} (u_k) \neq 1$.   □

As the above example shows, even though an apparently obvious value for $u\,|_\Gamma$ may be assigned $(u(0) = u(1) = 1)$, there is still ambiguity which results from the fact that γ is not a continuous operator. This has far-reaching consequences, as can well be appreciated by the following considerations: if $\gamma : L_2(\Omega) \to L_2(\Gamma)$ were a continuous operator then we

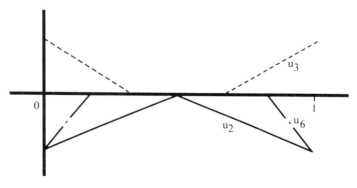

Figure 7.10

would have

$$\|\gamma(u)\|_{L_2(\Gamma)} \leqslant C \|u\|_{L_2(\Omega)} \tag{4}$$

for some constant $C > 0$. Thus if u and v are two functions in $L_2(\Omega)$ (they could be continuous functions) which are close in the sense that $\|u - v\|_{L_2(\Omega)} < \varepsilon$ for some small $\varepsilon > 0$, then (4) gives immediately

$$\|\gamma(u) - \gamma(v)\|_{L_2(\Gamma)} < C\varepsilon \tag{5}$$

so that $u|_\Gamma$ and $v|_\Gamma$ are correspondingly close. However, if γ is not continuous there is no guarantee that this situation would obtain. This is obviously untenable if we are to develop a coherent theory of boundary-value problems.

All is not lost, however; if a function u belongs to $C^1(\bar{\Omega})$ then it can be shown that the operator γ mapping u to its value on Γ is a *continuous* operator from $C^1(\bar{\Omega})$ to $C(\Gamma)$, with respect to the norms $\|\cdot\|_{H^1(\Omega)}$ and $\|\cdot\|_{L_2(\Gamma)}$. That is,

$$\gamma : C^1(\bar{\Omega}) \to C(\Gamma), \qquad \gamma(u) = u|_\Gamma$$

satisfies

$$\|\gamma(u)\|_{L_2(\Gamma)} \leqslant C \|u\|_{H^1(\Omega)} \tag{6}$$

for some constant $C > 0$ (note the norms used). The proof of this result is lengthy and will be omitted, but we concentrate on the consequences of (6), embodied in the following theorem.

Theorem 1 (The trace theorem).   *Let Ω be a bounded subset of R^n with a Lipschitz boundary Γ. (i) There exists a unique bounded linear operator γ which maps $H^1(\Omega)$ into $L_2(\Gamma)$:*

$$\gamma : H^1(\Omega) \to L_2(\Gamma), \qquad \|\gamma(u)\|_{L_2(\Gamma)} \leqslant C \|u\|_{H^1(\Omega)}. \tag{7}$$

If $u \in C^1(\bar{\Omega})$ then $\gamma(u) = u|_\Gamma$ in the conventional sense.
(ii) The range of γ is dense in $L_2(\Gamma)$.

Proof.   The proof of (i) follows immediately from (6) and the fact that $H^1(\Omega)$ and $L_2(\Gamma)$ are the completions of $C^1(\Omega)$ and $C(\Gamma)$, respectively, in the appropriate norms. Indeed, for any $u \in H^1(\Omega)$ we can set up a Cauchy sequence $\{u_k\}$ in $C^1(\bar{\Omega})$ which converges to u in the H^1-norm. Thus

$$\lim_{k \to \infty} u_k = u \quad \text{or} \quad \lim_{k \to \infty} \|u_k - u\|_{H^1} = 0,$$

and so from (6) we find, for $u \in H^1(\Omega)$, that

$$\|\gamma(u)\|_{L_2(\Gamma)} = \left\|\gamma\left(\lim_{k \to \infty} u_k\right)\right\| = \lim_{k \to \infty} \|\gamma(u_k)\|$$

$$\leq C \lim_{k \to \infty} \|u_k\|_{H^1(\Omega)} = C \|u\|_{H^1(\Omega)},$$

using Exercise 10.5. We remark that part (ii) of Theorem 1 implies that, while the range of γ is not all of $L_2(\Gamma)$, any member of $L_2(\Gamma)$ can be approximated arbitrarily closely by a function lying in the range of γ. ∎

The trace theorem enables us to define unambiguously $\gamma(u)$ or $u\mid_\Gamma$, provided that u is smooth enough to be in $H^1(\Omega)$. Now suppose that u is even smoother, so that u belongs to $H^2(\Omega)$. Then u is a member of $H^1(\Omega)$ and so in fact is $D^\alpha u$ for $|\alpha| = 1$:

$$u, \partial u/\partial x_1, \ldots, \partial u/\partial x_n \in H^1(\Omega).$$

This means that we can also describe the boundary values of the first derivatives of u, using the trace theorem.

The argument can be generalized to the space $H^m(\Omega)$; indeed, when $m > 1$ then for any $u \in H^m(\Omega)$ we have $D^\alpha u \in H^1(\Omega)$, for $|\alpha| \leq m - 1$. By the trace theorem the value of $D^\alpha u$ on the boundary is well-defined and belongs to $L_2(\Gamma)$:

$$\gamma(D^\alpha u) \in L_2(\Gamma), \qquad |\alpha| \leq m - 1. \tag{8}$$

Furthermore, if u is in fact m-times continuously differentiable then $D^\alpha u$ is at least continuously differentiable for $|\alpha| \leq m - 1$ and

$$\gamma(D^\alpha u) = (D^\alpha u)\mid_\Gamma. \tag{9}$$

We introduce the notation γ_α to denote the operator which, when applied to a member u of $H^m(\Omega)$, gives the trace or boundary value of $D^\alpha u$, for $|\alpha| \leq m - 1$:

$$\gamma_\alpha : H^m(\Omega) \to L_2(\Gamma), \qquad |\alpha| \leq m - 1,$$

$$\gamma_\alpha(u) = \gamma(D^\alpha u). \tag{10}$$

If $u \in C^m(\bar\Omega)$, then $\gamma_\alpha(u) = \gamma(D^\alpha u) = D^\alpha u\mid_\Gamma$. Clearly γ_α is a bounded operator.

A word about notation is in order at this point. Henceforth we shall deal with boundary values of a function only if these boundary values can be defined unambiguously, in the sense of Theorem 1 (or its extension to (10)); when referring to the value of a function u or that of its derivatives on the boundary we shall simply write $u, \partial u/\partial x, \ldots$, instead of $\gamma(u)$, $\gamma_{(1,0,\ldots)}(u), \ldots$, *it being understood that the boundary values are to be interpreted in the sense of the trace theorem.* So if we see, for example,

$$u = u_0 \quad \text{on} \quad \Gamma, \tag{11}$$

this means that $\gamma(u)$ takes on the value u_0 on Γ. Sometimes, in order to make this clearer, we may write "$u = u_0$ in the sense of traces".

Now that we know how to deal with boundary values of functions in $H^m(\Omega)$, it is fairly straightforward to extend Green's theorem, equation (25.1), to functions in $H^1(\Omega)$ (see Exercise 27.1): given functions u, $v \in H^1(\Omega)$, the identity

$$\int_\Omega u \frac{\partial v}{\partial x_i} \, dx = \int_\Gamma uvv_i \, ds - \int_\Omega \frac{\partial u}{\partial x_i} v \, dx \tag{12}$$

holds for $i = 1, 2, \ldots, n$. From this identity we can deduce higher-order identities; for example, if we replace u by $\partial u / \partial x_i$ (assuming now that $u \in H^2(\Omega)$) and sum over i from 1 to n then we find that

$$\int_\Omega \nabla u \cdot \nabla v \, dx = \int_\Gamma \frac{\partial u}{\partial v} v \, dx - \int_\Omega (\nabla^2 u) v \, dx \tag{13}$$

for $u \in H^2(\Omega)$, $v \in H^1(\Omega)$.

§28.   The spaces $H_0^m(\Omega)$ and $H^{-m}(\Omega)$

The space $H_0^m(\Omega)$ is a subspace of $H^m(\Omega)$ that arises frequently in boundary-value problems because members of $H_0^m(\Omega)$ are distinguished by the fact that certain of their derivatives vanish on the boundary. We define $H_0^m(\Omega)$ to be the completion, in the Sobolev norm $\|\cdot\|_{H^m}$, of the space $\hat{C}_0^m(\Omega)$ of functions with continuous derivatives of order $\leq m$, all of which have compact support on Ω and whose norms $\|u\|_{H^m}$ are finite. In other words, $H_0^m(\Omega)$ is formed by taking the union of $\hat{C}_0^m(\Omega)$ and all those limits of Cauchy sequences in $\hat{C}_0^m(\Omega)$ which are not in $\hat{C}_0^m(\Omega)$.

Since $D^\alpha u_k = 0$ on Γ ($|\alpha| \leq m$) for each member of a Cauchy sequence $\{u_k\}$ in $\hat{C}_0^m(\Omega)$, it suggests that the limit of such a Cauchy sequence, which of course belongs to $H_0^m(\Omega)$, also satisfies $D^\alpha u = 0$ on the boundary. This is borne out by the following theorem, which also gives other properties of $H_0^m(\Omega)$.

Theorem 1.   Let Ω be an open bounded domain in R^n with a sufficiently smooth boundary Γ and let $H_0^m(\Omega)$ be the completion of $\hat{C}_0^m(\Omega)$ in the norm $\|\cdot\|_{H^m}$. Then
(a) $H_0^m(\Omega)$ *is also the completion of $\hat{C}_0^\infty(\Omega)$ in the norm $\|\cdot\|_{H^m}$;*
(b) $H_0^m(\Omega) \subset H^m(\Omega)$;

(c)   *a function $u \in H^m(\Omega)$ belongs to $H_0^m(\Omega)$ if and only if*

$$D^\alpha u = 0 \quad \text{on} \quad \Gamma, \qquad |\alpha| \leqslant m - 1.$$

Proof.   The proof of (a) is similar to that of Theorem 26.1 (iv). Part (b) is obvious. To prove (c), we use the *continuity* of the trace operator: let $\{u_k\}$ be a Cauchy sequence in $\hat{C}_0^m(\Omega)$ with limit u in $H_0^m(\Omega)$. Then from the definition of γ_α we have

$$\gamma_\alpha(u_k) = 0 \qquad \text{for} \qquad |\alpha| \leqslant m - 1.$$

Hence

$$\lim_{k \to \infty} \gamma_\alpha(u_k) = \gamma_\alpha \left(\lim_{k \to \infty} u_k \right) = \gamma_\alpha(u) = D^\alpha u = 0. \qquad \blacksquare$$

Part (c) of Theorem 1 is particularly useful in characterizing members of $H_0^m(\Omega)$, as the following example shows.

Example

The function u defined by

$$u(x) = \begin{cases} x^2, & 0 \leqslant x \leqslant \tfrac{1}{2} \\ -x^2 + 2x - \tfrac{1}{2}, & \tfrac{1}{2} \leqslant x \leqslant \tfrac{3}{2}, \\ (2-x)^2, & \tfrac{3}{2} \leqslant x \leqslant 2 \end{cases}$$

is a member of $H^2(0, 2)$, as Figure 7.11 shows. Also, u and du/dx are equal to zero on the boundary $x = 0$, $x = 2$. Hence $u \in H_0^2(0, 2)$. $\square$

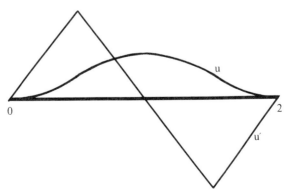

Figure 7.11

The space $H^{-m}(\Omega)$.   In Section 18 we discovered that the space $L_2(\Omega)$ is self-dual. The question now arises: how can we characterize

$H^m(\Omega)'$, the space of bounded linear functionals on $H^m(\Omega)$? There is a complication here: we would hope to find out about $H^m(\Omega)'$ by considering functionals f on $C_0^\infty(\Omega)$ (that is, distributions) and by looking at the limits of $f(\phi_k)$ as $k \to \infty$, where $\{\phi_k\}$ is a Cauchy sequence in $C_0^\infty(\Omega)$. But $C_0^\infty(\Omega)$ is *not* dense in $H^m(\Omega)$, so that not every $u \in H^m(\Omega)$ is the limit of a Cauchy sequence $\{\phi_k\}$ in $C_0^\infty(\Omega)$. We resolve this dilemma by restricting attention instead to the dual of $H_0^m(\Omega): C_0^\infty(\Omega)$ is dense in $H_0^m(\Omega)$, by Theorem 1(a), and we use this fact to define $H_0^m(\Omega)'$ in the following theorem. Before stating the theorem we introduce the convention whereby the dual of $H_0^m(\Omega)$ is denoted by $H^{-m}(\Omega)$:

$$H_0^m(\Omega)' \equiv H^{-m}(\Omega).$$

As we shall see, this notation makes complete sense.

Theorem 2.   *A distribution q is in the dual space $H^{-m}(\Omega)$ of $H_0^m(\Omega)$ if and only if it can be expressed in the form*

$$q = \sum_{|\alpha| \leq m} D^\alpha q_\alpha \tag{1}$$

where q_α are functions in $L_2(\Omega)$.

Proof.   Let f be any function in $L_2(\Omega)$: then, for any $\phi \in C_0^\infty(\Omega)$

$$|D^\alpha f(\phi)| = |f(D^\alpha \phi)|$$

$$= \left| \int_\Omega f(D^\alpha \phi)\, dx \right|$$

$$\leq \|f\|_{L_2} \|D^\alpha \phi\|_{L_2} \leq \|f\|_{L_2} \|\phi\|_{H^m} \tag{2}$$

using the Schwarz inequality. If $\{\phi_k\}$ is any Cauchy sequence in $C_0^\infty(\Omega)$ with limit u in $H_0^m(\Omega)$, then by replacing ϕ with ϕ_k in (2) and taking the limit as $k \to \infty$, we see that $D^\alpha f$ is a bounded linear functional on $H_0^m(\Omega)$ for $f \in L_2(\Omega)$ and $|\alpha| \leq m$. That is, $D^\alpha f$ belongs to $H_0^m(\Omega)' = H^{-m}(\Omega)$.

Conversely, if q belongs to $H^{-m}(\Omega)$, then by the Riesz representation theorem there is a $u \in H_0^m(\Omega)$ such that

$$q(\phi) = (u, \phi)_{H^m} \qquad \text{for all } \phi \in C_0^\infty(\Omega).$$

Now let $\{\phi_k\} \subset C_0^\infty(\Omega)$ with $\lim_{k \to \infty} \phi_k = v$ in $H_0^m(\Omega)$. Then

$$q(\phi_k) = (u, \phi_k)_{H^m}$$

$$= \int_\Omega \sum_{|\alpha| \leq m} (D^\alpha u)(D^\alpha \phi_k)$$

$$= \sum_{|\alpha| \leq m} (-1)^{|\alpha|} \int_\Omega D^\alpha(D^\alpha u)\phi_k.$$

Hence, as $k \to \infty$ we have

$$q(v) = \sum_{|\alpha| \le m} (-1)^{|\alpha|} D^\alpha (D^\alpha u)(v)$$

so that q is of the form

$$q = \sum_{|\alpha| \le m} (-1)^{|\alpha|} D^\alpha (D^\alpha u)$$

which gives the desired result since $D^\alpha u \in L_2(\Omega)$.   ∎

Example

Theorem 2 gives a useful way of characterizing the negative Sobolev spaces $H^{-m}(\Omega)$: indeed, (1) indicates that if we differentiate a member of $L_2(\Omega)$ up to m times, we get a functional q on $H_0^m(\Omega)$. For example, take

$$H(x) = \begin{cases} 0, & -1 < x < 0, \\ 1, & 0 < x < 1 \end{cases}$$

which belongs to $L_2(\Omega)$. We know that $H' = \delta$, the Dirac delta, so we conclude that

$$\delta \in H^{-1}(-1, 1).$$ □

This should give some idea of the nature of members of $H^m(\Omega)$; as m gets larger, we find progressively more irregular distributions in $H^{-m}(\Omega)$. We note here that

$$H^m(\Omega) \subset H^{m-1}(\Omega) \subset \cdots \subset H^0(\Omega) = L_2(\Omega) \subset H^{-1}(\Omega) \subset H^{-m}(\Omega)$$

(observe that $H^0(\Omega)' = H^0(\Omega)$, as we discovered in Section 18).

Bibliographical remarks

The definition of a distribution given in Section 24 is a simplified one which avoids a number of complicated topological considerations, but which is adequate for our purposes. The classical reference to distributions is Schwartz [42], who was responsible for developing much of the theory. The book [43] by Schwartz on mathematical methods also contains an account of distributions. Other references worth consulting are Oden and Reddy [34], Roman [41] and Showalter [44].

Generally, Sobolev spaces are defined to be subspaces of $L_p(\Omega)$, and are denoted by $W^{m,p}(\Omega)$, where

$$W^{m,p}(\Omega) = \{u \in L_p(\Omega) : D^\alpha u \in L_p(\Omega), \quad |\alpha| \le m\}.$$

We have restricted attention to the Hilbert space $W^{m,2}(\Omega) \equiv H^m(\Omega)$. A comprehensive treatment of Sobolev spaces may be found in the books by Adams [1], Nečas [30] and Lions and Magenes [27]. Oden and Reddy [34] give a good self-contained account of the theory, and Showalter [44] also gives a readable, though compact, account of the Hilbert spaces $H^m(\Omega)$. An alternative treatment is given by Rektorys [39], who defines $H^m(\Omega)$ to be the completion of $C^m(\Omega)$ in the Sobolev norm.

Exercises

24.1 If α is a multi-index in Z_+^n, define $\alpha!$ and $\mathbf{x}^\alpha$ by

$$\alpha! = \alpha_1! \, \alpha_2! \cdots \alpha_n!, \qquad \mathbf{x}^\alpha = x_1^{\alpha_1} x_2^{\alpha_2} \cdots x_n^{\alpha_n} \quad \text{for} \quad \mathbf{x} \in R^n.$$

Verify that the conventional Taylor expansion of a function f about the origin takes the form

$$f(\mathbf{x}) = \sum_{|\alpha|=0}^{\infty} \frac{\mathbf{x}^\alpha}{\alpha!} D^\alpha f(\mathbf{0}). \tag{1}$$

[Expand (1) for the case $n = 2$; work out the first few terms on the right-hand side of (1).]

24.2 Show that the Dirac delta δ is not generated by a locally integrable function $\delta(x)$ as follows: let $\phi_a(x)$ be the test function defined by

$$\phi_a(x) = \begin{cases} \exp\left[\dfrac{a^2}{x^2 - a^2}\right], & |x| < a, \\ 0, & b > |x| \geq a. \end{cases} \qquad \text{(here } b > a > 0\text{)}$$

Assume that a function $\delta(x)$ exists, and show that

$$\left| \int_{-b}^{b} \delta(x)\phi(x)\,dx \right| \leq \frac{1}{e} \int_{-a}^{a} \delta(x)\,dx.$$

Consider the limit as $a \to 0$ and obtain a contradiction.

24.3 Prove that the only continuous function f for which $f(\phi) = \int f(x)\phi(x)\,dx = 0$, is the zero function. [Use the result in Exercise 4.6.]

25.1 Use the standard form of Green's theorem to show that

$$\int_\Omega (D^\alpha u)v\,dx = (-1)^{|\alpha|} \int_\Omega u(D^\alpha v)\,dx + \int_\Gamma H(u,\,v)\,ds$$

for $|\alpha| = m$, where $u,\,v \in C^m(\bar{\Omega})$.

25.2 Show that $(\mathrm{sgn})' = 2\delta$ on $(-1, 1)$, where

$$\mathrm{sgn}\,(x) = \begin{cases} x/|x| & \text{for } x \neq 0, \\ 0 & \text{for } x = 0. \end{cases}$$

25.3 Show that $f'' = a\delta - a^2(\sin ax)H$ on $(-1, 1)$ where f is (the distribution generated by) the function $f(x) = \sin axH(x)$, and a is a constant.

25.4 Show that the function

$$f(x) = \begin{cases} x, & -1 < x \leq 0, \\ x + c, & 0 < x \leq 1, \end{cases}$$

has a generalized derivative given by $f' = df/dx + c\delta = 1 + c\delta$.

25.5 Let f be defined on $\Omega = (-1, 1) \times (-1, 1) \subset R^2$ by

$$f(\mathbf{x}) = \begin{cases} xy & \text{if } xy \geq 0 \\ 0 & \text{if } xy \leq 0 \end{cases}.$$

Show that $D^{(1,1)}f \equiv \partial^2 f/\partial x\,\partial y = g(\mathbf{x}) = \begin{cases} +1, & x \geq 0,\ y \geq 0, \\ -1, & x \leq 0,\ y \leq 0, \\ 0 & \text{otherwise.} \end{cases}$

25.6 Find the solution of the distributional differential equation

$$u' + u = e^{-x}\delta.$$

[Try u in the form $u = Hf$ where H is the step function and $f \in C^2(-1, 1)$.]

26.1 To which spaces $H^m(\Omega)$ do the following functions u belong?

(a) $u'(x) = \begin{cases} 0, & 0 < x < 1, \\ x - 1, & 1 \leq x < 2, \\ x^3 - x^2 - 3, & 2 \leq x < 3, \end{cases}$

(b) $u(x, y) = \begin{cases} xy, & 0 \leq x \leq 1,\ 0 \leq y < 1, \\ x(2 - y), & 0 \leq x \leq 1,\ 1 < y \leq 2, \\ x, & 0 \leq x < \frac{1}{2},\ y = 1, \\ 3\pi, & x = \frac{1}{2},\ y = 1, \\ x, & \frac{1}{2} < x \leq 1,\ y = 1. \end{cases}$

26.2 Show that the functions

$$u(x) = \begin{cases} x, & 0 \leqslant x \leqslant 1, \\ 2-x, & 1 \leqslant x \leqslant 2, \end{cases} \qquad v(x) = \sin \pi x,$$

are orthogonal in $L_2(0, 2)$. Investigate whether they are orthogonal in $H^1(0, 2)$. Find the distance between u and v in $L_2(0, 2)$ and in $H^1(0, 2)$.

26.3 For the functions u and $u - v$ in Exercise 26.2, verify the Schwarz inequality for $L_2(0, 2)$ and $H^1(0, 2)$.

26.4 Use the Sobolev embedding theorem to show that the function

$$u(\mathbf{x}) = \begin{cases} x^2 y^2 & x > 0, \, y > 0, \\ 0 & \text{otherwise}, \end{cases}$$

is continuous on $\Omega = (-1, 1) \times (-1, 1)$.

27.1 Starting with (25.1), derive Green's theorem (27.12) for functions in $H^1(\Omega)$. [Apply (25.1) to sequences u_n, v_n in $C^1(\bar{\Omega})$, and use the fact that $D^\alpha u_n \to D^\alpha u$ in L_2, together with the continuity of the inner product and of the trace operator.]

27.2 Derive the Green's formula

$$\int_\Omega \nabla^2 u \nabla^2 v \, dx = \int_\Omega (\nabla^4 u) v \, dx + \int_\Gamma \left[(\nabla^2 u) \frac{\partial v}{\partial \nu} - \frac{\partial}{\partial \nu} (\nabla^2 u) v \right] dx$$

for $u \in H^4(\Omega)$, $v \in H^2(\Omega)$.

28.1 Show that the semi-norm $|\cdot|_{H^m}$, defined by

$$|u|_{H^m} = \int_\Omega \sum_{|\alpha|=m} (D^\alpha u)^2 \, dx,$$

is a norm on $H_0^m(\Omega)$.

28.2 Use Green's theorem to show that

$$\|\nabla^2 v\|_{L_2} = |v|_{H^2} \qquad \text{for all } v \in H_0^2(\Omega).$$

28.3 Since $H^{-m}(\Omega)$ consists of continuous linear functionals on $H_0^m(\Omega)$, the norm on $H^{-m}(\Omega)$ is defined by (see Section 18)

$$\|f\|_{H^{-m}} = \sup \frac{|f(v)|}{\|v\|_{H^m}}, \qquad v \in H_0^m(\Omega).$$

Under what conditions is the Dirac delta a member of $H^{-m}(\Omega)$?

28.4 In the space $H^1(\Omega)$ show that the orthogonal complement of $H_0^1(\Omega)$ is the subspace of $u \in H^1(\Omega)$ for which $\nabla^2 u = u$ (distributionally). Find a basis for $H_0^1(\Omega)^\perp$ for the case $\Omega = (0, 1) \subset R$.

Part II
Boundary-value problems

8

Elliptic boundary-value problems

In this chapter we make our first acquaintance with boundary-value problems. Section 29 sets the stage by describing the various kinds of problems involving differential equations: here we can see how boundary-value problems fit into the general scheme of things.

In the remaining four sections we build up toward a general theory for the existence, uniqueness and regularity of solutions to elliptic boundary-value problems. We pose the problem as one of an elliptic operator from one Sobolev space to another. To the uninitiated, the ideas discussed here may seem rather esoteric: rather than discuss techniques for *solving* boundary-value problems, we obtain results of a *qualitative* nature. It will become evident, though, especially in our analysis of approximate methods of solution, that these qualitative aspects are indispensable to a proper understanding of the problem.

§29. Differential equations, boundary conditions and initial conditions

Differential equations. So far we have been dealing exclusively with mathematical ideas, with a view to their application to the theory of partial differential equations. At this stage we start our discussion of differential equations.

We deal with a function u, which could be a function of one or more independent variables $x_1, x_2, \ldots, x_n, t$. The variables $x_1, x_2, \ldots, x_n$, of which there are invariably three or less, usually refer to coordinates of a point in space. As before, we will write, for example, $\mathbf{x} = (x, y, z) = (x_1, x_2, x_3)$ for a point in R^3, the particular notation depending on circumstances. The variable t usually refers in a physical context to time.

A differential equation (DE) is any equation involving the independent variables $x_1, \ldots, x_n, t$, a function u of these variables, and some of the derivatives of u with respect to these variables. If there is only *one* independent variable then the DE is called an *ordinary differential equation* (ODE); on the other hand, if there are two or more independent variables the DE is a *partial differential equation* (PDE).

The *order* of a DE is defined to be the order of the highest derivative appearing in the equation.

Examples

(a) The equation

$$\frac{du}{dt} = [b(u) - d(u)]u \tag{1}$$

is a *first-order* ODE; physically, it describes the fact that a biological population $u(t)$ increases or decreases at a rate that depends on the current population u, and on the difference $b(u) - d(u)$ of the birth and death rates per capita. Here $b(u)$ and $d(u)$ are given functions that also depend on the current population.

(b) The equation

$$\frac{\partial u}{\partial t} - \frac{\partial}{\partial x}\left[k(x, y, u)\frac{\partial u}{\partial x}\right] - \frac{\partial}{\partial y}\left[k(x, y, u)\frac{\partial u}{\partial y}\right] = f(x, y) \tag{2}$$

is a *second-order* PDE in the function $u(x, y, t)$. This equation characterizes the temperature distribution in a material that has a thermal conductivity $k(x, y, u)$ depending on position and temperature. The function $f(x, y)$ represents a heat source. There is no dependence here on z. If the thermal conductivity is a function of position only, then (2) becomes

$$\frac{\partial u}{\partial t} - \frac{\partial}{\partial x}\left[k(x, y)\frac{\partial u}{\partial x}\right] - \frac{\partial}{\partial y}\left[k(x, y)\frac{\partial u}{\partial y}\right] = f(x, y). \tag{3}$$

Finally, if the temperature distribution does not depend on time—the so-called steady-state situation—and if, furthermore, the conductivity k is a constant, then (3) reduces to the *Poisson equation*

$$-\nabla^2 u = g(x, y) \tag{4}$$

where $g(x, y) = f(x, y)/k$ and ∇^2 is the (two-dimensional) *Laplacian* operator:

$$\nabla^2 = \frac{\partial^2}{\partial x^2} + \frac{\partial^2}{\partial y^2}. \qquad\qquad \square$$

A DE is *linear* if it can written as $Au = f$ where A is a *linear operator*.

Otherwise it is a *nonlinear* DE. For example, equation (1) is equivalent
to

$$Au = \frac{du}{dt} - [b(u) - d(u)]u = 0,$$

and this is a *nonlinear* equation. Equation (2) is also nonlinear, but
equations (3) and (4) are linear: it is easily verified in these last two cases
that the operators $A = \partial/\partial t - \partial/\partial x \, [k(x, y) \, \partial/\partial x] - \partial/\partial y \, [k(x, y) \, \partial/\partial y]$
and $A = -\nabla$ are both linear operators.

Physical considerations invariably dictate that a DE is required to be
satisfied only on a subset Ω of R^n and, if time is present as a variable,
over a prescribed length of time. It follows that a proper description of
the physical system must include, in addition to the DE, a statement
indicating the spatial and temporal range of interest. For example,
equation (1) needs to be supplemented by a statement to the effect that
we require $u(t)$ for t lying in the range $0 \leqslant t \leqslant T$ or $[0, T]$ where $t = 0$
represents some datum and T is the longest time of interest. If we take
$t = 0$ as the present and specify that a solution is required for all time in
the future, then the range of t is $(0, \infty)$. Similarly, in the heat conduction
problem corresponding to (2) or (3), we are required to specify the
spatial domain Ω as well as the temporal range $[0, T]$. For example, if the
problem has to do with heat conduction in a square slab occupying the
region $(0, 1) \times (0, 1)$, and if we require a solution for all time, then (2)
and (3) have to be supplemented by the statements

$$x \in \Omega = (0, 1) \times (0, 1), \qquad t \in (0, \infty).$$

Boundary conditions and initial conditions.   Once the domain of
interest has been specified, the next stage in the formulation of the
problem involves the specification of the unknown function and possibly
some of its derivatives on the boundary Γ (if the domain of interest is
spatial), and at the initial time $t = 0$ (if there is also a temporal
component of the domain). The former are known as *boundary conditions* (BCs) while the latter are called *initial conditions* (ICs); once again,
these are normally dictated by physical considerations.

If the domain of definition of a DE is purely spatial and denoted by Ω,
then only boundary conditions will be specified on the boundary of Ω,
and the DE together with the BCs is called a *boundary-value problem*
(BVP). A special kind of BVP is one defined on an interval $[a, b]$ of the
real line; then $\Omega = (a, b)$, $\Gamma = \{a, b\}$ and boundary conditions are given
at $x = a$ and $x = b$. This kind of BVP is called, for obvious reasons, a
two-point boundary-value problem.

When the domain is purely temporal, the problem consists of an ODE defined for $t \in (0, T)$—T may be infinity—and one or more initial conditions which specify the unknown function and possibly some of its derivatives at $t = 0$. This kind of problem is called an *initial-value problem* (IVP).

Finally, when the domain is both spatial and temporal, the problem will consist of a PDE together with boundary conditions and initial conditions. This problem is called an *initial-boundary-value problem* (IBVP).

We illustrate the above concepts with some examples.

Examples

(a) Consider equation (1); the complete problem is: find $u(t)$ satisfying

> ODE: $\mathrm{d}u/\mathrm{d}t = [b(u) - d(u)]u, \qquad t \in (0, \infty),$
>
> IC: $u(0) = u_0.$

The initial condition specifies that u must be equal to a prescribed value u_0 at time $t = 0$. This is an *initial-value problem*.

(b) Equation (4) is a PDE; suppose that we complete the specification of the problem by requiring that a function $u(x, y)$ be found which satisfies

> PDE: $-\nabla^2 u = g(x, y), \qquad \mathbf{x} \in \Omega = (0, 1) \times (0, 1),$
>
> BCs: $u(0, y) = u(1, y) = h(y), \qquad 0 \leq y \leq 1,$
>
> $\dfrac{\partial u}{\partial y}(x, 0) = \dfrac{\partial u}{\partial y}(x, 1) = 0, \qquad 0 \leq x \leq 1.$

This is an example of a *boundary-value problem*.

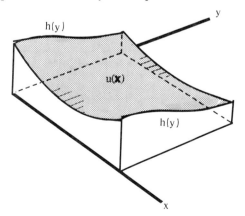

Figure 8.1

(c) The problem of finding a function $v(x)$ that satisfies

ODE: $\dfrac{d^2v}{dx^2} + a(x)\dfrac{dv}{dx} + b(x)v = c(x)$, $x \in (0, 1)$,

BCs: $v(0) = v(1) = 0$,

is an example of a *two-point boundary-value problem.*

(d) Consider the PDE, equation (2), defined on the unit square $\Omega = (0, 1) \times (0, 1)$ and on the time interval $(0, \infty)$, and suppose that this equation is supplemented by

BCs: $u(0, y, t) = u(1, y, t) = h(y, t)$, $0 \leqslant y \leqslant 1$, $0 \leqslant t < \infty$,

$\dfrac{\partial u}{\partial y}(x, 0, t) = \dfrac{\partial u}{\partial y}(x, 1, t) = 0$, $0 \leqslant x \leqslant 1$, $0 \leqslant t < \infty$.

IC: $u(x, y, 0) = g(x, y)$, $\mathbf{x} \in \Omega$.

This problem is an *initial-boundary-value problem.*   □

Though initial-value and initial-boundary-value problems are also important in their own right, we shall limit subsequent discussions to *boundary-value* problems in order to keep the scope of this work within reasonable limits. The sections that follow may, however, be regarded as a suitable prerequisite to the study of time-dependent problems.

We begin in earnest our study of BVPs in the following section, in which we discuss an important class of (ordinary or partial) differential operators called *linear elliptic operators.* The corresponding DE together with the appropriate boundary conditions is called an *elliptic boundary-value problem.*

§30. Linear elliptic operators

Let A be a linear partial differential operator of order k in n variables, defined by

$$Au = \sum_{|\alpha| \leqslant k} a_\alpha(\mathbf{x})D^\alpha u, \qquad \mathbf{x} \in \Omega \subset R^n, \tag{1}$$

where Ω is an open bounded set in R^n (recall the discussion of multi-index notation in Section 24). The coefficients $a_\alpha(\mathbf{x})$ are functions of $\mathbf{x}$ and D^α represents a partial differential operator of order $|\alpha|$, i.e.

$$D^\alpha = \dfrac{\partial^{|\alpha|}}{\partial x_1^{\alpha_1} \cdots \partial x_n^{\alpha_n}}. \tag{2}$$

The operator A is assumed to occur in a PDE of the form

$$Au = f \tag{3}$$

where f lies in the range of A.

The classification of A depends only on the coefficients of the highest-order derivatives, that is, derivatives of order k; the terms involving derivatives of order k are said to constitute the *principal part* of A.

Let $\boldsymbol{\xi}$ be a vector in R^n, and let

$$\boldsymbol{\xi}^\alpha = \xi_1^{\alpha_1} \xi_2^{\alpha_2} \cdots \xi_n^{\alpha_n}, \qquad \alpha \in Z_n^+. \tag{4}$$

Then we say that

(i) *A is elliptic at* $\mathbf{x}_0 \in \Omega$ *if*

$$\sum_{|\alpha|=k} a_\alpha(\mathbf{x}_0)\boldsymbol{\xi}^\alpha \neq 0 \qquad \text{for all} \qquad \boldsymbol{\xi} \neq \mathbf{0}; \tag{5}$$

(ii) *A is elliptic* if it is elliptic at $\mathbf{x}_0$ for all points $\mathbf{x}_0$ in Ω;

(iii) *A is strongly elliptic* if A is of *even order* (i.e. $k = 2m$ for some integer m) and if there exists a number $\mu > 0$ such that

$$\left| \sum_{|\alpha|=k} a_\alpha(\mathbf{x}_0)\boldsymbol{\xi}^\alpha \right| \geq \mu \, |\boldsymbol{\xi}|^{2m} \tag{6}$$

holds at every point $\mathbf{x}_0$ in Ω, and for $\boldsymbol{\xi} \in R^n$. Here $|\boldsymbol{\xi}| = (\xi_1^2 + \cdots + \xi_n^2)^{1/2}$ is the length of the vector $\boldsymbol{\xi}$.

These ideas are best appreciated by looking at a few examples.

Examples

(a)   Consider the second-order operator in R^2, given by

$$A = a_{20} \frac{\partial^2}{\partial x^2} + a_{11} \frac{\partial^2}{\partial x \, \partial y} + a_{02} \frac{\partial^2}{\partial y^2} + b \frac{\partial}{\partial x}$$

where $a_{20} = a_{(2,0)}$, etc. Here $k = 2$ ($m = 1$) and so

$$\sum_{|\alpha|=2} a_\alpha \boldsymbol{\xi}^\alpha = a_{20}\xi^2 + a_{11}\xi\eta + a_{02}\eta^2 \tag{7}$$

where $\boldsymbol{\xi} = (\xi, \eta)$. If $A = \nabla^2 = \partial^2/\partial x^2 + \partial^2/\partial y^2$, the Laplacian operator, then (7) becomes

$$\xi^2 + \eta^2 = |\boldsymbol{\xi}|^2$$

and so ∇^2 is elliptic; indeed, ∇^2 is strongly elliptic with $\mu = 1$ in (6).

(b)   The *biharmonic operator* $\nabla^4 = (\partial^4/\partial x^4) + (2\partial^4/\partial x^2 \, \partial y^2) + (\partial^4/\partial y^4)$ is strongly elliptic since, for $\Omega \subset R^2$,

$$\sum_{|\alpha|=4} a_\alpha \xi^\alpha = \xi^4 + 2\xi^2\eta^2 + \eta^4 = (\xi^2 + \eta^2)^2 = |\xi|^4.$$

(c) The operator

$$A = (1 - x)\frac{\partial^2}{\partial x^2} + 3\frac{\partial^2}{\partial y^2} - y\frac{\partial}{\partial x}$$

is elliptic only in the half-plane $x < 1$: to see this, consider

$$\sum_{|\alpha|=2} a_\alpha \xi^\alpha = (1 - x)\xi^2 + 3\eta^2;$$

this expression is non-zero for all non-zero vectors (ξ, η) provided that $x < 1$. However, for any point (x_0, y_0) in the half-plane $x \geqslant 1$ this expression is zero for all vectors of the form $\xi = (\sqrt{3}, \sqrt{(x_0 - 1)})$. □

We point out at this stage that it is possible to show that *all elliptic operators in R^n are of even order when $n \geqslant 2$.*

§31. Normal boundary conditions

Boundary conditions cannot be specified arbitrarily; there must be restrictions on their number, the order of the differential operators appearing in them, and so on, if the boundary-value problem is to admit a solution. For example, if two boundary conditions are identical or, in any case, not independent of each other, the formulation is defective. Similar considerations apply if two boundary conditions are contradictory; for example, suppose that we have a domain $\Omega \subset R^2$ with boundary Γ, and let the two boundary conditions be

$$u = g \qquad \text{on } \Gamma, \tag{1}$$

$$\nabla u \cdot \mathbf{s} = du/ds = h \qquad \text{on } \Gamma, \tag{2}$$

where du/ds is the tangential derivative, $\mathbf{s}$ being the unit tangent vector to the boundary curve. The condition (1) implies that the tangential derivative is $du/ds = dg/ds$, which contradicts (2). Hence (1) and (2) are inadmissible as boundary conditions when specified together, though either one of them on its own is admissible.

In order to avoid situations such as these, we restrict the manner in which boundary conditions may be specified. First, we shall be exclusively concerned with boundary-value problems involving differential

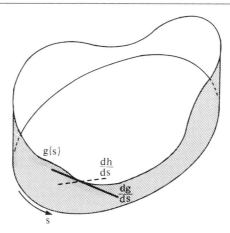

Figure 8.2

equations of *even* order, the order being equal to $2m$ $(m = 1, 2 \ldots)$, and we assume the boundary Γ to be of class C^∞.

Then we impose the following restrictions on the boundary conditions:

(i) a total of m boundary conditions must be specified at each point of the boundary. We write these as

$$B_0 u = g_0,$$
$$B_1 u = g_1,$$
$$\vdots \tag{3}$$
$$B_{m-1} u = g_{m-1},$$

where $g_0, \ldots, g_{m-1}$ are given functions and $B_0, \ldots, B_{m-1}$ are a set of linear differential operators called *boundary operators*. (The boundary conditions are numbered $0, 1, 2, \ldots$ rather than $1, 2, 3, \ldots$ for reasons that will become evident). The jth boundary operator $(j = 0, \ldots, m-1)$ is of the form

$$B_j u = \sum_{|\alpha| \leqslant q_j} b_\alpha^{(j)} D^\alpha u, \tag{4}$$

that is, it is a linear operator of order q_j. The coefficients $b_\alpha^{(j)}$ are given functions of $\mathbf{x}$ for $\mathbf{x} \in \Gamma$. We assume that $b_\alpha^{(j)}$ and g_j are infinitely differentiable functions.

(ii) the order of the highest derivative appearing in each boundary condition must be less than the order of the PDE: in other words,

$$0 \leqslant q_j \leqslant 2m - 1 \qquad \text{for } j = 0, \ldots, m-1;$$

(iii) $q_i \neq q_j$ for $i \neq j$—in other words, no two boundary conditions should have the same order of differential operators;

(iv) the final requirement is a restriction on the coefficients of the highest order derivatives, the *principal part* of B_j: we require

$$\sum_{|\alpha|=q_j} b_\alpha^{(j)} \mathbf{v}^\alpha \neq 0 \qquad \text{for all } \mathbf{x} \in \Gamma \tag{5}$$

where $\mathbf{v}$ is the unit normal to the boundary and $\mathbf{v}^\alpha = v_1^{\alpha_1} v_2^{\alpha_2} \cdots v_n^{\alpha_n}$.

Requirements (i)–(iii) are self-explanatory but the fourth requirement needs some explanation, which is best done by means of a simple example. Suppose that we have a boundary condition specified on $\Gamma \subset R^2$, which is

$$\nabla u \cdot \mathbf{a} = h$$

where $\mathbf{a}$ is an arbitrary unit vector: $\nabla u \cdot \mathbf{a}$ is the *directional derivative* in the direction of $\mathbf{a}$, and is equal to $a_x \, \partial u/\partial x + a_y \, \partial u/\partial y$. The principal part is given by (5) with $|\alpha| = 1$, $b_{(1,0)} = a_x$ and $b_{(0,1)} = a_y$, and so (5) becomes

$$b_{(1,0)} v_x^1 v_y^0 + b_{(0,1)} v_x^0 v_y^1 \neq 0$$

or $a_x v_x + a_y v_y \neq 0$ which implies $\mathbf{a} \cdot \mathbf{v} \neq 0$.

Thus (5) requires that the vector $\mathbf{a}$ should not be orthogonal to $\mathbf{v}$; this condition ensures that we do not have a situation such as that which occurred with the pair of boundary conditions (1) and (2) discussed earlier. There, $\mathbf{a} = \mathbf{s}$ in (2) and the two conditions are contradictory.

When conditions (i) to (iv) are satisfied the set $\{B_0, \ldots, B_{m-1}\}$ is said to be a *normal set of boundary operators*, and (3) is called a set of *normal boundary conditions*. An important special case of a set of normal boundary conditions arises when the order q_j of the highest derivative in the jth boundary condition is equal to j, i.e. $q_j = j$ for $j = 0, 1, \ldots, m-1$. This set of boundary conditions is called a *Dirichlet system* of order m.

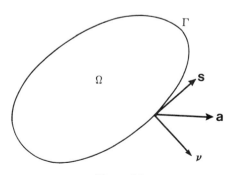

Figure 8.3

Examples

(a)   Consider the fourth-order elliptic PDE

$$\sum_{|\alpha|\leq 4} a_\alpha D^\alpha u = f \quad \text{in } \Omega \subset R^2; \tag{6}$$

this equation requires a total of two boundary conditions to be specified on the boundary Γ. One possibility is

$$\left.\begin{array}{l} B_0 u \equiv u = g_0, \\[2mm] B_1 u \equiv \nabla u \cdot \mathbf{v} = \dfrac{\partial u}{\partial v} = g_1, \end{array}\right\} \quad \text{on } \Gamma$$

which is a Dirichlet system of order 2 since $q_0 = 0$ and $q_1 = 1$. The system

$$\left.\begin{array}{l} \dfrac{\partial u}{\partial x} = g_0, \\[3mm] \dfrac{\partial u}{\partial y} = g_1, \end{array}\right\} \quad \text{on } \Gamma$$

on the other hand, violates requirement (ii) since $q_0 = q_1 = 1$.

(b)   It is not necessary that the total of m boundary conditions has to be in the form of m equations, each of which applies to the whole of Γ. The requirement is that m boundary conditions must be specified at each point $\mathbf{x}$ in Γ. So, for example, it is quite in order for (6) to have associated with it the boundary conditions

$$u = g \quad \text{on } \Gamma_1, \tag{7}$$

$$\frac{\partial u}{\partial v} = h \quad \text{on } \Gamma_2, \tag{8}$$

and

$$\frac{\partial^2 u}{\partial x^2} + u = f \quad \text{on } \Gamma \tag{9}$$

where Γ_1 and Γ_2 are disjoint subsets of Γ such that $\Gamma_1 \cup \Gamma_2 = \Gamma$. Clearly we have specified a total of two boundary conditions at each point of Γ: (7) and (9) for points on Γ_1 and (8) and (9) for points on Γ_2 (Figure 8.4).

(c)   Consider the two-point BVP

$$\frac{d^4 u}{dx^4} + 2\frac{d^2 u}{dx^2} + 3\frac{du}{dx} + u = f \quad \text{on } \Omega = (0, 1),$$

$$u(0) = 0, \qquad u(1) = 0,$$

$$u'(0) = 1, \qquad u'(1) = 2.$$

The set of BCs is a normal set (note that requirement (iv) is trivial for ODEs); in fact, the BCs constitute a Dirichlet system of order 2. We

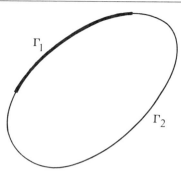

Figure 8.4

observe also that the BCs can be written in the format (3) if we define B_0 and B_1 by

$$B_0 u = (u(0), u(1)), \qquad B_1 u = (u'(0), u'(1));$$

then we have

$$B_0 u = (0, 0), \qquad B_1 u = (1, 2). \qquad\qquad \square$$

Having ensured that the boundary conditions are consistent and contain no ambiguities, we must now ensure that they are compatible with the partial differential equation $Au = f$. Intuitively it is clear that one cannot expect an arbitrary set of BCs to be compatible with the PDE; further restrictions must be placed on them if we are to ensure that the problem is well posed.

Let τ be a tangent vector to Γ at the point $\mathbf{x}$, and let $\mathbf{v}$ be an outward normal vector at this point. Now consider the pair of equations

$$\sum_{|\alpha|=2m} a_\alpha(\mathbf{x}) \left[\tau - i\mathbf{v} \frac{d}{ds} \right]^\alpha u(s) = 0, \qquad s > 0, \tag{10}$$

$$\sum_{|\alpha|=q_j} b_\alpha^{(j)}(\mathbf{x}) \left[\tau - i\mathbf{v} \frac{d}{ds} \right]^\alpha u(s) \,|_{s=0} = 0, \qquad j = 0, \ldots, m-1, \tag{11}$$

which involve only the principal parts of A and B_j (recall that $\xi^\alpha = \xi_1^{\alpha_1} \xi_2^{\alpha_2} \cdots \xi_n^{\alpha_n}$ for any vector ξ). The set $\{B_j\}$ of boundary operators is compatible with A, and is said to *cover* A at $\mathbf{x}$, if the only solution of (10), (11) is $u(s) = 0$. We will require of $\{B_j\}$ that it covers A at every point $\mathbf{x}$ in Γ.

Precisely why a condition such as the covering condition should ensure compatibility between B_j and A is not obvious at first sight. We do not pursue this point here (the reader is referred to Kellogg [22] and to Oden and Reddy [34] for a discussion and elucidation of this point); instead, we illustrate the covering condition with an example.

Example

Consider the Poisson equation

$$-\nabla^2 u = f \qquad \text{in } \Omega \subset R^2;$$

the most general boundary condition is of the form

$$B_0 u = a\frac{\partial u}{\partial x} + b\frac{\partial u}{\partial y} + cu = g \qquad \text{on } \Gamma, \tag{12}$$

and so we must investigate the restrictions on a, b and c. At a point $\mathbf{x}$ on Γ with normal $\mathbf{v} = (0, 1)$ and tangent $\boldsymbol{\tau} = (\sigma, 0)$, equation (10) gives (with $a_{20} = a_{02} = -1$)

$$-\sigma^2 u + \frac{\mathrm{d}^2 u}{\mathrm{d}s^2} = 0 \tag{13}$$

while (11) gives

$$a\sigma u(0) - ibu'(0) = 0. \tag{14}$$

A general solution of (13) is

$$u(s) = c_1 e^{\sigma s} + c_2 e^{-\sigma s};$$

since we require $u(s)$ to be finite as $s \to \infty$, we must have $c_1 = 0$. The use of (14) now gives

$$(a + ib)\sigma c_2 = 0,$$

so that $c_2 = 0$ and hence $u(s) = 0$ provided that the complex number $a + ib$ has non-zero modulus: $\sqrt{(a^2 + b^2)} \neq 0$.

This will be true for $a \neq 0$ or $b \neq 0$, so that (12) covers A at $\mathbf{x}$ for any values of a, b and c. In order to investigate the covering condition at other points on Γ, we simply introduce new axes $\bar{x}$, $\bar{y}$ so that $\mathbf{v} = (0, 1)$ relative to these axes, at the point under consideration. □

§32. Green's formulas and adjoint problems

In this and the following sections we concern ourselves with boundary-value problems of the form

$$Au = f \qquad \text{in } \Omega \subset R^n,$$

$$\left.\begin{array}{l} B_0 u = g_0 \\ B_1 u = g_1 \\ \quad\vdots \\ B_{m-1} u = g_{m-1} \end{array}\right\} \qquad \text{on } \Gamma \tag{1}$$

where A is a linear elliptic partial differential operator of order $2m$, that is,

$$Au = \sum_{|\alpha| \leqslant 2m} a_\alpha D^\alpha u; \qquad (2)$$

the coefficients a_α are functions of $\mathbf{x}$ and are infinitely differentiable. The set $B_0, \ldots, B_{m-1}$ of boundary operators are of the form

$$B_j u = \sum_{|\alpha| \leqslant q_j} b_\alpha^{(j)} D^\alpha u \qquad (3)$$

and they constitute a set of normal boundary conditions which cover A, the coefficients $b_\alpha^{(j)}$ also being infinitely differentiable functions on Γ. We call (1) a *regularly elliptic boundary-value problem of order $2m$*.

A central question of the theory of elliptic boundary-value problems relates to the conditions under which one may expect a *unique* solution of (1) to exist. In other words, given data in the form of the functions f, $g_0, \ldots, g_{m-1}$, under what circumstances can we expect to find a unique solution? Furthermore, if a solution exists then it is equally important to know something about the *regularity* or *smoothness* of this solution: if, for example, f belongs to $H^l(\Omega)$ and the functions g_j are members of spaces $H^{l_j}(\Gamma)$, we would like to know the smallest Sobolev space $H^\sigma(\Omega)$ to which the solution u belongs, since this tells us exactly how smooth u is. As one would expect, the regularity of u depends very much on that of the data: the smoother f and g_j are, the smoother u can be expected to be.

Before we can discuss questions of existence and uniqueness in any detail it is necessary to introduce the concept of a Green's formula associated with an operator A, as well as the concept of the adjoint operator A^* corresponding to A. These concepts are crucial to an understanding of the conditions under which a unique solution will exist.

Green's formula and the adjoint operator. Consider an elliptic operator A such as that defined by (2). Using Green's theorem (27.12) repeatedly, it is possible to transform the integral $\int_\Omega vAu \, dx$ so that it takes the form

$$\int_\Omega vAu \, dx = \int_\Omega uA^*v \, dx + \int_\Gamma F(u, v) \, ds, \qquad (4)$$

where A^* is also an operator of order $2m$. This identity is called a *Green's formula* for the operator A. The operator A^* that results from this manipulation is called the *formal adjoint* of A, and when A is given by (2) then A^* can be shown to have the form

$$A^*u = \sum_{|\alpha| \leqslant 2m} (-1)^{|\alpha|} D^\alpha(a_\alpha u). \qquad (5)$$

If $A^* = A$ then A is said to be *self-adjoint*. The boundary integral in (4) consists of products of various derivatives of u and v. We return to this integral in a moment, after looking at some simple examples.

Examples

(a)   Consider the second-order ordinary differential operator

$$A = \frac{d^2}{dx^2} + 1;$$

using integration by parts we have, for $\Omega = (0, 1)$,

$$\int_0^1 vAu \, dx = \int_0^1 \left(v \frac{d^2 u}{dx^2} + vu \right)$$

$$= \left[v \frac{du}{dx} \right]_0^1 - \int_0^1 \left(\frac{dv}{dx} \frac{du}{dx} - vu \right) dx$$

$$= \left[v \frac{du}{dx} \right]_0^1 - \left[\frac{dv}{dx} u \right]_0^1 + \int_0^1 \underbrace{\left(\frac{d^2 v}{dx^2} + v \right)}_{A^* v} u \, dx$$

The Green's formula is thus

$$\int_0^1 v \left(\frac{d^2 u}{dx^2} + u \right) dx = \underbrace{\left[v \frac{du}{dx} - \frac{dv}{dx} u \right]_0^1}_{F(u, v)} + \int_0^1 u \underbrace{\left(\frac{d^2 v}{dx^2} + v \right)}_{A^* v} dx \qquad (6)$$

and since $A^* = A$, A is self-adjoint.

(b)   Let A be the second-order differential operator defined by

$$Au = \frac{\partial^2 u}{\partial x^2} + 2 \frac{\partial^2 u}{\partial x \, \partial y} + \frac{\partial^2 u}{\partial y^2}. \qquad (7)$$

Using Green's theorem we have

$$\int_\Omega v \frac{\partial^2 u}{\partial x^2} \, dx = \int_\Gamma v \frac{\partial u}{\partial x} v_x \, ds - \int_\Omega \frac{\partial v}{\partial x} \frac{\partial u}{\partial x} \, dx$$

$$= \int_\Gamma \left(v \frac{\partial u}{\partial x} v_x - u \frac{\partial v}{\partial x} v_x \right) ds + \int_\Omega u \frac{\partial^2 v}{\partial x^2} \, dx.$$

Similarly,

$$\int_\Omega v \frac{\partial^2 u}{\partial y^2} \, dx = \int_\Gamma \left(v \frac{\partial u}{\partial y} v_y - u \frac{\partial v}{\partial y} v_y \right) ds + \int_\Omega u \frac{\partial^2 v}{\partial y^2} \, dx.$$

Finally,

$$2\int_{\Omega} v\frac{\partial^2 u}{\partial x\,\partial y}\,dx = \int_{\Omega}\left(v\frac{\partial^2 u}{\partial x\,\partial y} + v\frac{\partial^2 u}{\partial y\,\partial x}\right)dx$$

$$= \int_{\Gamma}\left(v\frac{\partial u}{\partial x}v_y + v\frac{\partial u}{\partial y}v_x - u\frac{\partial v}{\partial x}v_y - u\frac{\partial v}{\partial y}v_x\right)ds$$

$$+ 2\int_{\Omega}\frac{\partial^2 v}{\partial x\,\partial y}u\,dx.$$

Collecting terms, we obtain the Green's formula

$$\int_{\Omega} vAu\,dx = \int_{\Gamma}\underbrace{\left[v\frac{\partial u}{\partial v} - u\frac{\partial v}{\partial v} + v\left(\frac{\partial u}{\partial x}v_y + \frac{\partial u}{\partial y}v_x\right) - u\left(\frac{\partial v}{\partial x}v_y + \frac{\partial v}{\partial y}v_x\right)\right]}_{F(u,\,v)}ds$$

$$+ \int_{\Omega}u\underbrace{\left(\frac{\partial^2 v}{\partial x^2} + 2\frac{\partial^2 v}{\partial x\,\partial y} + \frac{\partial^2 v}{\partial y^2}\right)}_{A^*v}dx. \tag{8}$$

Hence $A^* = A$ so that A is self-adjoint. $\qquad\square$

It turns out that the boundary integral that appears in a Green's formula can be expressed very concisely in terms of four sets of boundary operators. One of these sets is B_j, which forms part of the description of the original BVP. The second set of boundary operators is denoted by S_j $(j = 0, \ldots, m - 1)$ and has the property that the $2m$ operators

$$B_0, B_1, \ldots, B_{m-1}, \qquad S_0, S_1, \ldots, S_{m-1} \tag{9}$$

form a *Dirichlet system of order* $2m$. Given these two sets of operators, it is possible to show that every Green's formula can then be written as

$$\int_{\Omega} vAu\,dx = \int_{\Omega} uA^*v + \sum_{j=0}^{m-1}\int_{\Gamma}(S_j uB_j^*v - B_j uS_j^*v)\,ds \tag{10}$$

where S_j and B_j are as defined above, while the operators B_j^* and S_j^* $(j = 0, \ldots, m - 1)$, which are uniquely defined, have the following properties:

(i) B_j^* is of order $2m - 1 - p_j$ (p_j is the order of S_j);
(ii) S_j^* is of order $2m - 1 - q_j$ (q_j is the order of B_j);
(iii) the system $\{B_0^*, B_1^*, \ldots, B_{m-1}^*, S_0^*, S_1^*, \ldots, S_{m-1}^*\}$ is a Dirichlet system of order $2m$ on Ω.

We return to the previous examples to illustrate these ideas.

Examples

(a)   In the Green's formula (6) we wish to express the boundary term in the form

$$[S_0uB_0^*v - B_0uS_0^*v]_0^1$$

(remember that $m = 1$ here). Exactly what form this integral takes depends of course on the boundary condition. Suppose that we have the boundary condition

$$B_0u \equiv (u(0), u(1)) = (0, 0).$$

Thus $q_0 = 0$, and so S_0 must be of order 1 in order for $\{B_0, S_0\}$ to be a Dirichlet system of order 2. Furthermore, S_0^* must be of order $2m - 1 - q_0 = 1$ and B_0^* must be of order $2m - 1 - p_0 = 0$. By inspection of (6) we have the following correspondence:

$$\left[v\frac{du}{dx} - u\frac{dv}{dx} \right]_0^1.$$

$$
\begin{array}{cc}
S_0u & S_0^*v \\
\\
B_0^*v & B_0u
\end{array}
$$

(b)   In the Green's formula (8) corresponding to the operator A given by (7), we have a boundary integral which, according to (10), ought to be expressible as

$$\int_\Gamma (S_0uB_0^*v - B_0uS_0^*v)\, ds.$$

Suppose that we are given $B_0 = -\partial/\partial v$: this is a first-order operator ($q_0 = 1$) so in order that $\{B_0, S_0\}$ forms a Dirichlet system of order 2 the operator S_0 must be of order zero. Hence we must have

$$S_0u = \alpha u \qquad (\text{i.e., } p_0 = 0)$$

for any function $\alpha(\mathbf{x})$. Next, S_0^* must be of order $2m - 1 - q_0 = 2 - 1 - 1 = 0$ and B_0^* must be of order $2m - 1 - p_0 = 2 - 1 - 0 = 1$. Thus we must have

$$S_0^*v = \beta v,$$

$$B_0^*v = \gamma\frac{\partial v}{\partial x} + \delta\frac{\partial v}{\partial y} + \zeta v,$$

from which it follows that

$$(\alpha u)\left(\gamma \frac{\partial v}{\partial x} + \delta \frac{\partial v}{\partial y} + \zeta v\right) + \frac{\partial u}{\partial v}(\beta v)$$

$$= v\frac{\partial u}{\partial v} - u\frac{\partial v}{\partial v} + v\left(\frac{\partial u}{\partial x}v_y + \frac{\partial u}{\partial y}v_x\right) - u\left(\frac{\partial v}{\partial x}v_y + \frac{\partial v}{\partial y}v_x\right).$$

Since $\dfrac{\partial u}{\partial v} = \dfrac{\partial u}{\partial x}v_x + \dfrac{\partial u}{\partial y}v_y$, we obtain, by equating coefficients,

$$u\frac{\partial v}{\partial x}: \qquad \alpha\gamma = -v_y - v_x, \qquad\qquad\qquad\qquad \text{(i)}$$

$$u\frac{\partial v}{\partial y}: \qquad \alpha\delta = -v_x - v_y, \qquad\qquad\qquad\qquad \text{(ii)}$$

$$uv: \qquad \alpha\zeta = 0, \qquad\qquad\qquad\qquad\qquad\quad \text{(iii)}$$

$$v\frac{\partial u}{\partial x}: \qquad \beta v_x = v_y + v_x, \qquad\qquad\qquad\qquad \text{(iv)}$$

$$v\frac{\partial u}{\partial y}: \qquad \beta v_y = v_x + v_y. \qquad\qquad\qquad\qquad \text{(v)}$$

v_x times (iv) $+$ v_y times (v) gives, using $v_x^2 + v_y^2 = 1$,

$$\beta = (1 + 2v_x v_y).$$

This leaves three equations with four unknowns α, γ, δ, ζ. We are thus free to choose one of these arbitrarily, so we set $\alpha = 1$. Then

$$\zeta = 0, \qquad \delta = \gamma = -v_y - v_x.$$

Hence the boundary integral in (10) can be written as

$$\int_\Gamma \underbrace{u}_{} \left[\underbrace{(-v_x - v_y)\left(\frac{\partial v}{\partial x} + \frac{\partial v}{\partial y}\right)}_{}\right] - \left(\underbrace{-\frac{\partial u}{\partial v}}_{}\right)\underbrace{(1 + 2v_x v_y)}_{}v \; ds. \qquad \square$$
$$\quad S_0 u \qquad\qquad B_0^* v \qquad\qquad\qquad B_0 u \qquad S_0^* v$$

Note that the boundary operators S_j have to be partial differential operators of such orders as to make (9) a Dirichlet system, but further than that they are not unique. Indeed, in the previous example we saw that α could be chosen arbitrarily. However, once S_j are fixed then so are the forms that the sets of operators B_j^* and S_j^* take.

With each regularly elliptic problem of the form (1) we may associate an *adjoint problem*

$$A^*u = f^* \qquad \text{in } \Omega,$$

$$\left.\begin{aligned}B_0^*u &= g_0^* \\ B_1^*u &= g_1^* \\ &\vdots \\ B_{m-1}^*u &= g_{m-1}^*\end{aligned}\right\} \qquad \text{on } \Gamma \qquad\qquad (11)$$

where $f^*, g_0^*, \ldots, g_{m-1}^*$ are given functions.

Like the original problem (1), the adjoint problem is also a regularly elliptic boundary-value problem of order $2m$ (note that A^* has the same principal part as A). We shall soon see that the adjoint problem plays a key role in discussions of existence and uniqueness of solutions to (1).

Example

Returning to the previous example (b) with the operator

$$A = (\partial^2/\partial x^2) + (2\partial^2/\partial x\, \partial y) + (\partial^2/\partial y^2),$$

we saw earlier that A is self-adjoint, so that $A^* = A$. Furthermore, B_0^* was found to be given by

$$B_0^*v = -(v_x + v_y)\left(\frac{\partial v}{\partial x} + \frac{\partial v}{\partial y}\right) = -\frac{\partial v}{\partial v} - v_x \frac{\partial v}{\partial y} - v_y \frac{\partial v}{\partial x}.$$

Hence the adjoint problem is

$$\frac{\partial^2 u}{\partial x^2} + 2\frac{\partial^2 u}{\partial x\, \partial y} + \frac{\partial^2 u}{\partial y^2} = f^* \qquad \text{in } \Omega,$$

$$-\frac{\partial u}{\partial v} - v_x \frac{\partial u}{\partial y} - v_y \frac{\partial u}{\partial x} = g_0^* \qquad \text{on } \Gamma. \qquad\qquad \square$$

§33. Existence, uniqueness and regularity of solutions

We come now to the main topic of this chapter, namely the discussion of existence, uniqueness and regularity of solutions to problems of the form (32.1). In order to keep the discussion as simple as possible we confine attention to problems having *homogeneous boundary conditions*, that is, problems for which $g_0, \ldots, g_{m-1}$ are all zero. This is no real restriction, since it is not difficult to show (see Exercise 33.1) that any problem with non-homogeneous boundary conditions can be converted to one with homogeneous boundary conditions in a fairly straightforward manner.

Thus we consider the problem

$$Au = f \qquad \text{in } \Omega \subset R^n,$$

$$\left.\begin{array}{l} B_0 u = 0 \\ B_1 u = 0 \\ \vdots \\ B_{m-1} u = 0 \end{array}\right\} \qquad \text{on } \Gamma. \tag{1}$$

Our primary aim is to determine when (1) has a solution u that belongs to $H^s(\Omega)$, s being an integer greater than or equal to $2m$, and to establish the largest value of s for which $u \in H^s(\Omega)$. This, of course, gives us a good idea of how smooth the solution is. Note that if u belongs to $H^s(\Omega)$, then Au is in $H^{s-2m}(\Omega)$ since A is a differential operator of order $2m$.

The question of existence and uniqueness of a solution to (1) is best approached by adopting the language of linear operator theory (Chapter 5). First, we define $N(B_j)$ to be the *null space* of the boundary operator B_j: that is, if we regard B_j as a linear operator from $H^s(\Omega)$ to $L_2(\Gamma)$, then

$$N(B_j) = \{u \in H^s(\Omega) : B_j u = 0 \quad \text{on } \Gamma\}, \qquad j = 0, \ldots, m-1.$$

It now follows that a solution of (1), if it exists, will belong to the subspace of $H^s(\Omega)$ consisting of all functions that are also in $N(B_j)$. We consequently take the domain of A to be the space $\tilde{H}^s(\Omega)$ defined by

$$\tilde{H}^s(\Omega) = H^s(\Omega) \cap N(B_0) \cap \cdots \cap N(B_{m-1})$$
$$= \{u \in H^s(\Omega) : B_0 u = \cdots = B_{m-1} u = 0 \quad \text{on } \Gamma\}, \tag{2}$$

and problem (1) now reads: find $u \in H^s(\Omega)$ which satisfies

$$A : \tilde{H}^s(\Omega) \to H^{s-2m}(\Omega), \qquad Au = f \qquad \text{in } \Omega. \tag{3}$$

Our first task is to determine the set of functions f in $H^{s-2m}(\Omega)$ for which (3) admits a solution. That is, we must identify $R(A)$, the range of A. This will enable us to solve the problem of the *existence* of a solution. We will find that $R(A)$ is not all of $H^{s-2m}(\Omega)$: there are functions f in $H^{s-2m}(\Omega)$ which do not lie in $R(A)$, and for which no solution exists. The situation is shown diagrammatically in Figure 8.5.

Our second task is to ascertain the conditions under which the solution is *unique*; in other words, we wish to know the conditions under which A is one-to-one. For this purpose we define the null space $N(A)$ of A by

$$N(A) = \{u \in \tilde{H}^s(\Omega) : Au = 0\}$$
$$= \{u \in H^s(\Omega) : Au = 0 \quad \text{in } \Omega, B_j u = 0 \quad \text{on } \Gamma\}. \tag{4}$$

Clearly, if $N(A) \neq \{0\}$ then we cannot have a unique solution since, if u_0

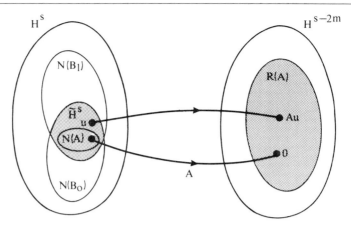

Figure 8.5

is a solution, then so is $u_0 + u$ for any $u \in N(A)$ because $A(u_0 + u) = Au_0 + Au = Au_0 = f$. We will obviously have to exclude elements of $N(A)$ from the domain of A if we are to ensure uniqueness. This is no problem: we first define the *orthogonal complement* $N(A)^\perp$ of $N(A)$ with respect to the L_2-inner product by

$$N(A)^\perp = \{v : v \in \bar{H}^s(\Omega), (v, u)_{L_2} = 0 \qquad \text{for all } u \in N(A)\}. \tag{5}$$

Now it can be shown that $N(A)$ is finite-dimensional, and hence complete, so that by the projection theorem (Theorem 14.2) we have

$$\bar{H}^s(\Omega) = N(A) \oplus N(A)^\perp; \tag{6}$$

in other words, every $u \in \bar{H}^s(\Omega)$ is of the form $u = v + w$ for $v \in N(A)$ and $w \in N(A)^\perp$, with $N(A) \cap N(A)^\perp = \{0\}$. Since $N(A)$ and $N(A)^\perp$ are disjoint (except for the zero element), we simply restrict the domain of A to $N(A)^\perp$ to ensure uniqueness.

Similar remarks of course apply to the adjoint problem

$$A^* u = f^* \qquad \text{in } \Omega,$$

$$\left.\begin{array}{l} B_0^* u = 0 \\ \qquad \vdots \\ B_{m-1}^* u = 0 \end{array}\right\} \qquad \text{on } \Gamma; \tag{7}$$

we define

$$\bar{H}^{*s}(\Omega) = \{u \in H^s(\Omega) : B_0^* u = \cdots = B_{m-1}^* u = 0 \quad \text{on } \Gamma\}$$

by analogy with (2), and rephrase (7) as

$$A^* : \bar{H}^{*s}(\Omega) \to H^{s-2m}(\Omega), \qquad A^* u = f \quad \text{in } \Omega. \tag{8}$$

The null space $N(A^*)$ of A^* and its orthogonal complement $N(A^*)^\perp$ are then

$$N(A^*) = \{u \in \tilde{H}^{*s}(\Omega) : A^*u = 0\},\tag{9}$$

$$N(A^*)^\perp = \{v \in \tilde{H}^{*s}(\Omega) : (v, u)_{L_2} = 0 \quad \text{for all} \quad u \in N(A^*)\}.\tag{10}$$

Like $N(A)$, the space $N(A^*)$ is finite-dimensional. Indeed, for most problems of practical interest,

$$\dim N(A) = \dim N(A^*).\tag{11}$$

We are not particularly concerned with solutions to the adjoint problem, but when discussing the existence of solutions to (3) we will need to call on the space $N(A^*)^\perp$. We now give a few examples.

Examples

(a)   Consider the problem

$$Au = u'' = f(x) \quad \text{in } \Omega = (0, 1),$$

$$B_0 u = (u(0), u(1)) = (0, 0).$$

We assume that $f \in L_2(0, 1)$ so we seek $u \in H^2(0, 1)$. Also,

$$N(B_0) = \{u \in H^2(0, 1) : u(0) = u(1) = 0\} = \tilde{H}^2(\Omega).$$

The null space of A is the set of solutions of

$$u'' = 0,$$

$$u(0) = u(1) = 0,$$

that is, $N(A) = \{0\}$, so a solution, if it exists, will be unique. Alternatively, suppose that

$$B_0 u = (u'(0), u'(1)) = (0, 0):$$

then $N(A) = \{u : u = \text{const.}\}$ so that

$$N(A)^\perp = \left\{ v : (v, c)_{L_2} = 0 \quad \text{or} \quad \int_0^1 v \, dx = 0 \right\}.$$

We can show that d^2/dx^2 is self-adjoint and $B_0^* = B_0$, so that $N(A^*) = N(A)$.

(b)   Consider the problem

$$Au = \frac{\partial^2 u}{\partial x^2} + 2\frac{\partial^2 u}{\partial x \, \partial y} + \frac{\partial^2 u}{\partial y^2} = f \quad \text{in } \Omega \subset R^2,$$

$$\frac{\partial u}{\partial v} = 0 \quad \text{on } \Gamma.$$

We saw in Section 32 that A is self-adjoint and that

$$B_0^* u = -(v_x + v_y)\left(\frac{\partial u}{\partial x} + \frac{\partial u}{\partial y}\right).$$

We thus have

$$N(A) = \{u : u = \text{const.}\}, \qquad N(A)^\perp = \left\{v : \int_\Omega v \, dx = 0\right\}.$$

Similarly, the null space of $A^* = A$ is the set of solutions of

$$A^* u = 0 \quad \text{in } \Omega, \qquad B_0^* u = 0 \quad \text{on } \Gamma,$$

and this is given by

$$N(A^*) = \{u : u(\mathbf{x}) = \alpha_1 + \alpha_2(x - y), \qquad \alpha_1, \alpha_2 \in R\}$$

and

$$N(A^*)^\perp = \left\{v : \int_\Omega \alpha_1 v + \alpha_2(x - y)v \, dx = 0\right\}$$

or, since α_1 and α_2 are arbitrary,

$$N(A^*)^\perp = \left\{v : \int_\Omega v \, dx = 0, \qquad \int_\Omega (x - y)v \, dx = 0\right\}. \qquad \square$$

A remark concerning the *regularity* of solutions is in order. If we can show that problem (3) has a solution, then we will, of course, conclude that u is in $H^s(\Omega)$, which establishes the degree of smoothness of u. Furthermore, it is very desirable to have a result which shows that, for some constant $C > 0$,

$$\|u\|_{H^s} \leq C \|f\|_{H^{s-2m}}. \tag{12}$$

This inequality has the following implication: suppose we consider (3) with two different sets of data f_1 and f_2, so that we seek functions u_1 and u_2 satisfying

$$Au_1 = f_1, \qquad Au_2 = f_2.$$

Since A is linear, we have

$$Au_2 - Au_1 = A\Delta u = \Delta f$$

where $\Delta u = u_2 - u_1$ and $\Delta f = f_2 - f_1$. The inequality (12) then gives

$$\|\Delta u\|_{H^s} \leq C \|\Delta f\|_{H^{s-2m}},$$

and from this we conclude that if f_1 and f_2 are close in the sense that $\|\Delta f\| < \varepsilon$ where ε is a small number, then $\|\Delta u\| < C\varepsilon$ so that u_1 and u_2 are correspondingly close. When this situation obtains we say that *the solution u depends continuously on the data.*

We are now in a position to state the main result of this section.

Theorem 1.   Consider the regularly elliptic boundary-value problem (1) *or* (3), *with* $s \geqslant 2m$. *Then*
(i) *(uniqueness) assuming the solution u exists, it is* unique *if* $u \in N(A)^{\perp}$, *that is, if*

$$(u, v)_{L_2} = 0 \qquad \text{for all } v \in N(A); \tag{13}$$

(ii) *(existence) there exists at least one solution to* (3) *if and only if* $f \in N(A^*)^{\perp}$, *that is, if*

$$(f, v)_{L_2} = 0 \qquad \text{for all } v \in N(A^*); \tag{14}$$

(iii) *if a unique solution exists then there is a constant* $C > 0$, *independent of u, such that*

$$\|u\|_{H^s} \leqslant C \|f\|_{H^{s-2m}}. \tag{15}$$

Remark.   The theorem states that A is a *surjective* operator from $\tilde{H}^s(\Omega)$ *onto the subspace* of functions in $H^{s-2m}(\Omega)$ that satisfy (14).
  Furthermore, A is *one-to-one* if we restrict its domain to the subspace of functions in $\tilde{H}^s(\Omega)$ which satisfy (13).

Proof.   (i) Take any $w \in N(A)$ and assume that there are two solutions u_1, u_2 satisfying

$$(u_1, w) = (u_2, w) = 0,$$

that is, u_1 and u_2 belong to $N(A)^{\perp}$. Since $Au_1 = Au_2 = f$, we have $A(u_1 - u_2) = 0$ so that $(u_1 - u_2) \in N(A)$. But we have $\tilde{H}^s(\Omega) = N(A) \oplus N(A)^{\perp}$ from Theorem 14.2, and since $N(A) \cap N(A)^{\perp} = \{0\}$ it follows that $u_1 - u_2 = 0$, or $u_1 = u_2$. Hence the solution is unique.
(ii) First assume that (3) has a solution u. Then for any $v \in N(A^*)$ we

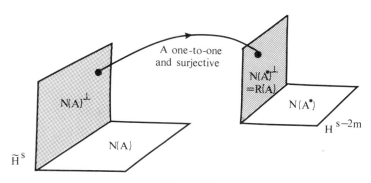

Figure 8.6

have, using Green's formula (32.10) and (9),

$$(v, f)_{L_2} = (v, Au)_{L_2} = (u, A^*v)_{L_2} + \sum_{j=0}^{m-1} \int_\Gamma (S_j u B_j^* v - B_j u S_j^* v)\, ds$$

$$= (u, 0)_{L_2} + \sum_{j=0}^{m-1} \int_\Gamma (S_j u \cdot 0 - 0 \cdot S_j^* v)\, ds = 0.$$

Hence $f \in N(A^*)^\perp$.

We sketch the proof of the converse and leave some of the details to the Exercises. We wish to prove that if $f \in N(A^*)^\perp$, then $f \in R(A)$, that is, that $N(A^*)^\perp \subset R(A)$. First, we note from (i) that, since A is one-to-one from $N(A)^\perp$ onto $R(A)$, we can define the inverse operator $A^{-1} : R(A) \to N(A)^\perp$. Secondly, we can show (Exercise 33.4) that both A and A^{-1} are bounded operators, and furthermore $R(A)$ is closed. It follows from Lemma 14.1 that $R(A)^{\perp\perp} = \overline{R(A)} = R(A)$.

Next, if $v \in R(A)^\perp$ and $u \in \tilde{H}^s(\Omega)$, then

$$(v, Au)_{L_2} = 0 = (u, A^*v)_{L_2} + \sum_{j=0}^{m-1} \int_\Gamma S_j u B_j^* v\, ds,$$

so that $v \in N(A^*)$ (since u is arbitrary, we must have $A^*v = 0$ and $B_j^* v = 0$). Hence $R(A)^\perp \subset N(A^*)$, which implies $N(A^*)^\perp \subset R(A)^{\perp\perp} = R(A)$ (see Exercise 14.4), which concludes the proof.

(iii) Once again we use the fact that A is a bounded, one-to-one linear operator from $N(A)^\perp$ onto $R(A)$: then (see Exercise 33.4(b)) there is a constant C such that

$$\|u\|_{H^s(\Omega)} \leqslant C \|Au\|_{H^{s-2m}(\Omega)} = C \|f\|_{H^{s-2m}}. \qquad \blacksquare$$

Part (ii) of the theorem expresses the fact that the data cannot be specified arbitrarily: it has to satisfy (14) if a solution is to exist. When (14) is satisfied we say that the data is *compatible* with the operator A, and (14) is called a *compatibility condition*.

Some examples follow; these should clarify the nature of all the ingredients of Theorem 1.

Examples

(a) Consider the problem

$$-\nabla^2 u = f \qquad \text{in } \Omega,$$
$$u = 0 \qquad \text{on } \Gamma.$$

In this case $A = A^* = -\nabla^2$, so A is self-adjoint. We assume that $f \in L_2(\Omega)$, so that $s = 2$ ($m = 1$). Thus

$$N(A) = N(A^*) = \{u \in H^2(\Omega): -\nabla^2 u = 0 \quad \text{on } \Omega, u = 0 \quad \text{on } \Gamma\} = \{0\}$$

so that (13) and (14) are satisfied identically. It follows that $-\nabla^2$ is one-to-one from $\bar{H}^2(\Omega)$ onto $L_2(\Omega)$. Furthermore, from (iii) there is a constant $C > 0$ such that

$$\|u\|_{H^2} \leqslant C \, \|f\|_{L_2}.$$

(b)   Consider now the problem

$$-\nabla^2 u = f \quad \text{in } \Omega,$$
$$\partial u / \partial v = 0 \quad \text{on } \Gamma.$$

In this case $N(A) = N(A^*) = \{c\}$, c being a constant function. From (i) of Theorem 1 we thus deduce that there exists a solution if and only if

$$(f, c) = 0 \rightarrow c \int_\Omega f \, dx = 0, \quad \text{or} \quad \int_\Omega f \, dx = 0.$$

Furthermore, from (i) the solution u is unique if we prescribe the condition

$$(u, c) = 0 \quad \text{or} \quad \int_\Omega u \, dx = 0.$$

(c)   Consider the problem

$$u'' = 12x^2 + 6x - 7 \quad \text{in } \Omega = (0, 1),$$
$$u'(0) = u'(1) = 0. \tag{16}$$

The spaces $N(A)$ and $N(A^*)$ are now

$$N(A) = N(A^*) = \{v : v(x) = \text{const.}\}$$

(A is self-adjoint), so for a solution to exist we require that

$$\int_0^1 cf(x) \, dx = 0 \quad \text{or} \quad \int_0^1 (12x^2 + 6x - 7) \, dx = 0,$$

which is satisfied. The solution is unique if

$$\int_0^1 u(x) \, dx = 0. \tag{17}$$

We easily verify that $u(x) = x^4 + x^3 - 5x^2/2 + d$ is a solution of (16); the constant d is fixed by (17), making the solution unique.   □

Bibliographical remarks

The concepts in Section 29 are elementary, and are normally encountered in beginning courses on differential equations. Further aspects, including

procedures for finding solutions, may be found in the classical text by
Courant and Hilbert [12].

The theory of elliptic boundary-value problems developed in Sections
30 to 33 draws heavily on the accounts given by Kellogg [22] and by Oden
and Reddy [34], wherein a much more detailed exposition can be found.
These accounts are in turn based on that given in the treatise by Lions
and Magenes [27]. Rektorys [39] also treats elliptic operators in detail.

We have deliberately focused attention on those aspects of the theory
that are most relevant to our primary objectives of variational boundary-
value problems (Chapter 9) and their approximation (Chapters 10 and
11), and have omitted more complex topics such as the question of
existence and uniqueness in the presence of non-homogeneous boundary
conditions and in the presence of data in H^{-r} for $r>0$. The latter would
cover problems such as $-\nabla^2 u = f$ in Ω where, for example, f is a Dirac
delta. Naturally the solution u is correspondingly irregular. These topics
require some knowledge of Sobolev spaces $H^s(\Omega)$ and $H^s(\Gamma)$ for which s
is real; the theory of such spaces is covered in the references to Sobolev
spaces given at the end of Chapter 7. Also, we have assumed the
boundary Γ to be of class C^∞; when the boundary is less smooth (for
example, Lipschitzian or polygonal) then the theory on regularity
becomes more complicated, though in many cases the results look similar
to those we have given here. For a comprehensive treatment of problems
in nonsmooth domains the text by Grisvard [18] is recommended.
Chapter 8 of the article by Babuška and Aziz [3] also discusses
nonsmooth domains.

Exercises

29.1 For each of the following differential equations specify the order of
the equation, state whether or not it is linear, and sketch the
spatial domain Ω:

(a) $\dfrac{\partial^2 u}{\partial x^2} + \dfrac{\partial u}{\partial x}\dfrac{\partial u}{\partial y} = y,$ in $\Omega = \{\mathbf{x} \in R^2 : x^2 + y^2 < 1, y > 0\};$

(b) $\dfrac{\partial^4 u}{\partial x^4} + 2\dfrac{\partial^4 u}{\partial x^2 \partial y^2} + \dfrac{\partial^4 u}{\partial y^4} - a\dfrac{\partial u}{\partial t} = 0$ in $\Omega = \{\mathbf{x} \in R^2 : x > 0,$

$y > 0, \quad x + y < 1\}, \qquad t \in (0, \infty).$

30.1 Find the regions in the xy plane in which the operator

$$A = (1-x)^2 \frac{\partial^4}{\partial x^4} + 2(1-x)(1-y)\frac{\partial^4}{\partial x^2 \partial y^2} + (1-y)^2\frac{\partial^4}{\partial y^4}$$

is (i) elliptic; (ii) strongly elliptic.

30.2 Show that the operator A defined by

$$A = -\frac{\partial}{\partial x}\left[(1 + x^2)\frac{\partial}{\partial x}\right] + 3\frac{\partial^2}{\partial y^2} + 2(1 + z^2)\frac{\partial^2}{\partial z^2}$$

is not elliptic anywhere in R^3.

31.1 Express the boundary condition

$$\frac{\partial^2 u}{\partial v^2} = g \qquad \text{on } \Gamma$$

in the form $\sum_{|\alpha|\leqslant 2} b_\alpha D^\alpha u = g$. Is it normal?

31.2 The boundary condition

$$-\frac{\partial}{\partial v}(\nabla^2 u) + (1 - \sigma)\frac{\partial}{\partial \tau}\left[\frac{\partial^2 u}{\partial x^2}v_1 v_2 - \frac{\partial^2 u}{\partial x\, \partial y}(v_1^2 - v_2^2) - \frac{\partial^2 u}{\partial y^2}v_1 v_2\right] = g$$

occurs in a particular fourth-order boundary-value problem. Here v is the outward unit normal, τ measures distance along the boundary and σ is a constant. Write the equation in the form $\sum b_\alpha D^\alpha u = g$ and investigate whether it fails to be a normal boundary condition for any values of σ.

31.3 Determine the conditions under which the boundary condition (31.2) covers the operator A given by

$$A = \alpha\frac{\partial^2 u}{\partial x^2} + 2\beta\frac{\partial^2 u}{\partial x\, \partial y} + \gamma\frac{\partial^2 u}{\partial y^2},$$

at a point on the boundary for which $v = (0, 1)$.

32.1 Show that the Green's formula for the operator A defined by

$$\frac{d^4 u}{dx^4} = f \qquad \text{in } \Omega = (0, 1)$$

is $\int_0^1 vu'''' \, dx = \int_0^1 uv'''' \, dx + [u'''v - u''v' + u'v'' - uv''']_0^1$. Given that $B_0 u = (u''(0), u''(1))$ and $B_1 u = (u'''(0), u'''(1))$, find the operators B_j^*, S_j and S_j^* $(j = 0, 1)$.

32.2 Show that the Green's formula for the operator A defined by

$Au = \nabla^2 u$ can be expressed in the form

$$\int_\Omega (\nabla^2 u)v \, dx = \int_\Omega u(\nabla^2 u) \, dx + \int_\Gamma (v\,\nabla u \cdot \mathbf{v} - u\,\nabla v \cdot \mathbf{v})\, ds.$$

Given that $B_0 = \partial/\partial v$, identify the boundary operators B_0^*, S_0, S_0^*.

32.3 If A is given by (32.2), verify that A^* is given by (32.5).

32.4 Consider the problem

$$Au = \frac{\partial^4 u}{\partial x^4} + 2\frac{\partial^4 u}{\partial x^2 \, \partial y^2} + \frac{\partial^4 u}{\partial y^2} = f \qquad \text{in } \Omega,$$

$$\left.\begin{array}{l} u = g_0 \\[2mm] \dfrac{\partial u}{\partial v} = g_1 \end{array}\right\} \qquad \text{on } \Gamma.$$

Derive the Green's formula for A, and express it in the form (32.10).

33.1 Consider the BVP

$$Au = f \qquad \text{in } \Omega,$$
$$B_j u = g_j \qquad \text{on } \Gamma \quad (j = 0, 1, \ldots, m-1)$$

where A is a $2m$th order operator. Let ϕ be a known function in $C^{2m}(\bar\Omega)$ such that $B_j\phi = g_j$ on Γ. Show that the BVP can be transformed to the problem

$$Aw = \hat f \qquad \text{in } \Omega,$$
$$B_j w = 0 \qquad \text{on } \Gamma,$$

where $w = u - \phi$ and $\hat f = f - A\phi$.

33.2 Investigate the existence, uniqueness and regularity of solutions to the problem

$$\frac{d^4 u}{dx^4} = f \quad \text{in} \quad (0, 1),$$

$$u''(0) = u''(1) = 0,$$
$$u'''(0) = u'''(1) = 0$$

(see Exercise 32.1).

33.3 Investigate the existence, uniqueness and regularity of solutions to the problem

$$\frac{\partial^2 u}{\partial x^2} + 2\frac{\partial^2 u}{\partial x\, \partial y} + \frac{\partial^2 u}{\partial y^2} = f \qquad \text{in } \Omega \subset R^2,$$

$$\frac{\partial u}{\partial v} = 0 \qquad \text{on } \Gamma$$

(see Examples in Sections 32 and 33). If $\Omega = (-1, 1) \times (-1, 1)$, show that any data f satisfying

$$f(x, y) = f(y, x), \qquad f \text{ odd in } x \text{ or } y,$$

is compatible.

33.4 The purpose of this exercise is to fill in some of the details of the proof of Theorem 33.1(ii).

(a)   Show that $A : H^s(\Omega) \to H^{s-2m}(\Omega)$ is a bounded operator, where

$$Au = \sum_{|\alpha| \leqslant 2m} a_\alpha(\mathbf{x}) D^\alpha u$$

and the coefficients a_α have bounded derivatives of all orders.

(b)   Use the fact that A is one-to-one from $N(A)^\perp$ onto its range, so that A has an inverse $A^{-1} : R(A) \to N(A)^\perp$. Now use the *Banach theorem*, Theorem 16.6, to conclude that A^{-1} is bounded. Use the boundedness of A^{-1} to show that $R(A)$ is closed.

9

Variational boundary-value problems

In the preceding few sections we have built up a theory of regularly elliptic BVPs, in which the typical problem involves finding a function u which satisfies an elliptic PDE of order $2m$ in a domain Ω, as well as a set of m normal boundary conditions:

PDE: $Au = f$ in Ω,

BCs: $B_0 u = g_0$

$\qquad\qquad \vdots \qquad\qquad$ on Γ.

$\qquad B_{m-1} u = g_{m-1}$

We have also settled the question of existence, uniqueness and regularity of solutions to elliptic BVPs, at least for the case of homogeneous BCs: provided certain conditions are satisfied, a unique solution exists. Furthermore, if $f \in H^l(\Omega)$ then u is smooth enough to belong to $H^{l+2m}(\Omega)$.

In this chapter we are going to broaden the concept of a boundary-value problem by introducing what is known as a *variational boundary-value problem* (VBVP). The variational formulation is a weaker one than the conventional formulation, as will be seen, since it requires fewer derivatives of u to be in $L_2(\Omega)$. Nevertheless, there is a VBVP corresponding to every BVP, and *vice versa*, so that we have the option of formulating a problem in either one of these two settings.

We start by examining a typical VBVP in Section 34; we take a simple example and show explicitly the relationship between the VBVP and a conventional BVP. Then in Section 35 we examine general features of VBVPs: how they are formulated and how they are related to BVPs. In Section 36 we consider the questions of existence and uniqueness of solutions to VBVPs. Finally, we show in Section 37 that certain VBVPs can be formulated alternatively as minimization problems in which it is required to find the function which minimizes a given functional.

§34.   A simple variational boundary-value problem

A variational boundary-value problem is a problem of the following form: find a function u which belongs to a Hilbert space V, and which satisfies the equation

$$B(u, v) = l(v) \tag{1}$$

for all functions v in V. Here B is a *bilinear form* and l a *linear functional*:

$$B : V \times V \to R \quad \text{and} \quad l : V \to R. \tag{2}$$

Before discussing general ideas, we consider a simple example of a VBVP: find $u \in H_0^1(\Omega)$ which satisfies

$$\int_\Omega \nabla u \cdot \nabla v \, dx = \int_\Omega fv \, dx \quad \text{for all } v \in H_0^1(\Omega), \quad \Omega \subset R^2. \tag{3}$$

Here $V = H_0^1(\Omega)$,

$$B(u, v) = \int_\Omega \nabla u \cdot \nabla v \, dx = \int_\Omega \left(\frac{\partial u}{\partial x} \frac{\partial v}{\partial x} + \frac{\partial u}{\partial y} \frac{\partial v}{\partial y} \right) dx$$

and

$$l(v) = \int_\Omega fv \, dx.$$

The first question we ask is: in what sense is (3) equivalent to a BVP, and what does this BVP look like? This we answer by observing first that since v in (3) is arbitrary, we can set $v = \phi \in C_0^\infty(\Omega)$ (note that $C_0^\infty(\Omega) \subset H_0^1(\Omega)$). This gives

$$B(u, \phi) \equiv \int_\Omega \left(\frac{\partial u}{\partial x} \frac{\partial \phi}{\partial x} + \frac{\partial u}{\partial y} \frac{\partial \phi}{\partial y} \right) dx = l(\phi). \tag{4}$$

Suppose for definiteness that f is in $L_2(\Omega)$; then f is locally integrable and generates a regular distribution, also denoted f, so that

$$l(\phi) = f(\phi) = \int_\Omega f\phi \, dx. \tag{5}$$

Now the functions $\partial u/\partial x$ and $\partial u/\partial y$ appearing in (4) belong to $L_2(\Omega)$ (since $u \in H_0^1(\Omega)$) and also generate regular distributions $\partial u/\partial x$ and $\partial u/\partial y$, from which it follows that

$$B(u, \phi) = \frac{\partial u}{\partial x}\left(\frac{\partial \phi}{\partial x} \right) + \frac{\partial u}{\partial y}\left(\frac{\partial \phi}{\partial y} \right), \tag{6}$$

the right-hand side indicating the action of the distributions $\partial u/\partial x_i$ on

$\partial\phi/\partial x_i$. From the definition of the generalized derivative of a distribution we have

$$\frac{\partial u}{\partial x_i}\left(\frac{\partial\phi}{\partial x_i}\right) = -\frac{\partial^2 u}{\partial x_i^2}(\phi), \tag{7}$$

$\partial^2 u/\partial x_i^2$ being a distribution (not necessarily regular). Bringing together (4)–(7) we thus obtain

$$(-\nabla^2 u - f)(\phi) = 0, \qquad \text{for all } \phi \in C_0^\infty(\Omega); \tag{8}$$

in other words, (3) is equivalent to the problem of finding $u \in H_0^1(\Omega)$ which satisfies

$$-\nabla^2 u = f \quad \text{in } \Omega \tag{9}$$

in the sense of distributions (see Section 25). Furthermore, since $u \in H_0^1(\Omega)$ it vanishes on the boundary, and we have

$$u = 0 \quad \text{on } \Gamma. \tag{10}$$

It is important to remember that by (9) we mean (8). That is, the PDE (9) may only make sense when viewed as a distributional differential equation. For example, if instead of (5) we had

$$l(v) = \delta(v) = v(0) \tag{11}$$

where δ is the Dirac singular distribution, then the same procedure leads to the equation

$$-\nabla^2 u = \delta \quad \text{in } \Omega \tag{12}$$

which, as we know very well, only has meaning in the distributional sense.

As (9), (10) stand, we seek a solution in the space $H_0^1(\Omega)$. Whether or not this solution coincides with a "classical" solution of the kind discussed in Chapter 8, will depend on the smoothness of f. If $f \in H^s(\Omega)$, then we known from Chapter 8 that $u \in H^{s+2}(\Omega)$, and so the solution to the VBVP is the same as that of the classical BVP.

So far we have shown that the VBVP (3) implies (9) (in the sense of distributions) and (10). What of the converse: suppose that we have the BVP

$$-\nabla^2 u = f \quad \text{in } \Omega, \tag{13}$$

$$u = 0 \quad \text{on } \Gamma, \tag{14}$$

with $f \in L_2(\Omega)$, and we wish to derive the corresponding VBVP? We first select V: here we choose V to be $H_0^1(\Omega)$ (the general procedure for selecting V will be made clear in a short while); next, we multiply (13) by an arbitrary function v in $H_0^1(\Omega)$ and integrate, to obtain

$$-\int_\Omega (\nabla^2 u)v \, dx = \int_\Omega fv \, dx. \tag{15}$$

Without bothering at this stage about which space u belongs to, we apply Green's theorem in the form (27.13) to the left-hand side of (15), to reduce this to

$$-\int_\Omega (\nabla^2 u)v \, dx = -\int_\Gamma \frac{\partial u}{\partial v} v \, ds + \int_\Omega \nabla u \cdot \nabla v \, dx. \tag{16}$$

Since $v \in H_0^1(\Omega)$, the boundary integral in (16) vanishes and we are left with

$$\int_\Omega \nabla u \cdot \nabla v \, dx = \int_\Omega fv \, dx. \tag{17}$$

We now observe that (17) makes sense even when u is in the larger space $H_0^1(\Omega)$, and so we simply pose the VBVP: find $u \in H_0^1(\Omega)$ such that (17) holds for all $v \in H_0^1(\Omega)$.

To summarize, then, the solution to the Dirichlet problem (13), (14) for the Poisson equation satisfies the VBVP (3). Conversely, the VBVP (3) is equivalent to the problem (9), (10) or (13), (14), *provided we interpret this problem in the broader sense of seeking $u \in H_0^1(\Omega)$ which satisfies* (8). Thus, the variational formulation contains all the information found in the conventional formulation *and more*, since we are able, when dealing with VBVPs, to work in a larger space and also to consider very irregular data such as that given by (11).

We proceed in the following section to discuss VBVPs corresponding to arbitrary regularly elliptic BVPs.

§35. Formulation of variational boundary-value problems

The ideas developed in the previous section are readily applicable to BVPs of arbitrary order. We confine attention to regularly elliptic BVPs of order $2m$, and now give details of the general procedure for formulating the corresponding variational boundary-value problems.

In anticipation of difficulties that may arise with boundary conditions of order $\geqslant m$ we start by partitioning the set of boundary conditions, of which there is a total of m, into two subsets:

(i) those of *order* $<m$, which are called *essential boundary conditions*;
(ii) those of *order* $\geqslant m$, which are called *natural boundary conditions*.

The reason for making the distinction is this: the aim is to formulate a VBVP in which the solution is required only to be in $H^m(\Omega)$ or a subspace of $H^m(\Omega)$. If this is so, then by the trace theorem it is possible to assign boundary values only to those derivatives of u which are of order less than m, that is, to those appearing in the *essential* boundary conditions. These boundary conditions may then be included in the description of the space in which we seek a solution (as with the inclusion of (34.14) in the problem description by choosing $V = H_0^1(\Omega)$) while the natural boundary conditions are catered for explicitly in the VBVP.

In order to simplify matters, attention is confined to problems with *homogeneous* essential boundary conditions. This assumption does not imply any restriction on the problem since it is a straightforward matter to convert any problem with nonhomogeneous boundary conditions to one whose boundary conditions are homogeneous (see Exercise 33.1). When necessary, for present purposes this conversion need be carried out only for the essential boundary conditions.

If we write down the BCs in the order of the highest derivatives appearing in each one, so that the first p BCs are essential BCs, then the BVP to be considered has the form

$$\text{PDE: } Au \equiv \sum_{|\alpha| \leqslant 2m} a_\alpha(\mathbf{x}) D^\alpha u = f \quad \text{in } \Omega, \tag{1}$$

$$\text{BCs: } \left.\begin{aligned} B_0 u &= 0 \\ &\vdots \\ B_{p-1} u &= 0 \end{aligned}\right\} \text{(essential)}$$

$$\left.\begin{aligned} B_p u &= g_p \\ &\vdots \\ B_{m-1} u &= g_{m-1} \end{aligned}\right\} \text{(natural)} \qquad\qquad \text{on } \Gamma. \tag{2}$$

The first step is to define a space V in which the solution to the VBVP is to be sought. This corresponds to the space $H_0^1(\Omega)$ in problem (34.3). Generally the space V, known as the *space of admissible functions*, is defined by

$$V = \{v \in H^m(\Omega) : v \text{ satisfies all essential boundary conditions}\}$$

or

$$V = \{v \in H^m(\Omega) : B_j v = 0 \text{ on } \Gamma, \quad j = 0, \dots, p-1\}. \tag{3}$$

As with the simple example worked through earlier, the next step is to multiply both sides of (1) by an arbitrary function from V, integrate and use Green's theorem to reduce the expression so obtained to one of the form

$$B(u, v) = l(v) \tag{4}$$

in which the bilinear form B consists of an integral over Ω involving derivatives of u and v up to and including those of order m, and possibly also an integral over Γ (see Example (a) below). While the essential boundary conditions are taken care of by the requirement that $u \in V$, the *natural* boundary conditions are applied directly in (4). Once the formulation (4) is arrived at, we may take (4) as it stands, without regard to any smoothness initially assumed of u, and pose the VBVP: find $u \in V$ which satisfies (4) for all $v \in V$. Since the VBVP is derived from the setting (1), (2), every solution of (1), (2) is a solution of the VBVP. Conversely, we can show that every solution of (4) satisfies the BCs (2) and also satisfies the PDE (1) in a distributional sense.

Examples

(a) Consider the problem

$$-\nabla^2 u + a(\mathbf{x})u = f \quad \text{in } \Omega,$$

$$\frac{\partial u}{\partial v} + b(\mathbf{x})u = g \quad \text{on } \Gamma. \tag{5}$$

Here $m = 1$, so that the boundary condition is a natural one. The space of admissible functions is thus $V = H^1(\Omega)$. Multiplying both sides of the PDE by $v \in H^1(\Omega)$, integrating and using Green's theorem, we get

$$\int_\Omega (\nabla u \cdot \nabla v + auv)\, dx - \int_\Gamma \left(\frac{\partial u}{\partial v}\right) v\, ds = \int_\Omega fv\, dx.$$

Making use of the boundary condition, this reduces to the VBVP of finding $u \in H^1(\Omega)$ which satisfies

$$\underbrace{\int_\Omega (\nabla u \cdot \nabla v + auv)\, dx + \int_\Gamma buv\, ds}_{B(u,\,v)} = \underbrace{\int_\Omega fv\, dx + \int_\Gamma gv\, ds}_{l(v)} \tag{6}$$

for all $v \in H^1(\Omega)$. Thus the problem of finding u which satisfies (5) for $f \in L_2(\Omega)$ is equivalent to the VBVP (6). Conversely, if u is a solution of (6), then upon setting $v = \phi \in C_0^\infty(\Omega)$ we get

$$(-\nabla^2 u + au - f)(\phi) = 0, \tag{7}$$

so that $(5)_1$ is satisfied distributionally. The interpretation of the boundary integrals in (6) is less straightforward, though, unless we assume that $u \in H^2(\Omega)$, in which case Green's theorem may be used to obtain

$$0 = \int_\Gamma \left(bu - g + \frac{\partial u}{\partial v} \right) v \ ds - \int_\Omega (\nabla^2 u - au + f) v \ dx$$

$$= \int_\Gamma \left(\frac{\partial u}{\partial v} + bu - g \right) v \ ds \tag{8}$$

using (7). The boundary value $\partial u / \partial v$ is, of course, well defined since $u \in H^2(\Omega)$ by assumption. From (8), then, we get the boundary condition $(5)_2$.

(b) Consider the BVP

$$\nabla^4 u = \frac{\partial^4 u}{\partial x^4} + 2 \frac{\partial^4 u}{\partial x^2 \, \partial y^2} + \frac{\partial^4 u}{\partial y^4} = f \qquad \text{in } \Omega, \tag{9}$$

$$\left. \begin{aligned} u &= 0 \\ \nabla^2 u - \partial u / \partial v &= 0 \end{aligned} \right\} \qquad \text{on } \Gamma. \tag{10}$$

Here $m = 2$, so there are two boundary conditions: $u = 0$ is an essential boundary condition while the second boundary condition is natural, since it is of order $2 \geqslant m$. Hence

$$V = \{ v \in H^2(\Omega) : v = 0 \text{ on } \Gamma \} = H^2(\Omega) \cap H_0^1(\Omega).$$

To obtain the bilinear form corresponding to this problem, we first observe that

$$\int_\Omega v \frac{\partial^4 u}{\partial x^4} \ dx = \int_\Gamma v \frac{\partial^3 u}{\partial x^3} v_x \ ds - \int_\Omega \frac{\partial v}{\partial x} \frac{\partial^3 u}{\partial x^3} \ dx$$

$$= \int_\Gamma \left(v \frac{\partial^3 u}{\partial x^3} - \frac{\partial v}{\partial x} \frac{\partial^2 u}{\partial x^2} \right) v_x \ ds + \int_\Omega \frac{\partial^2 v}{\partial x^2} \frac{\partial^2 u}{\partial x^2} \ dx \tag{i}$$

after two applications of Green's theorem. Similarly,

$$\int_\Omega v \frac{\partial^4 u}{\partial y^4} \ dx = \int_\Gamma \left(v \frac{\partial^3 u}{\partial y^3} - \frac{\partial v}{\partial y} \frac{\partial^2 u}{\partial y^2} \right) v_y \ ds + \int_\Omega \frac{\partial^2 v}{\partial y^2} \frac{\partial^2 u}{\partial y^2} \ dx. \tag{ii}$$

Finally, we consider

$$2 \int_\Omega v \frac{\partial^4 u}{\partial x^2 \, \partial y^2} \ dx = \int_\Omega \left(v \frac{\partial^4 u}{\partial x \, \partial y \, \partial x \, \partial y} + v \frac{\partial^4 u}{\partial x \, \partial y \, \partial y \, \partial x} \right) dx$$

(this decomposition is done in order to preserve the symmetry inherent in the PDE)

$$= \int_\Gamma \left[v \left(v_y \frac{\partial^3 u}{\partial x^2 \, \partial y} + v_x \frac{\partial^3 u}{\partial x \, \partial y^2} \right) - \frac{\partial v}{\partial x} \frac{\partial^2 u}{\partial y^2} v_x - \frac{\partial v}{\partial y} \frac{\partial^2 u}{\partial x^2} v_y \right] ds$$

$$+ \int_\Omega \left[\frac{\partial^2 v}{\partial y^2} \frac{\partial^2 u}{\partial x^2} + \frac{\partial^2 v}{\partial x^2} \frac{\partial^2 u}{\partial y^2} \right] dx, \tag{iii}$$

again after two applications of Green's theorem. Adding together (i), (ii) and (iii) we find that

$$\int_\Omega v\nabla^4 u \, dx = \int_\Omega \nabla^2 u \nabla^2 v \, dx + \int_\Gamma \left(v \frac{\partial}{\partial \nu}(\nabla^2 u) + \nabla^2 u \frac{\partial v}{\partial \nu} \right) ds.$$

But $v = 0$ on Γ and the natural boundary condition is $\nabla^2 u = \partial u / \partial \nu$; using this information, we obtain finally

$$\underbrace{\int_\Omega \nabla^2 u \nabla^2 v \, dx + \int_\Gamma \frac{\partial u}{\partial \nu} \frac{\partial v}{\partial \nu} \, ds}_{B(u, v)} = \underbrace{\int_\Omega fv \, dx}_{= \quad l(v).} \tag{11}$$

The VBVP corresponding to (9), (10) is thus: find $u \in V$ which satisfies (11) for all $v \in V$. In much the same way as was done for Example (a) above, we may show that (11) is equivalent to the distributional differential equation (9), together with the boundary conditions (10). $\square$

§36. Existence, uniqueness and regularity of solutions

Existence and uniqueness of solutions to VBVPs. Earlier, in Section 33, we discussed the conditions under which solutions to regularly elliptic BVPs exist and are unique. The results there apply, of course, to what we have referred to as the classical formulation, which consists of a PDE and a collection of boundary conditions (see (33.1)), the latter assumed homogeneous.

Now in much the same way we wish to know the conditions under which a unique solution to the corresponding variational boundary-value problem may be found. Just as the issues of existence and uniqueness of the solution to (33.1) depend on various properties of the differential operators A and $B_0, \ldots, B_{m-1}$, in the case of VBVPs we can expect these issues to be closely tied to properties of the bilinear form $B(.,.)$ and the linear functional $l(\cdot)$. It turns out that there is exactly one solution to a VBVP of the form (35.4) provided that l is continuous and provided that B is *continuous* and *V-elliptic*: recall from Section 19 that a bilinear operator B is *V-elliptic* if there is a constant $\alpha > 0$ such that

$$B(v, v) \geq \alpha \|v\|_V^2 \qquad \text{for all } v \in V, \tag{1}$$

V being the space of admissible functions and $\|\cdot\|_V$ the norm on this space. Without further ado we present the basic existence and uniqueness theorem for VBVPs, after which a few specific examples will be considered.

Theorem 1. *Let V be a Hilbert space and let $B(.,.): V \times V \to R$ be a continuous, V-elliptic bilinear form on V. Furthermore, let $l: V \to R$ be a* continuous *linear functional on V. Then*

(i)   *the VBVP of finding* $u \in V$ *which satisfies*

$$B(u, v) = l(v) \qquad \text{for all } v \in V \tag{2}$$

has one and only one solution;

(ii)   *the solution depends continuously on the data, in the sense that*

$$\|u\|_V \leqslant \frac{1}{\alpha} \|l\|_{V'}, \tag{3}$$

where $\|\cdot\|_{V'}$ *is the norm in the dual space* V' *of* V *and* α *is defined in* (1).

Proof. The proof of this theorem follows from the Lax–Milgram theorem (Theorem 19.1). Since B is continuous and V-elliptic, every bounded linear functional, and in particular the functional $l(\cdot)$, can be expressed in the form

$$l(v) = B(u, v)$$

where u is unique. This proves part (i); part (ii) follows by setting $v = u$ in (1) and using (2): this gives

$$\alpha \|u\|_V^2 \leqslant B(u, u) = l(u) \leqslant \|l\|_{V'} \|u\|_V,$$

the last inequality coming from the fact that l is bounded. Dividing throughout by $\|u\|$, we obtain (3).   ∎

Recall from the discussion in Section 33 the significance of a result like (3): this inequality assures us that a small change in the functional l leads to a correspondingly small change in the solution.

The inequality (3) may be expressed in an alternative form if l is given by

$$l(v) = \int_\Omega fv \, dx + \int_\Gamma gv \, ds$$

where f is in $L_2(\Omega)$ and $g \in L_2(\Gamma)$ (as in (35.6) above); for then we have

$$\begin{aligned}
\alpha \|u\|_V^2 \leqslant B(u, u) = l(u) &= (f, u)_{L_2(\Omega)} + (g, u)_{L_2(\Gamma)} \\
&\leqslant \|f\|_{L_2(\Omega)} \|u\|_{L_2(\Omega)} + (g, u)_{L_2(\Gamma)} \\
&\leqslant \|f\|_{L_2(\Omega)} \|u\|_{L_2(\Omega)} + \|g\|_{L_2(\Gamma)} \|u\|_{L_2(\Gamma)} \\
&\leqslant k \|u\|_{L_2(\Omega)} \{\|f\|_{L_2(\Omega)} + \|g\|_{L_2(\Gamma)}\}
\end{aligned}$$

using the Schwarz inequality and the trace theorem. Since V is a subspace of a Sobolev space $H^m(\Omega)$, the norm $\|\cdot\|_V$ will be the H^m-norm, and, of course $\|u\|_{L_2} \leqslant \|u\|_V$. Hence

$$\|u\|_V \leqslant \frac{k}{\alpha} (\|f\|_{L_2(\Omega)} + \|g\|_{L_2(\Gamma)}). \tag{4}$$

We now show how Theorem 1 is applied to actual problems.

Examples

(a) Consider the BVP

$$-\nabla^2 u + au = f \quad \text{in } \Omega,$$

$$u = 0 \quad \text{on } \Gamma, \tag{5}$$

where $a(\mathbf{x})$ is continuous on Ω. We assume first of all that

$$M_2 \geqslant a(\mathbf{x}) \geqslant M_1 \tag{6}$$

for some positive constants M_1, M_2. The VBVP corresponding to (5) is: find $u \in H_0^1(\Omega)$ which satisfies

$$\underbrace{\int_\Omega (\nabla u \cdot \nabla v + auv)\, dx}_{B(u,\, v)} = \underbrace{\int_\Omega fv\, dx}_{l(v)}, \quad v \in H_0^1(\Omega). \tag{7}$$

If we can show that l is continuous and that B is both continuous and H_0^1-elliptic, then we are guaranteed the existence of a unique solution to (7). First, l is continuous since

$$|l(v)| = \left| \int_\Omega fv\, dx \right| \leqslant \|f\|_{L_2} \|v\|_{L_2} \quad \text{(Schwarz inequalilty)}$$

$$\leqslant \|f\|_{L_2} \|v\|_{H^1} \quad (\|v\|_{L_2} \leqslant \|v\|_{H^m});$$

if we set $\|f\|_{L_2} = K$, then $|l(v)| \leqslant K \|v\|_{H^1}$ and so l is bounded, and hence continuous.

Next, B is continuous since

$$|B(u, v)| = \left| \int_\Omega (\nabla u \cdot \nabla v + auv)\, dx \right|$$

$$\leqslant \left| \int_\Omega \nabla u \cdot \nabla v\, dx \right| + \left| \int_\Omega auv\, dx \right| \quad \text{(triangle inequality)}$$

$$\leqslant \sum_{i=1}^n \left| \left(\frac{\partial u}{\partial x_i}, \frac{\partial v}{\partial x_i} \right)_{L_2} \right| + M_2 |(u, v)_{L_2}| \quad \text{(from (6))}$$

$$\leqslant \sum_{i=1}^n \left\| \frac{\partial u}{\partial x_i} \right\|_{L_2} \left\| \frac{\partial v}{\partial x_i} \right\|_{L_2} + M_2 \|u\|_{L_2} \|v\|_{L_2}$$

$$\leqslant \sum_{i=1}^n \|u\|_{H^1} \|v\|_{H^1} + M_2 \|u\|_{H^1} \|v\|_{H^1}$$

$$\text{(since } \|\partial u/\partial x_i\|_{L_2} \leqslant \|u\|_{H^1})$$

$$= K \|u\|_{H^1} \|v\|_{H^1} \quad \text{where } K = M_2 + n$$

$$\text{(for a problem in } R^n).$$

Thus B is continuous. Finally, to show that B is H_0^1-elliptic, consider

$$B(v, v) = \int_\Omega (\nabla v \cdot \nabla v + av^2)\, dx$$

$$\geq \int_\Omega (\nabla v \cdot \nabla v + M_1 v^2)\, dx \qquad \text{(from (6))}$$

$$\geq \begin{cases} \displaystyle\int_\Omega (\nabla v \cdot \nabla v + v^2)\, dx & \text{if} \quad M_1 \geq 1, \\[2ex] \displaystyle M_1 \left[\int_\Omega (\nabla v \cdot \nabla v + v^2)\, dx \right] & \text{if} \quad M_1 \leq 1. \end{cases}$$

In other words,

$$B(v, v) \geq \alpha \int_\Omega (\nabla v \cdot \nabla v + v^2)\, dx = \alpha \, \|v\|_{H^1}^2$$

where $\alpha = \min(1, M_1)$. Thus B is H_0^1-elliptic. All requirements are met, and so a unique solution to (7) exists.

(b)   Consider once again the BVP (5), but this time assume that we only know that $a(\mathbf{x})$ is non-negative and bounded above:

$$0 \leq a(\mathbf{x}) \leq M_2.$$

(Note that the Poisson equation $-\nabla^2 u = f$ is a particular example of this situation.) Continuity of l and B follow from the same arguments as those used in Example (a), but the proof of H_0^1-ellipticity in (a) does not apply here since we made use in (a) of the condition $a(\mathbf{x}) \geq M_1$. In order to show that B is H_0^1-elliptic, we have to use an important inequality which is valid for all functions v in $H_0^1(\Omega)$. This is called the *Poincaré–Friedrichs inequality*, and it states that a constant $C > 0$ can always be found such that

$$\int_\Omega v^2\, dx \leq C \int_\Omega \nabla v \cdot \nabla v\, dx \qquad \left(= C \int_\Omega \sum_{i=1}^n \frac{\partial v}{\partial x_i} \frac{\partial v}{\partial x_i}\, dx \right) \tag{8}$$

or

$$|v|_{L_2}^2 \leq C \, |v|_{H^1}^2$$

for all $v \in H_0^1(\Omega)$, ($|\cdot|_{H^m}$ is the *semi-norm* $|v|_{H^m} = [\sum_{|\alpha|=m} \int (D^\alpha v)^2\, dx]^{1/2}$). The proof of the Poincaré–Friedrichs inequality is outlined in Exercise 36.1.

From (7) we have

$$B(v, v) = \int_\Omega (\nabla v \cdot \nabla v + av^2)\, dx \geq \int_\Omega \nabla v \cdot \nabla v\, dx \tag{9}$$

and from (8),

$$(C+1)\int_\Omega \nabla v \cdot \nabla v \, dx \geq \int_\Omega (v^2 + \nabla v \cdot \nabla v) \, dx = \|v\|_{H^1}^2 \qquad (10)$$

so that

$$B(v, v) \geq \frac{1}{C+1} \|v\|_{H^1}^2$$

and so B is H_0^1-elliptic.

(c) An example of a problem which does not have a unique solution is the BVP

$$\begin{aligned} -\nabla^2 u &= f \quad &\text{in } \Omega, \\ \partial u/\partial v &= 0 \quad &\text{on } \Gamma. \end{aligned} \qquad (11)$$

The corresponding VBVP is: find $u \in H^1(\Omega)$ such that

$$\int_\Omega \nabla u \cdot \nabla v \, dx = \int_\Omega fv \, dx \qquad (12)$$

for all $v \in H^1(\Omega)$.

Thus (12) is similar in all respects to Example (b), with the exception that the space of admissible functions is $H^1(\Omega)$ and *not* $H_0^1(\Omega)$. As before, l and B are continuous, but B is *not* H^1-elliptic: indeed, the inequality

$$B(v, v) \geq \alpha \|v\|^2$$

ceases to hold for any function v which is constant (for which case $B(v, v) = 0$). Hence, while the lack of H^1-ellipticity does not necessarily mean that a unique solution does not exist (Theorem 1 gives *sufficient* conditions for the existence of a solution; if these conditions are not satisfied, it does not imply non-existence or non-uniqueness), we are unable to guarantee the existence of a unique solution. □

Recall from Section 33 that in order for a unique solution of the BVP (11) to exist, it was necessary and sufficient that the data f and the solution satisfy

$$\int_\Omega f \, dx = 0 \quad \text{and} \quad \int_\Omega u \, dx = 0. \qquad (13)$$

The reason why we cannot prove existence of a unique solution to the corresponding VBVP (12) is essentially that the conditions (13) are not satisfied in the statement of (12) as it stands. First of all, the space $H^1(\Omega)$

in which u is sought is too large: for uniqueness we must restrict attention
to the subspace of $H^1(\Omega)$ consisting of elements orthogonal to constants,
i.e. elements satisfying $(13)_2$. Secondly, the compatibility condition $(13)_1$
is a *necessary* condition for existence of a solution, since we get from
(12), setting $v = c$, a constant,

$$0 = \int_\Omega fc \, dx \quad \text{or} \quad \int_\Omega f \, dx = 0.$$

We will show that $(13)_1$ is in fact also a *sufficient* condition for existence,
as in Section 33. Without further ado we present the extension of
Theorem 1.

Theorem 2. *Let V be a closed subspace of $H^m(\Omega)$, and define a
continuous bilinear form $B(.,.)$ on $V \times V$ and a continuous linear
functional $l(\cdot)$ on V. Let P be a closed subspace of V such that*

$$B(u + p, v + \bar{p}) = B(u, v) \qquad \text{for all } u, v \in V \text{ and } p, \bar{p} \in P. \tag{14}$$

*Also, denote by Q the subspace of V consisting of functions orthogonal
to P in the L_2-norm:*

$$Q = \left\{ v \in V : \int_\Omega vp \, dx = 0 \qquad \text{for all } p \in P \right\}, \tag{15}$$

and assume that $B(.,.)$ is Q-elliptic: there is a constant $\alpha > 0$ such that

$$B(q, q) \geqslant \alpha \|q\|_Q^2 \qquad \text{for } q \in Q, \tag{16}$$

the norm on Q being the same as that on V. Then
(i) there exists a unique solution *to the problem of finding $u \in Q$ such
that*

$$B(u, v) = l(v) \qquad \text{for all } v \in V \tag{17}$$

if and only if the compatibility condition

$$l(p) = 0 \qquad \text{for } p \in P \tag{18}$$

holds;
(ii) (Continuous dependence on the data) *the solution u satisfies*

$$\|u\|_Q \leqslant \alpha^{-1} \|l\|_{Q'}. \tag{19}$$

Remark. Note that Theorem 1 is a special case for which $P = \{0\}$, so
that $Q = V$. Also, observe that for Example (c) above we have
$V = H^1(\Omega)$, $P = P_0$, the set of constant functions, and Q consists of functions satisfying $(13)_2$, while (18) reduces to $(13)_1$.

Proof.   First we show the necessity of (18). Assuming that (17) holds, we have from (14)

$$B(u, v + p) = B(u, v) = l(v + p)$$

so that, subtracting this from (17) and using the linearity of $l(.)$, (18) follows.

Conversely, assume that (18) holds: we want to show that (17) has exactly one solution. This we do by showing first that (17) is equivalent to the problem of finding $u \in Q$ such that

$$B(u, q) = l(q) \qquad \text{for all } q \in Q. \tag{20}$$

Once this has been done, the proof follows that in Theorem 1, since $B(., .)$ is continuous and Q-elliptic from Exercise 14.2, it is known that Q is complete in the H^m-norm.

Now since P is closed in V, which is in turned closed in $H^m(\Omega)$ and hence also in $L_2(\Omega)$, we have by Theorem 14.2 that

$$V = P \oplus Q.$$

Hence we can write any $v \in V$ uniquely as $v = p + q$ for $p \in P$ and $q \in Q$. Since (18) is assumed to hold we have from (14) and (18)

$$B(u, p) = 0 \qquad \text{and} \qquad l(p) = 0$$

so that, if $u \in Q$ is a solution of (20) then

$$B(u, q) + B(u, p) = l(q) + l(p)$$

or, with $p + q = v$,

$$B(u, v) = l(v) \qquad \text{for all } v \in V.$$

Hence a solution of (20) is also a solution of (17). Conversely, (17) holds *a fortiori* for all $v \in Q$ since Q is in V. Hence a solution of (17) is also a solution of (20), so that problems (17) and (20) are equivalent.

The remainder of the proof of part (i), as well as the proof of part (ii), now follow in much the same way as the proof of Theorem 1.   ∎

Example

We return to (11), but consider instead the non-homogeneous boundary condition

$$\frac{\partial u}{\partial v} = g \qquad \text{in } \Gamma.$$

The VBVP is now: find $u \in H^1(\Omega)$ such that

$$\int_\Omega \nabla u \cdot \nabla v \, dx = \int_\Omega fv \, dx + \int_\Gamma gv \, ds \qquad \text{for all } v \in H^1(\Omega). \tag{21}$$

Here $V = H^1(\Omega)$. Of course, (21) does not have a unique solution as it stands; we note that

$$B(u + p, v + \bar{p}) = \int_\Omega \nabla(u + p) \cdot \nabla(v + \bar{p}) \, dx = B(u, v)$$

$$\text{for all } p, \bar{p} \in P_0,$$

where P_0 is the space of constant functions. Thus $P = P_0$ and the solution must be sought in the space

$$Q = \left\{ q \in H^1(\Omega) : \int_\Omega q \, dx = 0 \right\}. \tag{22}$$

We know that $B(. , .)$ and $l(.)$ are continuous; to show that B is Q-elliptic we use the Poincaré inequality (26.10) with (22) to show that

$$\int_\Omega \nabla q \cdot \nabla q \geqslant \alpha \|q\|_{H^1}^2 = \alpha \|q\|_Q^2.$$

Hence there exists a *unique* solution u in Q to (21) if and only if

$$l(p) = 0 \Rightarrow \int_Q f \, dx + \int_\Omega g \, ds = 0.$$

Furthermore, we have

$$\|u\|_Q = \|u\|_{H^1} \leqslant \alpha^{-1} \|l\|_{Q'}. \tag{23}$$

Since $l(v)$ is given by the right-hand side of (21) we can in fact express (23) in the alternative form (4).   □

Regularity of solutions. A few words are in order regarding the regularity or degree of smoothness of the solution to a VBVP. Recall that we discussed this issue in Section 33 in the context of the conventional or classical formulation, and we showed there that the solution u belongs to $H^s(\Omega)$ (or rather, a subspace $\bar{H}^s(\Omega)$ of $H^s(\Omega)$) if the data f is in $H^{s-2m}(\Omega)$. Here $s \geqslant 2m$.

In the context of variational boundary-value problems, we also showed earlier that, when the linear functional $l(.)$ is of the form

$$l(v) = \int_\Omega fv \, dx \tag{24}$$

with $f \in L_2(\Omega)$, then the problem of finding $u \in V$ which satisfies

$$B(u, v) = l(v), \qquad v \in V \tag{25}$$

is equivalent to the problem of $u \in V$ which satisfies

$$Au = f \qquad \text{in } L_2(\Omega). \tag{26}$$

Thus, for a differential operator of order $2m$ we conclude that u is "almost" in $H^{2m}(\Omega)$, in the sense that u lies in the domain of an operator of order $2m$. If A contains *all* derivatives of order $2m$, then u in fact belongs to $H^{2m}(\Omega)$.

The above characterization of the smoothness of solutions of VBVPs can be strengthened considerably with the aid of the existence and uniqueness results given earlier in this section, and the theory of Section 33. Indeed, suppose that (24)–(26) hold, and that B and l satisfy the requirements for existence and uniqueness; since the solution u is unique, we may apply the theory of Section 33 to equation (26) and conclude from Theorem 33.1 that u in fact belongs to $H^{2m}(\Omega)$, or rather the subspace of $H^{2m}(\Omega)$ consisting of functions that satisfy all of the homogeneous boundary conditions. Going one step further, if f is even more smooth so that it lies in $H^s(\Omega)$, say, with $s \geqslant 0$, then we are assured that u is $H^{s+2m}(\Omega)$. We summarize these observations in the following theorem.

Theorem 3. *Let Ω be a smooth domain and let $u \in V$ be a solution of the* VBVP

$$B(u, v) = \int_\Omega fv \, dx, \qquad v \in V,$$

where $V \subset H^m(\Omega)$. If $f \in H^s(\Omega)$ then we have $u \in H^{s+2m}(\Omega)$, and the estimate

$$\|u\|_{H^{s+2m}} \leqslant C \, \|f\|_{H^s}$$

holds. ∎

§37. Minimization of functionals

We have seen how an elliptic boundary-value problem of the form

$$Au = f \qquad \text{in } \Omega,$$
$$B_j u = g_j \qquad \text{on } \Gamma \qquad (j = 0, \ldots, m-1) \tag{1}$$

can be posed in the alternative form of a variational boundary-value problem

$$u \in V; \qquad B(u, v) = l(v), \qquad v \in V. \tag{2}$$

We have also seen that the formulation (2) has certain advantages over (1). It turns out that if the bilinear form in (2) is *symmetric*, that is, if

$$B(u, v) = B(v, u) \qquad \text{for all } u, v \in V,$$

then a third formulation is possible: the variational boundary-value problem is equivalent to a *minimization problem* in which we are required to find a function u in V which makes the value of a functional $J : V \to R$ a minimum, that is,

$$J(u) \leqslant J(v) \qquad \text{for all } v \in V, \tag{3}$$

where J is defined by

$$J(v) = \tfrac{1}{2} B(v, v) - l(v). \tag{4}$$

The formulation (4) is often taken as a starting point, particularly in physical problems for which $J(v)$ represents the energy of a system, and (3) implies that the solution u is that vector or function which minimizes the energy of the system. For example, the problem of finding the displaced shape of a thin membrane under its own weight is equivalent to solving (Figure 9.1)

$$-\nabla^2 u = f \qquad \text{in } \Omega \subset R^2, \tag{5}$$
$$u = 0 \qquad \text{on } \Gamma.$$

The VBVP corresponding to (5) is: find $u \in H_0^1(\Omega)$ such that

$$\int_\Omega \nabla u \cdot \nabla v \, dx = \int_\Omega fv \, dx \qquad \text{for all } v \in H_0^1(\Omega), \tag{6}$$

and the equivalent minimization problem is: solve (3) where

$$J(v) = \tfrac{1}{2} \int_\Omega |\nabla v|^2 \, dx - \int_\Omega fv \, dx, \tag{7}$$

$J(v)$ being the total energy when the displacement is v. The statement that the solution of (5) or (6) is given by the function that minimizes (7) is known as *Dirichlet's principle* in the literature on boundary-value problems.

Our aim in this section, then, is to investigate the consequences of formulating problems such as (3), (4), and to show how these relate to VBVPs. Before embarking on this task, it is instructive to review the situation for functions of a single variable, wherein many of the ideas for general functionals are also present.

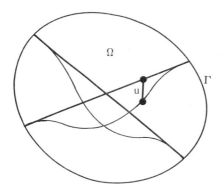

Figure 9.1

Consider a function $f: R \to R$, and suppose that we wish to locate the minimum value of $f(x)$, as well as the point x_0 at which this minimum occurs. That is, we wish to find x_0 such that

$$f(x_0) \leq f(x) \qquad \text{for all } x \in R. \tag{8}$$

From elementary calculus we know that this minimum occurs at x_0 which satisfies

$$f'(x_0) = 0. \tag{9}$$

We are assured that $f(x_0)$ is a minimum, and not a maximum or an inflection point, if the function f is *convex*, that is, if a straight line drawn between any two points on the curve of $f(x)$ lies on or above the graph of the function. Mathematically, f is convex if, for $0 < \theta < 1$,

$$f(y + \theta(x - y)) \leq f(y) + \theta(f(x) - f(y))$$

or

$$f(\theta x + (1 - \theta)y) \leq \theta f(x) + (1 - \theta)f(y). \tag{10}$$

The minimum may not be unique: for example, consider the function g shown in Figure 9.2; the minimum value of $g(x)$ occurs at all points between a and b. But we can guarantee a *unique* minimum of the *strictly convex* function $f(x)$ in Figure 9.2: that is, a function is strictly convex if

$$f(\theta x + (1 - \theta)y) < \theta f(x) + (1 - \theta)f(y), \qquad 0 < \theta < 1. \tag{11}$$

Then f is shaped like a bowl with no flats, and we have a unique minimum.

Remarkably, all of these ideas extend in a very simple way to functionals defined on arbitrary spaces. In order to see how this is done,

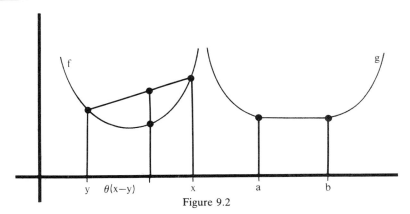

Figure 9.2

we first define a convex functional, after which we need to introduce a generalization of the idea of a derivative.

Convex functionals. Let $J: V \to R$ be a functional defined on a subspace V of a linear space U. Then J is said to be *convex* if

$$J(\theta u + (1 - \theta)v) \leqslant \theta J(u) + (1 - \theta)J(v) \tag{12}$$

for all $u, v \in V$ and for $0 < \theta < 1$. Furthermore, J is *strictly convex* if we replace "$\leqslant$" by "$<$" in (12).

We now introduce the concept of the Gateaux derivative of a functional.

Gateaux derivative. A functional $J: V \to R$ is said to be Gateaux-differentiable, or simply differentiable, at $u \in V$ if there exists an operator $DJ: V \to V'$ defined by

$$DJ(u)(v) = \lim_{\theta \to 0} \frac{[J(u + \theta v) - J(v)]}{\theta} \tag{13}$$

for all $v \in V$. Equivalently, DJ is defined by

$$DJ(u)(v) = \frac{\mathrm{d}}{\mathrm{d}\theta} [J(u + \theta v)] \bigg|_{\theta=0} \tag{14}$$

(see Exercise 37.1). The operator DJ is called the *gradient of J* and $DJ(u): V \to R$ is the (Gateaux) derivative of J at u. Observe from (13) or (14) that DJ maps V to its dual space V', so that $DJ(u) \in V'$; that is, $DJ(u)$ is a *continuous linear functional on V*. Note that the Gateaux derivative does not always exist: you may verify, for example, that if J is defined by $J: R^2 \to R$,

$$J(\mathbf{u}) = \begin{cases} u_1^2(1 + 1/u_2), & u_2 \neq 0, \\ 0, & u_2 = 0 \end{cases}$$

then

$$\lim_{\theta \to 0} \theta^{-1}[J(\mathbf{u} + \theta \mathbf{v}) - J(\mathbf{v})] = v_1^2/v_2,$$

which is not linear in $\mathbf{v}$.

Examples

(a) If V is an interval in R then we see that DJ reduces to the conventional derivative: $DJ(x) \equiv dJ/dx$. Furthermore, if $V \subset R^n$, then according to (13) or (14) we have (note that $(R^n)' = R^n$)

$$DJ : R^n \to R^n, \qquad DJ(\mathbf{x})\mathbf{y} = \sum_{i=1}^{n} \frac{\partial J}{\partial x_i} y_i = \nabla J \cdot \mathbf{y},$$

that is, the Gateaux derivative is the directional derivative (see Exercise 37.3).

(b) Let $J : H^1(\Omega) \to R$ be defined by

$$J(v) = \frac{1}{2} \int_\Omega \nabla v \cdot \nabla v \, dx - \int_\Omega fv \, dv.$$

Then J is strictly convex: indeed,

$$J(\theta u + (1 - \theta)v) = \tfrac{1}{2}\theta^2 \int_\Omega |\nabla u|^2 \, dx + \tfrac{1}{2}(1 - \theta)^2 \int_\Omega |\nabla v|^2 \, dx$$

$$+ \theta(1 - \theta) \int_\Omega \nabla u \cdot \nabla v \, dx$$

$$- \theta \int_\Omega fu \, dx - (1 - \theta) \int_\Omega fv \, dx.$$

Now $\int_\Omega (\nabla u - \nabla v) \cdot (\nabla u - \nabla v) \, dx > 0$, so that

$$2 \int_\Omega \nabla u \cdot \nabla v \, dx < \int_\Omega (|\nabla u|^2 + |\nabla v|^2) \, dx.$$

Hence

$$J(\theta u + (1 - \theta)v) < \tfrac{1}{2}\theta^2 \int_\Omega |\nabla u|^2 \, dx + \tfrac{1}{2}(1 - \theta)^2 \int_\Omega |\nabla v|^2 \, dx$$

$$+ \tfrac{1}{2}\theta(1 - \theta) \int_\Omega (|\nabla u|^2 + |\nabla v|^2) \, dx$$

$$- \theta \int_\Omega fu \, dx - (1 - \theta) \int_\Omega fv \, dx$$

$$= \theta J(u) + (1 - \theta)J(v).$$

To find the derivative of J, we use

$$DJ(u)v = \frac{d}{d\theta} \frac{1}{2} \int_\Omega (|\nabla u|^2 + 2\theta \nabla u \cdot \nabla v + \theta^2 |\nabla v|^2 - 2f(u + \theta v)) \, dx \big|_{\theta=0}$$

$$= \int_\Omega (\nabla u \cdot \nabla v + \theta |\nabla v|^2 - fv) \, dx \big|_{\theta=0}$$

or

$$DJ(u)v = \int_\Omega (\nabla u \cdot \nabla v - fv) \, dx. \qquad \square$$

Note that DJ is an operator from $H^1(\Omega)$ to $H^1(\Omega)'$, and so $DJ(u)$ is a member of $H^1(\Omega)'$, that is, a functional on $H^1(\Omega)$.

We are now in a position to demonstrate the relationship between minimization problems and VBVPs, and we start with the following fundamental result.

Theorem 1. *Let J be a convex functional defined on a subspace V of a linear space U. An element $u \in V$ is a solution of the minimization problem*

$$J(u) \leq J(v) \qquad \text{for all } v \in V \tag{15}$$

if and only if u is a solution of the VBVP

$$DJ(u)v = 0 \qquad \text{for all } v \in V. \tag{16}$$

Proof. We show first that (15) implies (16). Assume that (15) holds: then, replacing v by $u + \theta v$, we have

$$J(u + \theta v) - J(u) \geq 0.$$

Dividing by θ and allowing θ to go to zero, we obtain

$$DJ(u)v \geq 0. \tag{17}$$

But v is arbitrary, so (17) holds if we replace v by $-v$. Using the linearity of $DJ(u)$ we get $DJ(u)v \leq 0$, and so $DJ(u)v = 0$.

To show that (16) implies (15), we start with

$$J(\theta v + (1 - \theta)u) = J(u + \theta(v - u)) \leq \theta J(v) + (1 - \theta)J(u)$$

$$= J(u) + \theta(J(v) - J(u))$$

by the convexity of J. Hence

$$J(v) - J(u) \geq \frac{J(u + \theta(v - u)) - J(u)}{\theta}$$

and, as $\theta \to 0$, we get

$$J(v) - J(u) \geqslant DJ(u)(v - u) = 0$$

so that (16) implies (15).   ∎

Example

Suppose that $U = H^1(\Omega)$, $V = H_0^1(\Omega)$ and

$$J(v) = \frac{1}{2} \int_\Omega \nabla v \cdot \nabla v \, dx - \int_\Omega fv \, dx. \tag{18}$$

We found $DJ(u)$ in the previous example, so it follows that the problem of finding $u \in H_0^1(\Omega)$ that minimizes (18) is equivalent to the problem of finding $u \in H_0^1(\Omega)$ that satisfies

$$DJ(u)v = 0 \quad \text{or} \quad \int_\Omega \nabla u \cdot \nabla u \, dx = \int_\Omega fv \, dx \quad \text{for all } v \in H_0^1(\Omega).$$
$$\tag{19}$$

We recognize (19) as a VBVP.   □

The above example is just a special case of the general minimization problem which involves *quadratic functionals* of the form

$$J : V \to R, \qquad J(v) = \tfrac{1}{2} B(v, v) - l(v) \tag{20}$$

in which $B(.\,,.)$ is a *symmetric bilinear form* on V and $l(\cdot)$ is a *linear functional* on V. Here V is a subspace of a Sobolev space $H^m(\Omega)$. When J takes the form (20) then we have

$$DJ(u)v = \lim_{\theta \to 0} \theta^{-1}[\tfrac{1}{2}B(u + \theta v, u + \theta v) - l(u + \theta v) - \tfrac{1}{2}B(u, u) + l(u)]$$

$$= \lim_{\theta \to 0} \theta^{-1}[\theta(B(u, v) - l(v)) + \tfrac{1}{2}\theta^2 B(v, v)]$$

$$= B(u, v) - l(v)$$

using the bilinearity and symmetry of $B(.\,,.)$ and the linearity of $l(\cdot)$. Hence the problem of minimizing (20) is equivalent to the VBVP of finding $u \in V$ satisfying

$$B(u, v) = l(v) \qquad \text{for all } v \in V, \tag{21}$$

assuming that $J(\cdot)$ is convex. This is not usually a problem: if $B(.\,,.)$ is V-elliptic then it is strictly convex, as is easily shown (see Exercise 37.4).

To summarize, then, any VBVP of the form (21) in which $B(.\,,.)$ is V-elliptic is equivalent to the problem of minimizing the functional (20), and *vice versa*. This equivalence, as a matter of interest, explains the reason for the terminology "variational" in the expression "variational boundary-value problem". The classical *calculus of variations* is concerned with the problem of minimizing functionals of a general nature, and the expression $DJ(u)v$ is known in that theory as the *first variation* of J. A necessary condition for a minimum is that the *first variation vanishes*, that is, $DJ(u)v = 0$, and this is what we call a variational BVP. It is important to note, though, that we refer to problems of the form (21) as VBVPs *even if $B(.\,,.)$ is not symmetric*, in which case there is *no* corresponding minimization problem.

We close this section with a theorem that gives conditions for the *existence* and *uniqueness* of solutions to minimization problems involving functionals of the form (20). Of course, existence and uniqueness could be discussed in terms of the equivalent VBVP (21), using the theory of Section 36. But for completeness we discuss problem (20) on its own, and show in fact that the requirements of Theorem 36.1 and Theorem 2 below coincide.

Theorem 2.   *Let $J : V \to R$ be the functional given by* (20), *in which V is a* closed subspace *of a Hilbert space H. Assume that $B(.\,,.)$ is bilinear, symmetric, continuous and V-elliptic, and that l is continuous and linear. Then the problem of finding $u \in V$ that minimizes $J(v)$ over all $v \in V$ has one and only one solution.*

Proof.   We start by observing that $B(.\,,.)$ defines an *inner product* on V: indeed, if we write $B(u, v) \equiv (u, v)_B$ then we have

$$(u, v)_B = (v, u)_B, \qquad (\alpha u + \beta v, w)_B = \alpha(u, w)_B + \beta(v, w)_B,$$

and the positive-definiteness of $(.\,,.)_B$ follows from the continuity and V-ellipticity of B:

$$K \|u\|_H^2 \geqslant (u, u)_B \geqslant \alpha \|u\|_H^2,$$

so that $(u, u)_B \geqslant 0$ and $(u, u)_B = 0$ if and only if $u = 0$. Hence $(.\,,.)_B$ is an inner product on the Hilbert space V, and in accordance with the Riesz representation theorem there exists $\bar{l} \in V$ such that $l(v) = (\bar{l}, v)_B$. Hence (20) reads

$$J(v) = \tfrac{1}{2} \|v - \bar{l}\|_B^2 - \tfrac{1}{2} \|\bar{l}\|_B^2, \tag{22}$$

where $\|\cdot\|_B$ is the norm generated by $(.\,,.)_B$. From (22) it is clear that our problem amounts to one of finding $u \in V$ such that

$$\|u - \bar{l}\|_B \leqslant \|\bar{l} - v\|_B \qquad \text{for all } v \in V.$$

By Theorem 14.1 such an element exists and is unique. Indeed, since $\bar{l} \in V$ we have $u = \bar{l}$. ∎

We remark in conclusion that Theorem 2 is equivalent to Theorem 36.1 when the bilinear form is *symmetric*, but that Theorem 36.1 alone is of use if $B(.,.)$ is non-symmetric.

Bibliographical remarks

Sections 34 to 36 follow to some extent the treatment given by Lions and Magenes [27], Oden and Reddy [34], [35], and Rektorys [39]. The discussion of non-V-elliptic problems which includes Theorem 36.2 is adapted from the treatment of this topic by Nečas [30] and Rektorys [39]. These two authors, as well as Gilbarg and Trudinger [15], should be consulted for an alternative treatment of the regularity of solutions to VBVPs. Grisvard [18] also discusses the issue of regularity in great detail, giving results for problems in nonsmooth domains.

The discussion in Section 37 of the minimization of convex functionals has focused only on those aspects pertinent to our main goals. For more details on the topic the texts by Blanchard and Brüning [9], Glowinski [16], Mikhlin [28], Oden and Reddy [35] and Reddy [37] are good sources.

We have avoided discussion of problems such as (37.15) when V is a *convex subset* but not a subspace. For example, consider the problem of finding a function that satisfies

$$-\nabla^2 u - f \geqslant 0, \qquad u \geqslant g \qquad \text{in } \Omega,$$

$$(u - g)(-\nabla^2 u - f) = 0 \qquad \text{in } \Omega,$$

$$u = 0 \qquad \text{on } \Gamma.$$

This corresponds to the problem of finding the shape of a membrane stretched over an obstacle, as shown in Figure 9.3.

It can be shown that the corresponding VBVP is the *variational inequality*: find $u \in K$ such that

$$\int_\Omega \nabla u \cdot \nabla(v - u) \, dx - \int_\Omega f(v - u) \, dx \geqslant 0 \qquad \text{for all } v \in K$$

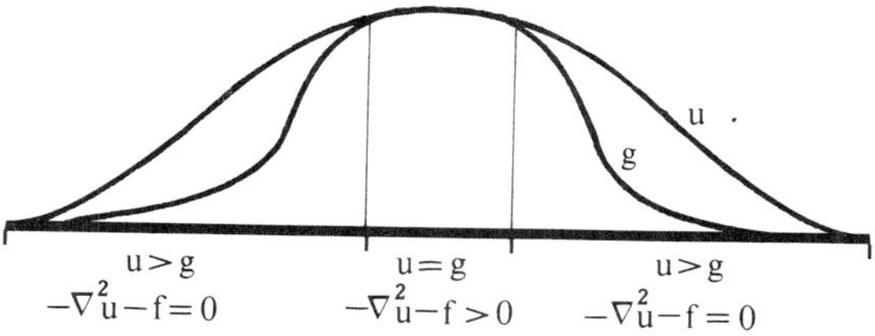

Figure 9.3

where K is the *convex subset*

$$K = \{v \in H_0^1(\Omega) : v \geq g \text{ a.e. in } \Omega\}$$

(note that K is not a subspace). The corresponding minimization problem is: find $u \in K$ such that

$$J(u) \leq J(v) = \frac{1}{2} \int_\Omega \nabla v \cdot \nabla v \, dx - \int_\Omega fv \, dx$$

for all $v \in K$. For a detailed account of variational inequalities see, for example, Baiocchi and Capelo [4], Fučik and Kufner [14] and Glowinski [16]. The book by Duvaut and Lions [13] is devoted to a thorough study of variational inequalities that arise in mechanics and physics.

Finally, we have restricted attention to linear problems. For general results on nonlinear variational problems, including the minimization of arbitrary functionals, see Oden and Reddy [35], Reddy [37] and Blanchard and Brüning [9].

Exercises

35.1 Derive the VBVP corresponding to the problem

$$-\frac{d}{dx}\left(p(x)\frac{du}{dx}\right) + r(x)u = f \qquad \text{in } \Omega = (0, 1),$$

$$u(0) = 0, \qquad u'(1) + u(1) = 0.$$

35.2 Find the VBVP corresponding to

$$-\nabla^2 u = f \qquad \text{in } \Omega \subset R^2$$

$$\frac{\partial u}{\partial \tau} = g \qquad \text{on } \Gamma,$$

in which $\partial u / \partial \tau \equiv \nabla u \cdot \boldsymbol{\tau}$ is the directional derivative in the direction of the unit vector $\boldsymbol{\tau}$ ($\boldsymbol{\tau}$ is not tangential to the boundary Γ).

35.3 Show that

$$\int v \nabla^4 u \, dx = \int_\Omega \left[\frac{\partial^2 u}{\partial x^2}\frac{\partial^2 v}{\partial x^2} + 2\frac{\partial^2 u}{\partial x \, \partial y}\frac{\partial^2 v}{\partial x \, \partial y} + \frac{\partial^2 u}{\partial y^2}\frac{\partial^2 v}{\partial y^2}\right] dx$$

$$+ \int_\Gamma \left[v\frac{\partial}{\partial \nu}(\nabla^2 u) - v\frac{\partial}{\partial s}\left[\nu_1\nu_2\left(\frac{\partial^2 u}{\partial x^2} - \frac{\partial^2 u}{\partial y^2}\right)\right.\right.$$

$$\left.\left. - (\nu_1^2 - \nu_2^2)\frac{\partial^2 u}{\partial x \, \partial y}\right] - \frac{\partial v}{\partial \nu}\frac{\partial^2 u}{\partial \nu^2}\right] ds.$$

Hence obtain the VBVP corresponding to

$$\nabla^4 u = f \quad \text{in } \Omega \subset R^2,$$

$$u = 0, \frac{\partial^2 u}{\partial v^2} = h \quad \text{on} \quad \Gamma.$$

36.1 Derive the Poincaré–Friedrichs inequality

$$\int_\Omega v^2 \, dx \leqslant C \int_\Omega \nabla v \cdot \nabla v \, dx \qquad \text{for } v \in H_0^1(\Omega).$$

[Consider $\int_\Omega v^2 \, dx$; use Green's theorem to transform this to $-2 \int_\Omega x_1 v \frac{\partial v}{\partial x_1} \, dx$. Then use the fact that Ω is bounded, and use the Schwarz inequality.]

36.2 Consider the BVP of finding u which satisfies

$$-\sum_{i,j} \frac{\partial}{\partial x_i} \left(a_{ij} \frac{\partial u}{\partial x_j} \right) + bu = f \quad \text{in} \quad \Omega,$$

$$u = 0 \quad \text{on} \quad \Gamma;$$

the coefficients a_{ij} are such that A is strongly elliptic. Derive the corresponding VBVP and show that the bilinear form is V-elliptic provided that $b(\mathbf{x}) \geqslant 0$. Show also that the bilinear form is continuous provided that $|a_{ij}(\mathbf{x})| \leqslant K$.

36.3 Show that the bilinear form

$$B(u, v) = \int_\Omega \left[\frac{\partial^2 u}{\partial x^2} \left(\frac{\partial^2 v}{\partial x^2} + \sigma \frac{\partial^2 v}{\partial y^2} \right) \right.$$

$$+ 2(1 - \sigma) \frac{\partial^2 u}{\partial x \, \partial y} \frac{\partial^2 v}{\partial x \, \partial y} + \left. \left(\frac{\partial^2 v}{\partial y^2} + \sigma \frac{\partial^2 v}{\partial x^2} \right) \frac{\partial^2 u}{\partial y^2} \right] dx$$

which comes from the BVP

$$\nabla^4 u = f \quad \text{in } \Omega,$$

$$u = 0 \quad \text{and} \quad \sigma \nabla^2 u + (1 - \sigma) \partial^2 u / \partial v^2 = 0 \qquad \text{on } \Gamma,$$

is V-elliptic provided that σ lies in the range $0 \leqslant \sigma < 1$. [Use the inequality

$$\|v\|_{H^2}^2 \leqslant C \sum_{|\alpha|=2} \int_\Omega (D^\alpha v)^2 \, dx + c \int_\Gamma u^2 \, ds,$$

which holds for all functions in $H^2(\Omega)$ (Rektorys [39], p. 341), after writing $B(v, v)$ in the form

$$B(v, v) = \int_\Omega \sigma\left[\frac{\partial^2 v}{\partial x^2} + \frac{\partial^2 v}{\partial y^2}\right]^2 dx + (1 - \sigma)\int_\Omega \cdots \bigg]$$

36.4   Show that the bilinear form associated with the BVP

$$-(pu')' + ru = f \quad \text{in } \Omega = (0, 1),$$
$$u(0) = 0, \; u'(1) + u(1) = 0,$$

is V-elliptic and continuous; here

$$p_1 \geqslant p(x) \geqslant p_0 > 0, \qquad r_1 \geqslant r(x) \geqslant r_0 > 0.$$

36.5   Investigate the existence and uniqueness of a solution to the VBVP corresponding to

$$\frac{d^4 u}{dx^4} = f \quad \text{in } (0, 1),$$

$$u''(0) = g_0, \; u''(1) = g_1,$$
$$u'''(0) = h_0, \; u'''(1) = h_1,$$

using Theorem 36.2.

37.1   Show that an equivalent definition of the Gateaux derivative is

$$DJ(u)v = \frac{d}{d\theta}[J(u + \theta v)]_{\theta=0}.$$

37.2   If $J : V \to R$ is a convex differentiable functional on a convex set V, show that

$$(DJ(u) - DJ(v))(u - v) \geqslant 0 \qquad \text{for all } u, v \in V.$$

37.3   Consider the functional $J : R^n \to R$; show that

$$DJ(\mathbf{x})\mathbf{y} = \sum_{i=1}^{n} \frac{\partial J}{\partial x_i} y_i.$$

37.4   If $B : V \times V \to R$ is a V-elliptic, symmetric bilinear form and $l : V \to R$ is a linear functional, show that

$$J(v) = \tfrac{1}{2}B(v, v) - l(v)$$

is strictly convex.

10

Approximate methods of solution

In the two preceding chapters we have devoted considerable attention to various aspects of boundary-value problems. The stage has now been reached where we can quite justifiably ask: how does one actually obtain solutions? The answer is rather disappointing, unfortunately: except for a few problems involving very simple PDEs and geometries, it is quite impossible (as far as is known) to obtain exact solutions to most BVPs in either the conventional or variational formulations. This state of affairs naturally leads us to ask next about the possibility of obtaining *approximate* solutions. Here things are far more encouraging; there are available many good methods for finding approximate solutions. Some of these are based on the conventional formulation, for example, finite difference methods, while others, such as the Galerkin method, are based on the variational formulation.

In the long run the approximate methods that make use of variational formulations are more versatile, and have enjoyed a great upsurge in popularity in the past two decades, particularly since the finite element method (a particular case of the Galerkin method) has become established. In the sections that follow we show how approximate solutions to VBVPs or, equivalently, to the corresponding minimization problems, can be obtained. The emphasis is on the Galerkin method and the finite element method, though we also give some indication of other related methods in Section 40.

§38. The Galerkin method

The basic idea behind the Galerkin method is an extremely simple one. Consider the VBVP of finding $u \in V$ that satisfies

$$B(u, v) = l(v) \qquad \text{for all } v \in V, \tag{1}$$

where V is a subspace of a Hilbert space. The difficulty in trying to solve (1) lies with the fact that V is a very large space (infinite-dimensional, in the language of Chapter 6), with the result that it is impossible to set up a logical procedure for finding the solution. But suppose that, instead of posing the problem in V, we pick a few linearly independent functions $\phi_1, \phi_2, \ldots, \phi_N$ in V and define the space V_h to be the finite-dimensional subspace of V spanned by the functions ϕ_i. That is,

$$V_h \subset V, \qquad \text{span } \{\phi_i\}_{i=1}^N = V_h. \tag{2}$$

The index h is a parameter that lies between 0 and 1, and whose magnitude gives some indication of how close V_h is to V; h is related to the dimension of V_h, and as the number N of basis functions chosen gets larger, h gets smaller (for example, we could set $h = 1/N$). In the limit, as $N \to \infty$, $h \to 0$ and we would like to choose $\{\phi_i\}$ in such a way that V_h will approach V.

Having defined the space V_h we now pose problem (1) in V_h instead of in V. That is, we try to find a function $u_h \in V_h$ that satisfies

$$B(u_h, v_h) = l(v_h) \qquad \text{for all } v_h \in V_h. \tag{3}$$

This is the essence of the Galerkin method. In order to solve for u_h, we simply note that both u_h and v_h must be linear combinations of the basis functions of V_h, so that

$$u_h = \sum_{i=1}^N a_i \phi_i, \qquad v_h = \sum_{j=1}^N b_j \phi_j. \tag{4}$$

Of course, since v_h is arbitrary, so are the coefficients b_k. We substitute (4) in (3) and use the fact that B is bilinear and l is linear to obtain

$$\sum_{i=1}^N \sum_{j=1}^N B(\phi_i, \phi_j) a_i b_j = \sum_{j=1}^N l(\phi_j) b_j$$

or, more concisely,

$$\sum_{j=1}^N b_j \left(\sum_{i=1}^N K_{ij} a_i - F_j \right) = 0 \tag{5}$$

where

$$K_{ij} = B(\phi_i, \phi_j) \quad \text{and} \quad F_j = l(\phi_j) \tag{6}$$

are, respectively, an $N \times N$ matrix and an N-tuple (vector). Note that K_{ij} and F_j can be evaluated in practice since the ϕ_i are known functions and the forms of B and l are also known.

Since the coefficients b_j are arbitrary, it follows that (5) will only hold if the term in brackets is zero. We have thus reduced the problem to one of solving the set of *simultaneous linear equations*

$$\sum_{i=1}^{N} K_{ij}a_i = F_j, \qquad j = 1, \ldots, N. \tag{7}$$

Once these equations are solved, the approximate solution u_h can be found from the first of equations (4).

Examples

(a)   Consider the BVP

$$-\frac{d^2u}{dx^2} = \sin\frac{\pi x}{2} \text{ in } \Omega = (0, 1), \qquad u(0) = u'(1) = 0.$$

The corresponding VBVP is: find $u \in V$ such that

$$\int_0^1 u'v'\, dx = \int_0^1 (\sin \pi x/2)v\, dx \qquad \text{for all } v \in V,$$

where $V = \{v \in H^1(0, 1) : u(0) = 0\}$ (note that $u'(1) = 0$ is a natural boundary condition). Define V_h to be the subspace of V spanned by

$$\phi_i(x) = x^i, \qquad i = 1, 2, \ldots, N.$$

We then have

$$K_{ij} = B(\phi_i, \phi_j) = \int_0^1 \phi_i' \phi_j'\, dx$$

and

$$F_i = l(\phi_i) = \int_0^1 (\sin \pi x/2)\phi_i\, dx.$$

Suppose that we take $N = 2$: then we obtain the set of simultaneous equations

$$\sum K_{ij}a_j = F_i \Leftrightarrow a_1 + a_2 = 0.405 \qquad a_1 = 0.738$$

$$\Rightarrow$$

$$a_1 + \tfrac{4}{3}a_2 = 0.295 \qquad a_2 = -0.33$$

Hence

$$u_h(x) = a_1\phi_1(x) + a_2\phi_2(x)$$
$$= 0.738x - 0.33x^2.$$

The exact solution to this problem is

$$u(x) = 0.405 \sin \pi x/2$$

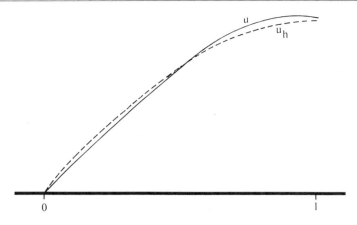

Figure 10.1

and is compared with the approximate solution in Figure 10.1. We see that even the crude approximation in a two-dimensional subspace produces in this case a solution barely distinguishable from the exact solution.

(b)   Consider the VBVP of finding $u \in H_0^1(\Omega)$ that satisfies

$$\int_\Omega \nabla u \cdot \nabla v \, \mathrm{d}x = \int_\Omega xyv \, \mathrm{d}x \qquad \forall \, v \in H_0^1(\Omega).$$

[The corresponding BVP is

$$-\nabla^2 u = xy \qquad \text{in } \Omega,$$
$$u = 0 \qquad \text{on } \Gamma.]$$

Here Ω is the unit square $(0, 1) \times (0, 1)$ in R^2. We now choose a set of basis functions in $H_0^1(\Omega)$: let

$$\phi_1 = \sin \pi x \sin \pi y, \qquad \phi_2 = \sin \pi x \sin 2\pi y,$$
$$\phi_3 = \sin 2\pi x \sin \pi y, \qquad \phi_4 = \sin 2\pi x \sin 2\pi y.$$

These four functions span a four-dimensional subspace of $H_0^1(\Omega)$. The next step is to evaluate

$$K_{ij} = B(\phi_i, \phi_j) \quad \text{and} \quad F_i = (f, \phi_i) = (xy, \phi_i),$$

which is straightforward if we make use of the identity

$$\int_0^1 \sin n\pi x \sin m\pi x \, \mathrm{d}x = \int_0^1 \cos n\pi x \cos m\pi x \, \mathrm{d}x = \begin{cases} 0 & \text{if } n \neq m, \\ \tfrac{1}{2} & \text{if } n = m. \end{cases} \qquad (8)$$

Then

$$K_{ij} = \int_0^1 \int_0^1 \left(\frac{\partial \phi_i}{\partial x} \frac{\partial \phi_j}{\partial x} + \frac{\partial \phi_i}{\partial y} \frac{\partial \phi_j}{\partial y} \right) dx\, dy, \qquad F_i = \int_0^1 \int_0^1 xy\phi_i\, dx\, dy.$$

Because of (8) we find that the only non-zero terms of K_{ij} are

$$K_{ii} = \int_0^1 \int_0^1 \left(\frac{\partial \phi_i}{\partial x} \right)^2 + \left(\frac{\partial \phi_i}{\partial y} \right)^2 dx\, dy$$

$$= \pi^2 \int_0^1 \int_0^1 (n^2 \cos^2 n\pi x \sin^2 m\pi y + m^2 \sin^2 n\pi x \cos^2 m\pi y)\, dx\, dy$$

where n and m take the following values:

i	1	2	3	4
n	1	1	2	2
m	1	2	1	2

After carrying out the integration we obtain

$$K_{ii} = \frac{\pi^2}{4}(n^2 + m^2) \qquad \text{or} \qquad \mathbf{K} = \begin{bmatrix} 2 & 0 & 0 & 0 \\ 0 & 5 & 0 & 0 \\ 0 & 0 & 5 & 0 \\ 0 & 0 & 0 & 8 \end{bmatrix} \frac{\pi^2}{4}.$$

Similarly,

$$F_i = \int_0^1 \int_0^1 xy \sin n\pi x \sin m\pi y\, dx\, dy$$

$$= \frac{1}{\pi^2}(1, -2, -2, 4).$$

Hence the solution is

$$\mathbf{a} = \frac{4}{\pi^4} (\tfrac{1}{2}, -\tfrac{2}{5}, -\tfrac{2}{5}, \tfrac{1}{2})$$

and so

$$u_h(x, y) = \frac{4}{\pi^4} [\tfrac{1}{2}(\sin \pi x \sin \pi y + \sin 2\pi x \sin 2\pi y)$$

$$- \tfrac{2}{5}(\sin \pi x \sin 2\pi y + \sin 2\pi x \sin \pi y)]. \qquad \square$$

We recall from Section 19 that the bilinear form $B(.\,,.)$ defines an *inner product* on V if B is symmetric and V-elliptic; indeed, the properties of linearity and symmetry are obvious, while the property of positive-definiteness comes from the V-ellipticity of B:

$$B(v, v) \geqslant \alpha \, \|v\|_V^2 > 0 \qquad \text{for all non-zero } v. \tag{9}$$

As before, we denote this inner product by $(.\,,.)_B$ and refer to it as the *energy inner product* (the name comes from the physical origins of certain elliptic BVPs, in which $B(u, u)$ is a measure of the amount of energy in a system); the corresponding norm is called the *energy norm*, and is denoted by $\|\cdot\|_B$.

Thus we have

$$(u, v)_B = B(u, v), \qquad \|u\|_B^2 = (u, u)_B. \tag{10}$$

Now if the set of basis functions $\{\phi_i\}_{i=1}^N$ is chosen in such a way that they are *orthogonal with respect to the energy inner product*, then the system of equations (7) simplifies considerably, since

$$K_{ij} = B(\phi_i, \phi_j) = (\phi_i, \phi_j)_B = 0 \quad \text{if } i \neq j,$$

and so

$$K_{ii}a_i = F_i \quad \text{or} \quad a_i = F_i / K_{ii}.$$

This is in fact the case in Example (b) above.

However, a word of warning is appropriate. While for the above example it was quite simple to find a basis that was orthogonal with respect to $(.\,,.)_B$, in general this is quite difficult. One could of course choose any non-orthogonal basis and use the Gram–Schmidt procedure of Section 21 to orthogonalize or even orthonormalize, but for all except the most trivial problems this is a laborious procedure, and little is to be gained from it.

The problem of how to construct a basis $\{\phi_i\}_{i=1}^N$ in such a way that V_h approaches V as $N \to \infty$, can be rather awkward. Remember that while orthonormal bases for spaces such as L_2 are well known (see, for example, Section 23), when using the Galerkin method we are required to find bases for spaces V that are subspaces of Sobolev spaces $H^m(\Omega)$, and that are defined on domains Ω which may be quite irregular in shape. A very simple and elegant method for constructing such bases is provided by the finite element method. We shall describe this procedure after first discussing, in the following section, a few important properties of Galerkin approximations.

The Rayleigh–Ritz method. The Rayleigh–Ritz method is very closely linked to the Galerkin method. It takes as its starting-point the minimization problem (37.15) and, as with the Galerkin method, this problem is posed in a finite-dimensional subspace. That is, we consider the problem of finding $u_h \in V_h$ such that

$$J(u_h) \leqslant J(v_h) \qquad \text{for all } v_h \in V_h, \tag{11}$$

where V_h is a finite-dimensional subspace of V. If $\{\phi_k\}_{k=1}^N$ is a basis for V_h then substitution of $V_h = \sum_{k=1}^N \alpha_k \phi_k$ in the expression for J yields the function

$$\hat{J}(\alpha_k) \equiv J(\textstyle\sum \alpha_k \phi_k), \tag{12}$$

which is a function of the N variables $\alpha_1, \ldots, \alpha_N$. In order to minimize $J(v_h)$, therefore, we require that

$$\frac{\partial \hat{J}}{\partial \alpha_k} = 0, \qquad k = 1, \ldots, N, \tag{13}$$

which yields a set of N simultaneous algebraic equations in the N unknowns $\alpha_1, \ldots, \alpha_N$. Solution of these equations yields the components α_k of u_h. In particular, if J is given by (37.20) then

$$\hat{J}(\alpha_k) = \tfrac{1}{2} \sum_{j,l=1}^N K_{jl} \alpha_j \alpha_l - \sum_{j=1}^N f_j \alpha_j$$

where

$$K_{jl} = B(\phi_j, \phi_l) \quad \text{and} \quad f_j = l(\phi_j)$$

and (13) yields the set of linear equations

$$\sum_{j=1}^N K_{jl} \alpha_j = f_l, \qquad l = 1, \ldots, N$$

which is precisely (7). Here, though, K_{jl} is always symmetric since $B(.,.)$ is symmetric.

§39. Properties of Galerkin approximations

In Section 38 we introduced the Galerkin method and illustrated how the method is used in practice. It is not very satisfactory, however, simply to leave things at that; we ought to know, first of all, whether the Galerkin method always works and, if so, how significantly the approximate solution differs from the exact solution. Also, we would like to be confident that as the number of functions ϕ_i in the basis of V_h is increased, V_h approaches in some sense the space V and u_h approaches the exact solution u. This last consideration is of course one of *convergence* of the approximate solution as $h \to 0$ (or as $N \to \infty$).

Existence and uniqueness. The question of existence and uniqueness of a solution is easily resolved, in view of the results in Section 36.
Since V_h is a finite-dimensional subspace of a Hilbert space, it is itself a
Hilbert space (see Theorem 21.1). Assuming then that B is a V-elliptic
continuous bilinear form on $V \times V$ and l is a continuous function on V,
the same obviously holds true when we restrict attention to V_h. Hence, if
we have already been able to show that a unique solution to (38.1) exists,
the same will hold true for the approximate problem (38.3). We record
this information in the following theorem.

Theorem 1. *Let V_h be a finite-dimensional subspace of a Hilbert
space V, and let $B : V_h \times V_h \to R$ and $l : V_h \to R$ be respectively a
continuous, V-elliptic bilinear form and a continuous linear functional.
Then there exists a unique function $u_h \in V_h$ that satisfies*

$$B(u_h, v_h) = l(v_h) \qquad \forall\, v_h \in V_h. \tag{1}$$

Furthermore, if l is of the form

$$l(v_h) = \int_\Omega f v_h \, dx,$$

then

$$\|u_h\|_V \leq \frac{1}{\alpha} \|f\|_{L_2}.$$

Errors in Galerkin approximations. Having established the conditions under which we can find an approximate solution, we proceed now
to characterize the *error e*, defined by

$$e = u - u_h. \tag{2}$$

We know that u satisfies

$$B(u, v) = l(v) \qquad \text{for all } v \in V, \tag{3}$$

and since any function $v_h \in V_h$ also belongs to V, we can write (3) as

$$B(u, v_h) = l(v_h) \qquad \text{for all } v_h \in V_h. \tag{4}$$

Furthermore, the Galerkin approximation u_h satisfies

$$B(u_h, v_h) = l(v_h) \qquad \text{for all } v_h \in V_h. \tag{5}$$

If we subtract (4) from (5) and use the bilinearity of B, we get

$$B(u_h, v_h) - B(u, v_h) = 0$$

or

$$B(u_h - u, v_h) = 0,$$

that is,

$$B(e, v_h) = 0. \tag{6}$$

This seemingly innocuous result has a useful geometrical interpretation in the event that B is *symmetric*; for then we have at our disposal the inner product $(.\,,.)_B$ defined in (38.9), and we can use the theory of Sections 17 and 23 on orthogonal projections. Indeed, according to Theorem 23.1, if $\{\phi_k\}_{k=1}^N$ is an orthonormal basis of V_h with respect to $(.\,,.)_B$ then the orthogonal projection onto V_h is defined by

$$Pv = \sum_{k=1}^N (v, \phi_k)_B \phi_k. \tag{7}$$

But from (4) and (5), if we set $v_h = \phi_k$ in these two equations we find

$$(u, \phi_k)_B = (u_h, \phi_k)_B$$

so that

$$Pu = \sum_{k=1}^N (u_h, \phi_k)_B \phi_k$$

$$= Pu_h = u_h. \tag{8}$$

Hence the orthogonal projection of the solution u onto V_h is the approximate solution u_h, using the inner product $(.\,,.)_B$. Clearly, then, the error $e = u - u_h = u - Pu \in N(P)$, that is,

$$(e, v_h)_B = 0,$$

which confirms (6).

In other words, relative to the inner product $(.\,,.)_B$ *the error is orthogonal to the subspace V_h,* a result which may be illustrated as in Figure 10.2.

The geometrical analogy may be carried a step further. Figure 10.2 strongly suggests that the distance $\|u - v_h\|$, when measured using the norm $\|\cdot\|_B$ defined in the usual way by $\|u\|_B^2 = (u, u)_B$, is a *minimum* when $v_h = u_h$. That this is indeed so is borne out by the following calculation: we have

$$\begin{aligned} B(u - v_h, u - v_h) &= B(u - u_h + u_h - v_h, u - u_h + u_h - v_h) \\ &= B(e + (u_h - v_h), e + (u_h - v_h)) \tag{9} \\ &= B(e, e) + 2B(e, u_h - v_h) + B(u_h - v_h, u_h - v_h), \end{aligned}$$

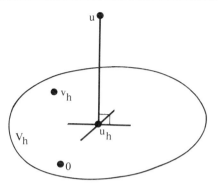

Figure 10.2

using the bilinearity of B. Now the second term on the right-hand side is zero, since e is orthogonal to all members of V_h using the inner product $(.\,,.)_B$. Thus

$$\|u - v_h\|_B^2 = \|e\|_B^2 + \|u_h - v_h\|_B^2$$

and for fixed u and u_h (and hence for fixed e) we conclude that $\|u - v_h\|$ is smallest when $v_h = u_h$, that is,

$$\|u - u_h\|_B \leqslant \|u - v_h\|_B \qquad \text{for all } v_h \in V_h. \tag{10}$$

In other words, the function in V_h that is closest to u is the Galerkin approximation. In this sense *the Galerkin approximation is the best approximation to u in V_h*. Of course we could have deduced (10) directly from Theorem 17.4 and (8). It is important to bear in mind, though, that (10) is only valid when B is symmetric.

Convergence of Galerkin approximations. As we mentioned earlier, each value of the parameter h defines a subspace V_h of V: the smaller h is, the larger the dimension of V_h will be. We could use for h the reciprocal of the number of basis functions that span V_h, though in Chapter 11 we will give h a more geometrical meaning, in the context of the finite element method. In any case, h lies in $(0, 1)$ and we can think of equation (5) as representing a *family* of Galerkin approximations, each value of h having associated with it a problem (5). Corresponding to this family of problems we have also a family of solutions u_h, and once again we associate with each value of h a particular solution u_h. Of course, if we define h by $h = 1/N$ where $N = \dim V_h$, then h cannot take on all values in $(0, 1)$, but only those of the form $1/N$ for integer N.

With these ideas at our disposal, it is quite simple to give a definition of the convergence of a family of Galerkin approximations: we say that the family of solutions u_h *converges to the exact solution* u if

$$\lim_{h \to 0} \|u_h - u\|_V = 0. \tag{11}$$

The task of proving convergence, once we have identified a basis or family of bases, is made considerably easier by a deceptively simple result, which has far-reaching implications. Since the bilinear form B is V-elliptic and continuous by assumption, we have

$$\alpha \|u - u_h\|^2 \leq B(u - u_h, u - u_h)$$
$$= B(u - u_h, u - v_h - u_h + v_h)$$
$$= B(u - u_h, u - v_h) - B(e, u_h - v_h).$$

The last term on the right hand side is zero so we have, using the continuity of B,

$$\alpha \|u - u_h\|^2 \leq B(u - u_h, u - v_h)$$
$$\leq K \|u - u_h\| \|u - v_h\|$$

or

$$\|u - u_h\| \leq \frac{K}{\alpha} \|u - v_h\|. \tag{12}$$

This result tells us that in order to get some idea of the size of the error $u - u_h$, it is not in fact necessary to work with u_h; it is good enough to pick any member v_h in V_h and to find the distance from u to v_h. Once this is known, (12) assures us that the error will be bounded above by $(K/\alpha) \|u - v_h\|$. The next question which of course arises is: which member of v_h do we use for this purpose? The most convenient choice is to take the *interpolate* of u: this is a function $\bar{u}_h$ in V_h whose value coincides with that of u at N points $x_1, x_2, \ldots, x_N$ in Ω. Since $\bar{u}_h$ has the representation

$$\bar{u}_h = \sum_{k=1}^{N} \bar{a}_k \phi_k \tag{13}$$

where $\{\phi_k\}$ is any basis of V_h, we can determine the coefficients $\bar{a}_k$ from the fact that

$$u(x_k) = \bar{u}_h(x_k), \qquad k = 1, \ldots, N, \tag{14}$$

for a given function u. That is, we solve for $\bar{a}_h$ the N simultaneous equations

$$\sum_{j=1}^{N} \bar{a}_j \phi_j(x_k) = u(x_k), \qquad k = 1, \ldots, N. \tag{15}$$

We observe that the operator $P : V \to V$ defined by

$$Pu = \bar{u}_h \qquad (16)$$

is a *projection operator*.

A simple example is illustrated below for the two-dimensional subspace of $H^1(0, 1)$ which consists of linear functions

$$\phi_1(x) = x, \qquad \phi_2(x) = 1 - x. \qquad (17)$$

Suppose that we require the interpolate u to be equal to u at the points $x_1 = 1/3$ and $x_2 = 2/3$: then (15) and (17) give

$$\bar{a}_1 \cdot \tfrac{1}{3} + \bar{a}_2 \cdot \tfrac{2}{3} = u(\tfrac{1}{3})$$
$$\bar{a}_1 \cdot \tfrac{2}{3} + \bar{a}_2 \cdot \tfrac{1}{3} = u(\tfrac{2}{3}).$$

Solving, we have

$$\bar{a}_1 = 2u(\tfrac{2}{3}) - u(\tfrac{1}{3}), \qquad \bar{a}_2 = 2u(\tfrac{1}{3}) - u(\tfrac{2}{3}).$$

With the choice (13) for v_h, (12) becomes

$$\|u - u_h\| \leqslant \frac{K}{\alpha} \|u - \bar{u}_h\| \qquad (18)$$

and so the problem of convergence reduces to one of finding out whether $\bar{u}_h \to u$ as $h \to 0$, and if so at what rate this occurs. We have thus reduced the problem of *convergence of Galerkin approximations* to one of *convergence of interpolates*.

In the case of the finite element method, which is distinguished by the fact that the basis functions are piecewise polynomials, we will see that the distance between u and its interpolate $\bar{u}_h$ satisfies an inequality of the form

$$\|u - \bar{u}_h\|_V \leqslant Ch^\beta \qquad (19)$$

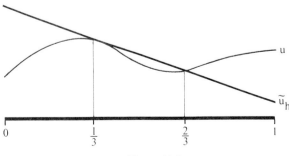

Figure 10.3

where the constant C is independent of h, and β is positive. Then (18) immediately tells us that

$$\|u - u_h\|_V \le \frac{CK}{\alpha} h^\beta. \tag{20}$$

Hence, as the approximation is progressively improved so that N gets larger and h gets smaller, we can expect u_h to converge to u at a rate that is determined by the magnitude of β. We write this as

$$\|u - u_h\| = 0(h^\beta) \tag{21}$$

and say that *the convergence of u_h to u is of order β.* Clearly we should like to have β as large as possible.

A result like (19) is called an *interpolation error estimate,* for obvious reasons, while (20) or (21) are called *Galerkin error estimates.* Our means of examining the convergence of Galerkin approximations will always be via estimates of the form (20).

§40.   Other methods of approximation

The Galerkin–Rayleigh–Ritz approach is very widely used, particularly in the context of finite-element methods. But a number of other approximate methods also exist, and while these are perhaps not as popular as the Galerkin method, they are nevertheless worth knowing about as they make interesting and viable alternative approaches. In this section we introduce some of the alternative methods, namely, the method of weighted residuals, the method of least squares, collocation methods and H^{-1}-methods.

We start by returning to the BVP

$$Au = f \qquad\qquad \text{in } \Omega \tag{1}$$
$$B_0 u = \cdots = B_{m-1}u = 0 \qquad \text{on } \Gamma \tag{2}$$

in which, as before, A is an elliptic operator of order $2m$. For convenience we will confine attention to problems with homogeneous boundary conditions. As in Chapter 9, we multiply both sides of (1) by a function v and integrate to obtain

$$(Au, v) = (f, v) \tag{3}$$

in which $(.\,,.)$ represents the L_2-inner product. Now in Chapter 9 we used Green's theorem to shift half the derivatives in Au over to v and then posed the VBVP

$$B(u, v) = (f, v) \tag{4}$$

for all $v \in V$ (V being a subspace of $H^m(\Omega)$). The other methods discussed here all rely on the observation that there are other ways besides (4) of formulating VBVPs. At one extreme, we could consider (3) as it stands and seek

$$u \in U, \qquad (Au, v) = (f, v) \qquad \text{for all } v \in L_2(\Omega) \tag{5}$$

in which, for example,

$$U = \{u \in H^{2m} : B_0 u = \cdots = B_{m-1} u = 0\}. \tag{6}$$

In this case no derivatives of u will have been shifted over to v. At the other extreme we could shift *all* derivatives over to v by repeated application of Green's theorem, and then pose the problem of finding $u \in L_2(\Omega)$ such that

$$(u, A^*v) = (f, v) \qquad \text{for all } v \in V \tag{7}$$

where A^* is the adjoint of A and $V = H_0^{2m}(\Omega)$, for example.

In all three cases (5), (4) and (7) we see that the solution u is sought in a space U and the equation is required to be satisfied for all functions v from a space V that is generally distinct from U. In the case of (4) we have $U = V$. The Galerkin method is of course based on (4), and we will find that weighted residual, least squares and collocation methods are based on (5) while H^{-1}-methods are based on (7).

Example

Consider the problem

$$-u'' + u = f \qquad \text{in } \Omega = (0, 1), \tag{8}$$
$$u(0) = u(1) = 0.$$

Equation (5) reads for this problem

$$\int_0^1 (-u'' + u)v \, dx = \int_0^1 fv \, dx$$

where $U = H^2(0, 1) \cap H_0^1(0, 1)$ and $V = L_2(0, 1)$. Integrating by parts once we obtain

$$\int_0^1 (u'v' + uv) \, dx = \int_0^1 fv \, dx$$

with $U = V = H_0^1(0, 1)$. Finally, by integrating by parts once more we may pose the problem of finding $u \in L_2(0, 1)$ such that

$$\int_0^1 (-uv'' + uv) \, dx = \int_0^1 fv \, dx$$

for all $v \in H_0^2(0, 1)$.   □

We now proceed to discuss the group of approximate methods based on the formulation (5). Suppose that we define finite-dimensional subspaces $U^h \subset U$ and $V^h \subset V$ where U is given by (6) and $V = L_2(\Omega)$. We pose the problem of finding a function $u_h \in U^h$ such that

$$(Au_h - f, v_h) = 0 \quad \text{or} \quad (r(u_h), v_h) = 0 \qquad (9)$$

for all $v_h \in V^h$. The expression $r(u_h) \equiv Au_h - f$ is called the *residual*; if u_h is the exact solution then of course the residual vanishes.

The method of weighted residuals. In this method we choose U^h and V^h above such that $\dim U^h = \dim V^h = N$, say, with bases

$$\{\phi_k\}_{k=1}^N \quad \text{for } U^h \qquad (10)$$

and

$$\{\psi_k\}_{k=1}^N \quad \text{for } V^h. \qquad (11)$$

Then we have

$$u_h = \sum_{k=1}^N a_k \phi_k \quad \text{and} \quad v_h = \sum_{k=1}^N b_k \psi_k \qquad (12)$$

where the coefficients b_k are arbitrary in view of the arbitrariness of v_h. Substitution of (12) in (9) yields a set of N simultaneous equations

$$\sum_{k=1}^N M_{kl} a_k = F_l, \quad l = 1, \ldots, N,$$

in which

$$M_{kl} = (A\phi_k, \psi_l) \quad \text{and} \quad F_l = (f, \psi_l).$$

The method of least squares. This method entails finding a function $u_h \in U^h$ that *minimizes* the residual, the magnitude of the residual being measured in the L_2-norm (hence the name of the method). That is, we define a functional

$$J: U^h \to R, \quad J(v_h) = \|r(v_h)\|_{L_2}^2$$

$$= \int_\Omega (Av_h - f)^2 \, dx$$

and we seek $u_h \in U^h$ such that

$$J(u_h) \leq J(v_h) \quad \text{for all } v_h \in U^h.$$

From Section 37 we know that this is equivalent to the problem of finding $u_h \in U^h$ such that

$$DJ(u_h)v_h = 0$$

or

$$\int_\Omega (Au_h - f)Av_h \, dx = 0 \qquad \text{for all } v_h \in U^h. \tag{13}$$

Comparison with (9) indicates that the method of least squares is equivalent to the method of weighted residuals if in the latter we choose the space V^h to be the span of the functions $\{A\phi_k\}_{k=1}^N$ where $\{\phi_k\}_{k=1}^N$ is the basis for U^h.

Collocation methods. The idea behind these methods is to force the residual to vanish at a finite number of points in the domain Ω. That is, if dim $U^h = N$ then we seek $u_h \in U^h$ such that

$$Au_h - f = 0 \qquad \text{at} \qquad \mathbf{x} = \mathbf{x}_k, \qquad k = 1, 2, \ldots, N.$$

The collocation method is also a variant of the method of weighted residuals, as we can appreciate from the following considerations. Corresponding to each v_h in V^h we can define a linear functional l_h such that

$$l_h(Au_h - f) = (Au_h - f, v_h).$$

Suppose for definiteness that u_h and f are smooth enough for the residual $r(u_h) = Au_h - f$ to belong to $H_0^1(\Omega)$: that is,

$$Au_h - f \in H \subset H_0^1(\Omega).$$

Then l_h is a linear functional on H, so that l_h lies in the dual space H' of H and we must find $u_h \in U^h$ such that

$$l_h(r(u_h)) = 0 \qquad \text{for all } l_h \in H'. \tag{14}$$

The collocation method arises from choosing the functionals l_h to be the N Dirac delta functionals $\delta_{\mathbf{x}_j}$, $j = 1, \ldots, N$, defined by

$$\delta_{\mathbf{x}_j} : H \to R, \qquad \delta_{\mathbf{x}_j}(w) = w(\mathbf{x}_j).$$

H^{-1}-methods. Unlike the methods discussed previously, the H^{-1}-method is based on the formulation (7) in which all derivatives are transferred by means of successive applications of Green's theorem. Working in finite-dimensional subspaces, we now seek $u_h \in U^h$ such that

$$(u_h, A^*v_h) = (f, v_h) \qquad \text{for all } v_h \in V^h, \tag{15}$$

where $U^h \subset L_2(\Omega)$ and $V^h \subset H_0^{2m}(\Omega)$. When using this approach we are free to choose U^h in such a way that some or all of its members are discontinuous, while at the same time the functions making up V^h have to belong to $H_0^{2m}(0, 1)$, so that we are demanding from them here a greater degree of smoothness than in the Galerkin method.

If we return to the earlier example (8) then the H^{-1}-method entails finding $u_h \in U^h \subset L_2(0, 1)$ such that

$$\int_0^1 (-u_h v_h'' + u_h v_h) \, dx = \int_0^1 f v_h \, dx$$

for all $v_h \in V^h \subset H_0^2(0, 1)$. So, for example, we could choose as a basis for the U^h a set of piecewise-constant functions while for V^h we need a basis of functions which are at least continuously differentiable and which, together with their first derivatives, vanish on the boundary.

In Exercise 40.2 the four methods introduced here are applied to a simple problem.

Bibliographical remarks

There is a large body of literature on the Galerkin and Rayleigh–Ritz methods, with texts ranging from those that deal mainly with computational aspects to those that also consider questions of convergence. The text by Vichnevetsky [49] falls into the former category, while Rektorys [39], Reddy [37] Mikhlin [28] and Strang and Fix [48] cover both aspects. The books by Becker, Carey and Oden [6] and Oden and Reddy [34, 35] all give full coverage of the Galerkin method, though they emphasize the use of the method as part of the finite element method. Further details on the alternative methods discussed in Section 40 may be found in Carey and Oden [10] and in Rektorys [39], among others.

Exercises

38.1   The BVP $(xu')' = x$ in $\Omega = (1, 2)$, $u(1) = u(2) = 0$, has the exact solution $u(x) = \frac{1}{4}x^2 - \frac{3}{4}\dfrac{\ln x}{\ln 2} - \frac{1}{4}$. Use the Galerkin method to find an approximate solution u_h in the subspace of $H_0^1(1, 2)$ spanned by $\phi_1(x) = (x - 1)(x - 2)$ and $\phi_2(x) = x(x - 1)(x - 2)$. Compare the exact and approximate solutions by (a) sketching graphs of u and u_h; (b) evaluating the errors $\|e\|_{L_\infty}$, $\|e\|_{L_2}$ and $\|e\|_{H^1}$, where $e = u - u_h$.

38.2   Use the Galerkin method with basis functions $\phi_1(x, y) = (-x^2 + x)(-y^2 + y)$ and $\phi_2(x, y) = (x^3 - \frac{3}{2}x^2 + \frac{1}{2}x)(y^3 - \frac{3}{2}y^2 + \frac{1}{2}y)$ to solve the BVP

$$-\nabla^2 u = xy \quad \text{on } \Omega = (0, 1) \times (0, 1),$$
$$u = 0 \quad \text{on } \Gamma.$$

38.3   Let $J: H \to R$ be a function on a Hilbert space H with

$$J(v) = \tfrac{1}{2}B(v, v) - l(v), \qquad v \in H$$

where $B(.,.)$ is continuous and H-elliptic and $l(\cdot)$ is continuous, and let $\{u_n\}$ be a sequence in H such that

$$\lim_{n \to \infty} J(u_n) = J(u) \leqslant J(v).$$

Then $\{u_n\}$ is called a *minimizing sequence*. The aim of this Exercise is to show how such a minimizing sequence can be generated, and secondly to show that $u_n \to u$ in the norm $\|\cdot\|_B$ (see (38.10)).
Let $\{\phi_k\}_{k=1}^{\infty}$ be an orthonormal basis for H with respect to the inner product $(.,.)_B = B(.,.)$, and let $H^{(n)} = \mathrm{span}\,\{\phi_1, \ldots, \phi_n\}$. Let u_n be the minimizer of J in the space $H^{(n)}$. Show that $u_n = \sum_{k=1}^{n} l(\phi_k)\phi_k$ and that $l(\phi_k) = (u, \phi_k)_B$ where u is the minimizer of J in H. Hence deduce that u_n is the nth partial sum of the Fourier series expansion for u, and conclude that $\|u_n - u\|_B \to 0$ and $\|u_n - u\|_H \to 0$.

38.4   Use the Rayleigh–Ritz method with basis function $\phi_1(x, y) = (x^2 - \alpha^2)(y^2 - \beta^2)$ to find an approximate solution to the problem of minimizing

$$J(v) = \frac{D}{2} \int_{-\alpha}^{\alpha} \int_{-\beta}^{\beta} \left[(\nabla^2 v)^2 - \frac{2qv}{D} \right] \mathrm{d}x\,\mathrm{d}y,$$

$$J: H_0^2(\Omega) \to R \quad \text{where} \quad \Omega = (-\alpha, \alpha) \times (-\beta, \beta).$$

The exact solution to the corresponding classical problem $\nabla^4 u = q$ in Ω, $u = \partial u/\partial v = 0$ on $\partial \Omega$ satisfies $u(0, 0) = 0.0202\, q\alpha^4/D$ at the origin, if $\alpha = \beta$.

39.1   Given the VBVP

$$B(u, v) = l(v), \qquad v \in V$$

in which B is symmetric and V-elliptic, show that the Galerkin approximation u_h satisfies

$$\|u - u_h\|_B^2 = \|u\|_B^2 - \|u_h\|_B^2,$$

that is, the error in the energy norm equals the error in the energy. From this deduce that

$$\|u_h\|_B \leqslant \|u\|_B.$$

39.2 If u minimizes the functional $J : V \to R$ given by

$$J(v) = \tfrac{1}{2} B(v, v) - l(v),$$

show that $J(u) = -\tfrac{1}{2} B(u, u)$.

39.3 Verify that the operator P defined by (39.16) is a projection.

39.4 Let V^h be the subspace of $H^1(-1, 1)$ spanned by the three functions

$$\phi_1(x) = \tfrac{1}{2} x(x - 1), \qquad \phi_2(x) = 1 - x^2, \qquad \phi_3(x) = \tfrac{1}{2} x(x + 1),$$

and define the interpolate u_h of $u \in H^1(-1, 1)$ by

$$P_h u = \bar{u}_h = \sum_{k=1}^{3} a_k \phi_k(x)$$

where $a_k = u(x_k)$ and $x_1 = -1$, $x_2 = 0$, $x_3 = +1$. Find $\bar{u}_h(x)$ if $u(x) = \sin \pi(x - \tfrac{1}{2})/2$, and evaluate $\|u - u_h\|_{H^1}$.

40.1 Use the Green's formula

$$(Au, v)_{L_2} = (u, A^*v) + G(u, v)$$

(see Section 32) to show that the least-squares problem (40.13) is equivalent to the higher-order problem.

$$A^*Au_h = A^*f \quad \text{in} \quad \Omega \tag{i}$$

plus appropriate boundary conditions. If $A = -\nabla^2$ (the Laplacian), show that (i) is equivalent to the problem

$$\nabla^4 u_h = \nabla^2 f$$

where ∇^4 is the biharmonic operator.

40.2 Consider the problem

$$-u'' + u = \sin x \quad \text{in} \ \Omega = (0, 1)$$
$$u(0) = u(1) = 0.$$

(a) Reformulate this problem in a manner suitable for solution using the method of weighted residuals. Then find an approximate solution in a two-dimensional subspace spanned by polynomials of appropriate order (that is, make suitable choices for U^h and V^h in (40.10) and (40.11)).

(b)   Find approximate solutions using the method of least squares and the method of collocation.

(c)   Apply integration by parts successively to transform the problem into one suitable for solution using H^{-1} methods. Again choose suitable subspaces of U^h and V^h, and find an approximate solution.

11

The finite element method

In practical situations the determination of suitable basis functions for use in the Galerkin method can be extremely difficult, especially in cases for which the domain Ω does not have a simple shape. The finite element method overcomes this difficulty by providing a systematic means for generating basis functions on domains of fairly arbitrary shape. What makes the method especially attractive is the fact that these basis functions are piecewise polynomials that are non-zero only on a relatively small part of Ω, with the result that evaluation of the integrals in (38.6) is very simple. Furthermore, as we will show a little later, the family of spaces V_h ($h \in (0, 1)$) defined by the finite element procedure possesses the property that V_h approaches V as h approaches zero, in the sense of (39.11). This is, of course, an indispensable property for convergence of the Galerkin method.

In Section 41 we outline the steps that lead to the construction of finite element bases for *second-order BVPs* defined on *domains in R and R^2*. This is the simplest and most commonly encountered class of problems, and most of the features of finite element approximations can be demonstrated in this context. The generalization to higher-order BVPs and to problems in R^n for $n \geq 3$ follows in a natural way, and is covered in great detail in all texts dealing with the finite element method.

Then in Sections 42 and 43 we describe in detail, for problems in R and R^2 respectively, some commonly used basis functions. We also show by means of examples how these are used to solve actual boundary-value problems in R and R^2.

The final three sections of this chapter form a self-contained account of the method for obtaining error estimates in H^m-norms. Section 44 discusses further the concept of a *master element* $\hat{\Omega}$: every element Ω_e in the domain is regarded as having been generated by a simple map F_e, so that Ω_e is the image of $\hat{\Omega}$ under this mapping. The master element and the collection of maps are then all the information we need in order to construct a finite element mesh. In Section 45 we derive estimates for the *interpolation error* on a single element and finally, in Section 46, we use this error estimate and the basic result (39.12) to derive *error estimates for finite-element approximations of VBVPs.*

§41. The finite element method for second-order problems

Suppose that we have a VBVP of the form: find $u \in V$ such that

$$B(u, v) = l(v) \qquad \text{for all } v \in V. \tag{1}$$

For a second-order BVP the space of admissible functions V consists of all those functions in $H^1(\Omega)$ that satisfy the essential boundary conditions.

A Galerkin approximation u_h to the solution of (1) may be sought by constructing a finite-dimensional subspace V_h of V, which is spanned by a finite number of basis functions ϕ_i. We then pose the problem of finding $u_h \in V_h$ that satisfies

$$B(u_h, v_h) = l(v_h) \qquad \text{for all } v_h \in V_h. \tag{2}$$

Our aim in this section then is to describe a method for constructing a suitable basis ϕ_i; once this has been achieved, the problem is reduced to one of solving

$$\sum_i K_{ij} a_i = F_j \tag{3}$$

where, as before,

$$K_{ij} = B(\phi_i, \phi_j) \quad \text{and} \quad F_j = l(\phi_j). \tag{4}$$

1. *The finite element mesh.*   We start by subdividing the domain Ω into a finite number E of subdomains $\Omega_1, \Omega_2, \ldots, \Omega_E$, called *finite elements.* These elements are non-overlapping and cover Ω, in the sense that

$$\Omega_e \cap \Omega_f = \varnothing \quad \text{for} \quad e \neq f, \qquad \bigcup_{e=1}^{E} \bar{\Omega}_e = \bar{\Omega}. \tag{5}$$

To avoid complicating matters unnecessarily, we assume that the domain Ω is *polygonal* if it is a subset of R^2. That is, the boundary $\Gamma \subset R^2$ of Ω is made up of straight segments. Under these conditions, it is easy to see that the entire domain can be covered exactly by polygonal elements, as shown in Figure 11.1.

We impose one more condition on the subdivision of Ω: every side of

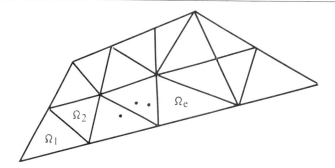

Figure 11.1

the boundary of an element in R^2 is *either* part of the boundary Γ, *or* it is a side of another element. This condition rules out a situation such as that shown in Figure 11.2, in which AB is a side of Ω_2 but not of Ω_1.

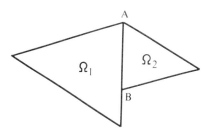

Figure 11.2

2. Nodal points. We next identify certain points called *nodes* or *nodal points* in the subdivided domain; these points play a key role in the finite element method, as will soon become evident. Nodes are allocated at least at the vertices of elements as shown in Figure 11.3(a), but in order to improve the approximation, further nodes may be introduced, for example at the midpoints of the sides of elements as shown in Figure 11.3(b). In any case, there is a total of G nodes, say, which are numbered $1, 2, \ldots, G$ and which have position vectors $\mathbf{x}_1, \mathbf{x}_2, \ldots, \mathbf{x}_G$. The set of elements and nodes that make up the domain Ω is called a *finite element mesh*.

3. Basis functions ϕ_i. We are now ready to describe how the finite element basis functions are formed. In carrying out this procedure we must bear in mind that the basis functions define a subspace of V, so that they must be functions in $H^1(\Omega)$ that satisfy the essential boundary

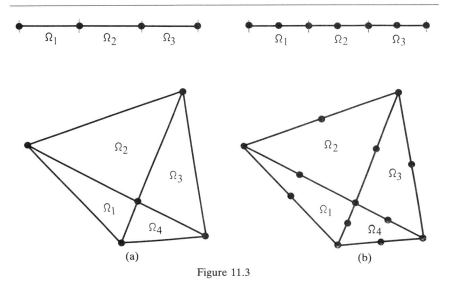

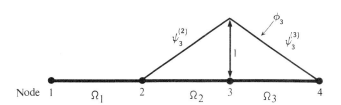

Figure 11.3

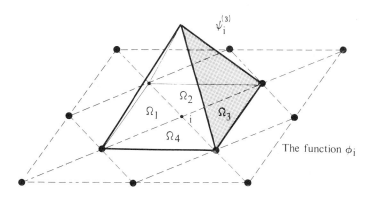

Figure 11.4

conditions. We leave aside for now the question of boundary conditions and proceed to construct a set of functions with the following properties:

(i)   The functions ϕ_i are *continuous*, that is, $\phi_i \in C(\bar{\Omega})$.

(ii) There is a total of G basis functions, and each function ϕ_i is non-zero only on those elements that are connected to node i.

(iii) ϕ_i is equal to 1 at node i, and equal to zero at other nodes:

$$\phi_i(\mathbf{x}_j) = \begin{cases} 1 & \text{if } i = j, \\ 0 & \text{otherwise.} \end{cases} \tag{6}$$

(iv) Let $\psi_i^{(e)}$ be the *restriction* of ϕ_i to Ω_e, in other words,

$$\phi_i|_{\Omega_e} = \psi_i^{(e)}; \tag{7}$$

then $\psi_i^{(e)}$ is a *polynomial function*.

From (iii) and (iv) it is clear that the function $\psi_i^{(e)}$ defined on element (e) will have the property that

$$\psi_i^{(e)}(\mathbf{x}_j) = \begin{cases} 1 & \text{if } i = j \\ 0 & \text{otherwise,} \end{cases} \tag{8}$$

i and j running over all nodes in Ω_e. We call $\psi_i^{(e)}$ a *local basis function*.

It is not difficult to show that conditions (i) and (iv) ensure that the functions ϕ_i belong to $H^1(\Omega)$, as required (see Exercise 41.1). We are thus going to set up basis functions which are *piecewise polynomials*, and which have *small supports*, in that they are non-zero only in a "small" region. We give specific examples in the following section, but in the meantime observe that if we write

$$v_h(\mathbf{x}) = \sum_{i=1}^{G} b_i \phi_i(\mathbf{x}),$$

then as a result of (6) we find that

$$v_h(\mathbf{x}_j) = \sum_{i=1}^{G} b_i \phi_i(\mathbf{x}_j) = b_j;$$

that is, the coefficient b_j is simply *the value of v_h at node j.*

We denote by X_h the space spanned by the basis functions $\{\phi_i\}_{i=1}^{G}$. Note that we have as yet said nothing about the essential boundary conditions, which form part of the definition of V_h. We simply set

$$V_h = \{v_h \in X_h : v_h \text{ satisfies essential BCs}\}$$

$$= \text{span } \{\phi_i : \phi_i \text{ satisfies essential BCs}\}. \tag{9}$$

In the same way we denote by X_e the space spanned by the functions

$\psi_i^{(e)}$. That is,

$$X_e = \text{span } \{\psi_i^{(e)}\} = \{v_h \mid_{\Omega_e}, \quad v_h \in X_h\},$$

that is, X_e is the space consisting of the restriction of all functions in X_h to Ω_e. In view of (iv) above we see that X_e consists of all polynomials up to a given degree.

4. The approximate solution. We go straight to (3) and (4) and note that, since $\psi_i^{(e)}$ is the restriction of ϕ_i to Ω_e, we can write (4) as

$$K_{ij} \equiv B(\phi_i, \phi_j) = \sum_{e=1}^{E} B^{(e)}(\phi_i, \phi_j) = \sum_{e=1}^{E} \underbrace{B^{(e)}(\psi_i^{(e)}, \psi_j^{(e)})}_{K_{ij}^{(e)}} \tag{10}$$

and

$$F_j \equiv l(\phi_j) = \sum_{e=1}^{E} l^{(e)}(\phi_j) = \sum_{e=1}^{E} \underbrace{l^{(e)}(\psi_j^{(e)})}_{F_j^{(e)}}. \tag{11}$$

Here we have used the fact, first of all, that $B(\phi_i, \phi_j)$ is of the form $\int_\Omega \ldots$, which we can write as a sum of integrals $\sum_{e=1}^{E} \int_{\Omega_e} \ldots$ over each element Ω_e. The final form comes simply from (7). Thus the actual evaluation of K_{ij} and F_j reduces to the evaluation of a series of matrices $K_{ij}^{(e)}$ and vectors $F_j^{(e)}$ for each element, and then summing these contributions over all elements. The condition (ii) in part 3 results in an additional simplifying feature: since $\phi_i = 0$ for all elements which do not have node i as a node, clearly $K_{ij}^{(e)} = 0$ *if nodes i and j do not belong to* Ω_e. It follows that a judicious numbering of nodes will result in the matrix **K** having a *banded* structure in which all non-zero entries are clustered around the main diagonal. From a computational viewpoint this represents a distinct advantage.

It should be clear from what has been discussed that we may regard a typical basis function ϕ_i as being built up by patching together the local basis functions $\psi_i^{(e)}$ associated with node i, as shown in Figure 11.5.

In the next section we give details of a systematic procedure for setting

Figure 11.5

up local basis functions $\psi_i^{(e)}$ and hence the basis functions ϕ_i, for the case of one-dimensional problems.

§42. One-dimensional problems

We show in this section how the local basis functions $\psi_i^{(e)}$ and the functions $\phi_i(x)$ are set up for one-dimensional problems, after which we work through an example.

Consider a second-order problem defined on a subset $\Omega = (a, b)$ of the real line. The domain is divided into elements $\Omega_1, \Omega_2, \ldots, \Omega_E$, each element Ω_e being a segment of length h_e, say.

Suppose now that we would like X_h to be the space of piecewise polynomials of degree 1, that is, piecewise straight lines. Then if we denote the set of all polynomials of degree $\leq k$ on Ω_e by $P_k(\Omega_e)$, X_e is the same as $P_1(\Omega_e)$. Furthermore, since every straight line is of the form $f(x) = a + bx$, it follows that a knowledge of the value of f at two points in Ω_e is sufficient to determine f uniquely on Ω_e. With this in mind we define nodal points at the ends of all elements, so that each element has *two* nodal points. As shown in Figure 11.6, in view of this simple arrangement we may number the nodes sequentially so that Ω_e will be connected to nodes e and $e + 1$.

The next step is to define two linear functions $\psi_i^{(e)}$ and $\psi_{i+1}^{(e)}$ that span X_e, and that satisfy (41.8). The only functions that fit all requirements are

$$\psi_i^{(e)}(x) = (x_{i+1} - x)/h_e, \qquad \psi_{i+1}^{(e)}(x) = (x - x_i)/h_e. \tag{1}$$

By patching together the functions in the manner outlined in the previous section we see that each basis function $\phi_i(x)$ is a piecewise linear "hat" function made up of the local basis functions associated with node i. Hence every function v_h in X_h is a *piecewise linear function* of the form

$$v_h(x) = \sum_{i=1}^{G} a_i \phi_i(x)$$

in which a_i is the value of v_h at node i (Figure 11.7).

Instead of defining local basis functions for each element as in (1), we can simplify matters considerably by setting up a *master element* $\hat{\Omega}$, say, which is isolated from the actual finite element mesh and which is referred to its own coordinate system $\hat{x}$. The master element extends

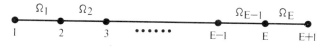

Figure 11.6

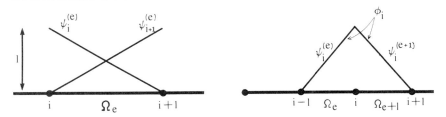

Figure 11.7

from $\hat{x} = -1$ to $\hat{x} = +1$ and has the same system of nodal points as the elements Ω_e in the actual mesh (two nodes in this case). This situation is shown in Figure 11.8. Each element Ω_e can now be thought of as having been generated by a map from $\hat{\Omega}$ to Ω_e, that is

$$x = \frac{h_e}{2}\hat{x} + x_e \quad \text{or} \quad \hat{x} = \frac{2}{h_e}(x - x_e) \tag{2}$$

where x_e is the coordinate of the *centre* of Ω_e ($x_e = \frac{1}{2}(x_i + x_{i+1})$ in Figure 11.8) and h_e is the length of Ω_e. Thus as $\hat{x}$ goes from -1 to $+1$, x goes from x_i to x_{i+1}.

The advantage of setting up a master element in this way is that we can define, once and for all, local basis functions $\hat{\psi}_1$ and $\hat{\psi}_2$ on $\hat{\Omega}$ which have the requisite properties, and having done this we simply use (2) to map $\hat{\psi}_1$ and $\hat{\psi}_2$ to $\psi_i^{(e)}$ and $\psi_{i+1}^{(e)}$, respectively, by defining $\psi_i^{(e)}$ and $\psi_{i+1}^{(e)}$ to be functions on Ω_e satisfying

$$\psi_i^{(e)}(x) = \hat{\psi}_1(\hat{x}), \qquad \psi_{i+1}^{(e)}(x) = \hat{\psi}_2(\hat{x}), \tag{3}$$

as shown in Figure 11.9.

For linear functions we thus define

$$\hat{\psi}_1(\hat{x}) = \tfrac{1}{2}(1 - \hat{x}), \qquad \hat{\psi}_2(\hat{x}) = \tfrac{1}{2}(1 + \hat{x}) \tag{4}$$

and from (3) and (2) we then recover (1) since, for example,

$$\psi_i^{(e)}(x) = \hat{\psi}_1(\hat{x}) = \frac{1}{2}\left(1 - \frac{2}{h_e}(x - x_e)\right) = \frac{1}{h_e}(x_{i+1} - x),$$

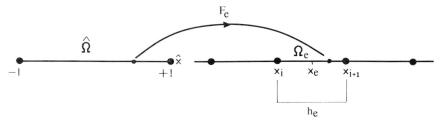

Figure 11.8

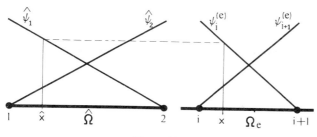

Figure 11.9

as is readily verified. In future we will define local basis functions on a
master element, with the assumption that the actual local basis functions
can be recovered by means of a relation such as (3).

The procedure is readily extended to higher order approximations. For
example, suppose that we wish to construct a space of *piecewise quadratic*
functions, with the restrictions to each element being a quadratic function
(i.e. a member of $P_2(\Omega_e)$). Every function $f \in X_e$ is thus of the form
$f(x) = a + bx + cx^2$ and so specification of f at *three* points in Ω_e
determines f uniquely in Ω_e. Hence we place nodes at the ends as well as
at the midpoints of elements, so that an arbitrary element will have
associated with it nodes i, $i + 1$ and $i + 2$.

We next set up a master element $\hat{\Omega}$ with nodes at the ends and at the
centre, and define quadratic basis functions

$$\hat{\psi}_1(\hat{x}) = \tfrac{1}{2}\hat{x}(\hat{x} - 1), \qquad \hat{\psi}_2(\hat{x}) = 1 - \hat{x}^2, \qquad \hat{\psi}_3(\hat{x}) = \tfrac{1}{2}\hat{x}(\hat{x} + 1) \qquad (5)$$

as shown in Figure 11.10. Naturally the functions in (5) and the local
basis functions $\psi_i^{(e)}$, $\psi_{i+1}^{(e)}$, $\psi_{i+2}^{(e)}$ that they generate using

$$\psi_i^{(e)}(x) = \hat{\psi}_1(\hat{x}), \qquad \psi_{i+1}^{(e)}(x) = \hat{\psi}_2(\hat{x}), \qquad \psi_{i+2}^{(e)}(x) = \hat{\psi}_3(\hat{x}) \qquad (6)$$

all satisfy (41.8). A few typical piecewise quadratic basis functions ϕ_i that
result from patching together the quadratic local basis functions are also
shown in Figure 11.10.

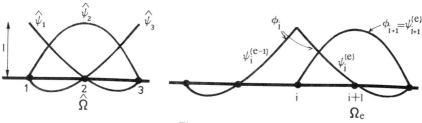

Figure 11.10

Extension to piecewise cubic and higher order approximations follows in a logical fashion, but these higher order approximations are seldom used in practice, so we omit details.

We now show how an approximate solution to a two-point BVP may be found, using piecewise-linear basis functions. In the Exercises you are asked to rework this example using a piecewise quadratic basis.

Example

Consider the BVP

$$-u'' + u = \sin \pi x, \qquad x \in \Omega = (0, 1), \tag{7}$$

$$u(0) = u(1) = 0. \tag{8}$$

The corresponding VBVP is: find $u \in V = H_0^1(0, 1)$ such that

$$\int_0^1 (u'v' + uv) \, dx = \int_0^1 v \sin \pi x \, dx \qquad \text{for all } v \in H_0^1(0, 1),$$

and the approximate problem is: find $u_h \in V_h$ such that

$$\int_0^1 (u_h' v_h' + u_h v_h) \, dx = \int_0^1 v_h \sin \pi x \, dx \qquad \text{for all } v_h \in V_h$$

where $V_h = \{v_h \in X_h : v_h(0) = v_h(1) = 0\}$.

Suppose that we use three elements for this problem (Figure 11.11). Since we require X_h to be spanned by *piecewise linear* basis functions, nodes are required at the ends of elements only, and the basis functions ϕ_i are formed by patching together the functions defined in (1). These functions span X_h.

Next, we construct V_h by requiring that (see (41.9))

$$V_h = \text{span} \{\phi_i \in X_h : \phi_i(0) = \phi_i(1) = 0\} = \text{span} \{\phi_2, \phi_3\},$$

since ϕ_1 and ϕ_4 do not satisfy (8). Hence every member of V_h is a linear combination of ϕ_2 and ϕ_3.

From (41.3) and (41.4) we require, for i and $j = 2, 3$,

$$B(\phi_i, \phi_j) = \int_0^1 (\phi_i' \phi_j' + \phi_i \phi_j) \, dx = \int_0^{1/3} \underbrace{[\psi_i^{(1)'} \psi_j^{(1)'} + \psi_i^{(1)} \psi_j^{(1)}] \, dx}_{K_{ij}^{(1)}}$$

$$+ \int_{1/3}^{2/3} \underbrace{[\psi_i^{(2)'} \psi_j^{(2)'} + \psi_i^{(2)} \psi_j^{(2)}] \, dx}_{K_{ij}^{(2)}} + \int_{2/3}^1 \underbrace{[\psi_i^{(3)'} \psi_j^{(3)'} + \psi_i^{(3)} \psi_j^{(3)}] \, dx}_{K_{ij}^{(3)}},$$

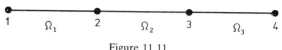

Figure 11.11

using the fact that the restriction of ϕ_i to element Ω_e is ψ_i^e. Of course, $\psi_i^e = 0$ if node i is not a node of Ω_e.

Now recall that $K_{ij}^{(e)} = 0$ if either node i or node j does not belong to Ω_e. Hence the only non-zero contributions which have to be calculated are $K_{22}^{(1)}$, $K_{22}^{(2)}$, $K_{23}^{(2)}$, $K_{33}^{(2)}$, $K_{33}^{(3)}$ (note that K_{ij} is symmetric).

In order to facilitate the computational work we evaluate integrals on the master element: thus

$$
\begin{aligned}
K_{22}^{(1)} &= \int_{\Omega_1} \left(\frac{d\psi_2^{(1)}}{dx}\frac{d\psi_2^{(1)}}{dx} + \psi_2^{(1)}\psi_2^{(1)} \right) dx \\
&= \int_{\hat{\Omega}} \left[\frac{d\hat{\psi}_2}{d\hat{x}}\frac{d\hat{\psi}_2}{d\hat{x}}\left(\frac{2}{h_1}\right)^2 + \hat{\psi}_2\hat{\psi}_2 \right] \frac{h_1}{2} d\hat{x}
\end{aligned}
\tag{9}
$$

using the fact that from (3)

$$
\frac{d\psi_2^{(1)}}{dx} = \frac{d\hat{\psi}_2}{d\hat{x}}\frac{d\hat{x}}{dx} \qquad \text{and} \qquad d\hat{x} = \frac{2}{h_1}\,dx
$$

(here $h_e = 1/3$ for all elements). We now substitute for $\hat{\psi}_2$ from (4) and integrate between $\hat{x} = -1$ and $\hat{x} = +1$ to get $K_{22}^{(1)} = 28/9$. After transforming to $\hat{\Omega}$, we find that $K_{33}^{(3)}$ is exactly the same as (9); $K_{22}^{(2)}$ and $K_{33}^{(3)}$ differ only in that $\hat{\psi}_2$ is replaced by $\hat{\psi}_1$ in (9), but since $(\hat{\psi}_1)^2$ and $(\hat{\psi}_2)^2$ have the same integrals, as do $(\hat{\psi}_1')^2$ and $(\hat{\psi}_2')^2$, we end up with the same answer. Hence

$$
K_{22}^{(2)} = K_{33}^{(2)} = K_{33}^{(3)} = \frac{28}{9}.
$$

The term $K_{23}^{(2)}$ is given by

$$
\begin{aligned}
K_{23}^{(2)} &= \int_{1/3}^{2/3} [\psi_2^{(2)'}\psi_3^{(2)'} + \psi_2^{(2)}\psi_3^{(2)}]\,dx \\
&= \int_{\hat{\Omega}} \left[\left(\hat{\psi}_1'\hat{\psi}_2'\right)\left(\frac{2}{h_2}\right)^2 + \hat{\psi}_1\hat{\psi}_2 \right] \frac{h_2}{2}\,d\hat{x} = -\frac{53}{18}.
\end{aligned}
$$

Collecting all terms, then, we have

$$
K_{ij} = \sum_{e=1}^{3} K_{ij}^{(e)} \Rightarrow K_{22} = \frac{28}{9} + \frac{28}{9} = \frac{56}{9} = K_{33},
$$

$$
K_{23} = -\frac{53}{18}.
$$

Next we have to evaluate

$$F_i = \int_0^1 \sin(\pi x)\phi_i \, dx. \tag{10}$$

Now, although evaluation of this integral over each element is a simple enough matter for this problem, it can be quite tedious in general, and, in the case of more complicated data, impossible to carry out exactly. We observe, however, that we are seeking a *piecewise linear* approximation to the exact solution, so it would make sense to replace $f(x) = \sin \pi x$ by its *linear interpolate* $\tilde{f}_h$ in X_h. This would enable evaluation of the terms F_i to be done very easily while preserving the piecewise linear nature of the approximation. The interpolate of f is, we recall, a function of the form

$$\tilde{f}_h(x) = \sum_{i=1}^4 c_i \phi_i(x), \qquad c_i = f(x_i)$$

with the property that $\tilde{f}_h$ is linear between nodes and equal to f at the nodes (Figure 11.12). When $f(x) = \sin \pi x$ we have $c_1 = c_4 = 0$ and $c_2 = c_3 = (\sqrt{3})/2$. We thus replace (10) by

$$\tilde{F}_i = \int_0^1 \left(\sum_{j=1}^4 c_j \phi_j \right) \phi_i \, dx = \sum_{e=1}^3 \int_{\Omega_e} \left(\sum_{j=1}^4 c_j \psi_j^{(e)} \right) \psi_i^{(e)} \, dx$$

$$= \left(\frac{\sqrt{3}}{2} \right) \Bigg[\underbrace{\int_0^{1/3} \psi_2^{(1)} \psi_i^{(1)} \, dx}_{\tilde{F}_i^{(1)}} + \underbrace{\int_{1/3}^{2/3} (\psi_2^{(2)} + \psi_3^{(2)}) \psi_i^{(2)} \, dx}_{\tilde{F}_i^{(2)}} + \underbrace{\int_{2/3}^1 \psi_3^{(3)} \psi_i^{(3)} \, dx}_{\tilde{F}_i^{(3)}} \Bigg].$$

Transforming each of these integrals to integrals on $\hat{\Omega}$ as before, and evaluating, we find that

$$\tilde{F}_1^{(1)} = \tilde{F}_4^{(3)} = (\sqrt{3})/36, \qquad \tilde{F}_2^{(1)} = \tilde{F}_3^{(3)} = (\sqrt{3})/18,$$
$$\tilde{F}_2^{(2)} = \tilde{F}_3^{(2)} = (\sqrt{3})/12,$$

the other components of $\tilde{F}_i^{(e)}$ being zero. Thus

$$\tilde{F}_i = \frac{\sqrt{3}}{18} [\underbrace{(\tfrac{1}{2}, 1, 0, 0)}_{\tilde{F}_i^{(1)}} + \underbrace{(0, \tfrac{3}{2}, \tfrac{3}{2}, 0)}_{\tilde{F}_i^{(2)}} + \underbrace{(0, 0, 1, \tfrac{1}{2})}_{\tilde{F}_i^{(3)}}]$$

$$= \frac{\sqrt{3}}{36} (1, 5, 5, 1).$$

Finally, then, we have to solve

$$K_{22}a_2 + K_{23}a_3 = \tilde{F}_2,$$
$$K_{32}a_2 + K_{33}a_3 = \tilde{F}_3,$$

Figure 11.12

which gives $a_2 = a_3 = 0.0734$. Hence the approximate solution is

$$u_h(x) = 0.0734(\phi_2(x) + \phi_3(x)).$$

We compare this with the *exact* solution $u(x) = (1 + \pi^2)^{-1} \sin \pi x$ in Figure 11.13 and we see that the finite element solution is a fair approximation of the exact solution, given the very small subspace V_h that we have used. Furthermore, the approximate solution, like the exact solution, is symmetric about $x = \frac{1}{2}$. The approximate solution could of course be improved in two ways: (i) by subdividing the domain into a larger number of elements, (ii) by using a higher order element, such as the quadratic element. Of course, either of these refinements will result in a greater amount of computational work, which must be taken into consideration.   □

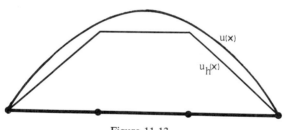

Figure 11.13

It is clearly of interest to know beforehand by how much an approximate solution will improve as a result of either of the two refinements mentioned above, so that one may decide whether the refinement is worth the effort. This is a question that we will address in Section 46. In the following section we apply the ideas developed here to second-order problems defined on domains Ω in R^2.

§43.   Two-dimensional elements

We have already given a rough indication in Section 41 of how a finite element mesh is constructed for problems defined on domains in R^2. We take up the subject again, and generalize the ideas in Section 42 to two-dimensional second-order problems.

Recall from Section 41 that the domain $\Omega \subset R^2$ is assumed *polygonal*, so that the boundary Γ is made up of straight segments. Next, Ω is divided into elements $\Omega_1, \Omega_2, \ldots, \Omega_E$, and we impose the condition that every side of an element is either part of the boundary Γ, or it is the side of another element. We proceed now to look at a few commonly used elements.

Triangular elements. Triangles are the simplest polygonal shapes in R^2, so it is not surprising that triangular elements possess features that make for very simple means of approximation, as we shall see.

Suppose that we want X_h to be the space of piecewise polynomials of degree 1, so that we require X_e to be the space $P_1(\Omega_e)$ (see the corresponding discussion at the beginning of Section 42). A *linear* function in two dimensions is of the form $f(x, y) = a + bx + cy$, so that if the values of the function are known at three points, then it is uniquely determined (in other words, we can then find a, b and c). We thus require *three* nodal points, which we place at the *vertices* of the triangle. This positioning of nodes ensures that if any of the sides ij, jk or ki of Ω_e is shared with an adjacent element Ω_f, say, *the piecewise linear function formed by patching together the functions defined on Ω_e and Ω_f will be continuous across the interface of these elements.*

Once again we set up a master element $\hat{\Omega}$ and a transformation from $\hat{\Omega}$ to each Ω_e, as shown in Figure 11.14. The master element is a right-angled isosceles triangle, defined in the xy plane: it is not difficult to verify that the *linear* transformation

$$x = x_i(1 - \hat{x} - \hat{y}) + x_j\hat{x} + x_k\hat{y}$$
$$y = y_i(1 - \hat{x} - \hat{y}) + y_j\hat{x} + y_k\hat{y} \quad \text{or} \quad \mathbf{x} = (1 - \hat{x} - \hat{y})\mathbf{x}_i + \hat{x}\mathbf{x}_j + \hat{y}\mathbf{x}_k$$

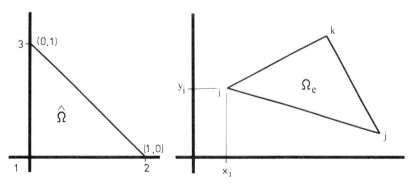

Figure 11.14

maps the nodal points $1, 2, 3$ of $\hat{\Omega}$ to nodal points i, j, k of Ω_e, and indeed maps each point $\hat{\mathbf{x}} \in \hat{\Omega}$ to a point $\mathbf{x} \in \Omega_e$. The inverse of this transformation is

$$
\begin{aligned}
\hat{x} &= \frac{1}{2A_e}[(y_k - y_i)(x - x_i) - (x_k - x_i)(y - y_i)], \\
\hat{y} &= \frac{1}{2A_e}[-(y_j - y_i)(x - x_i) + (x_j - x_i)(y - y_i)],
\end{aligned}
\tag{1}
$$

where A_e is the area of Ω_e.

Our next step is to define local basis functions $\hat{\psi}_1, \hat{\psi}_2, \hat{\psi}_3$ on $\hat{\Omega}$, and then to obtain the basis functions on Ω_e from

$$
\psi_i^{(e)}(\mathbf{x}) = \hat{\psi}_1(\hat{\mathbf{x}}), \qquad \psi_j^{(e)}(\mathbf{x}) = \hat{\psi}_2(\hat{\mathbf{x}}), \qquad \psi_k^{(e)}(\mathbf{x}) = \hat{\psi}_3(\hat{\mathbf{x}}).
$$

The local basis functions on $\hat{\Omega}$ are

$$
\hat{\psi}_1(\hat{\mathbf{x}}) = 1 - \hat{x} - \hat{y}, \qquad \hat{\psi}_2(\hat{\mathbf{x}}) = \hat{x}, \qquad \hat{\psi}_3(\hat{\mathbf{x}}) = \hat{y};
$$

these are shown in Figure 11.15, as are their images on Ω_e and the basis function $\phi_i(\mathbf{x})$ that results from patching together all local basis functions associated with node i. Clearly the condition (41.8) is satisfied.

The basis function ϕ_i formed by patching together all the local functions $\psi_i^{(e)}$ associated with node i is the two-dimensional counterpart of the "hat" function in one dimension, and is pyramidal in shape. Naturally ϕ_i is piecewise linear, and is non-zero only on those elements that have node i as a node.

Piecewise quadratic triangular elements are obtained by adding a further three nodes to an element, at the midpoints of the sides, as in Figure 11.16. A quadratic function $f(\hat{x}, \hat{y}) = a_1 + a_2\hat{x} + a_3\hat{y} + a_4\hat{x}^2 + a_5\hat{x}\hat{y} + a_6\hat{y}^2$ is then uniquely determined on $\hat{\Omega}$ (and hence on Ω_e) by its

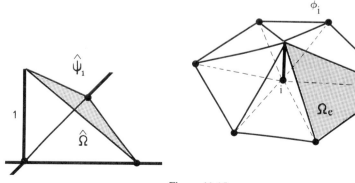

Figure 11.15

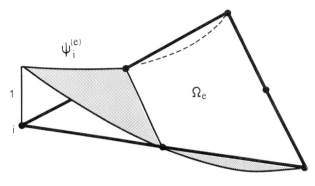

Figure 11.16

values at the six nodes. Furthermore, the basis functions ϕ_i formed by patching together all local basis functions $\psi_i^{(e)}$ are *piecewise quadratic functions*, and it is important to note once again that ϕ_i are continuous (see Exercise 43.1).

Rectangular elements.   We turn now to a second category of finite elements, namely those that are rectangular in shape. If we are to adhere to the policy of having nodal points at least at the vertices of elements, then clearly the simplest rectangular element will be one with *four* nodes, one node at each corner.

The question now arises: what kind of space of polynomials X_e can be defined on Ω_e so that any function in X_e is uniquely determined by its values at the four vertices. Linear functions are completely determined by *three* nodal values, so they will not do. On the other hand, quadratic functions require *six* nodal values, which is more than we have at our disposal. The solution to the problem is first to write

$$f(x) = a_1 + a_2 x + a_3 y + a_4 xy + a_5 x^2 + a_6 y^2 + \cdots$$

for an arbitrary polynomial, and to resolve that four terms be retained.

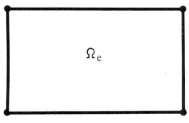

Figure 11.17

The constant and linear terms are obviously retained, and the question arises as to which one of the remaining terms should be used. It is inadvisable to retain the terms involving x^2 or y^2, since this would result in a lopsided approximation in which a quadratic term appears for only one of the coordinates. However, there is no objection to retaining the term involving xy: this ensures that the coordinates x and y are equally represented, for then the approximation is of the form

$$f(x, y) = a_1 + a_2 x + a_3 y + a_4 xy. \tag{2}$$

We call $f(x)$ a *bilinear polynomial*; in general, the space of polynomials containing terms of degree $\leqslant k$ *in each of the variables* is denoted by $Q_k(\Omega)$, so that $f(x, y)$ in (2) is a member of $Q_1(\Omega)$. Note that the inclusions $P_k(\Omega) \subset Q_k(\Omega) \subset P_{2k}(\Omega)$ hold for $\Omega \subset R^2$. The situation now is that $X_e = Q_1(\Omega_e)$, so that X_h will consist of *piecewise bilinear polynomials*.

As before, we set up a master element $\hat{\Omega}$ which this time is a square, as in Figure 11.18. The master element is mapped onto an arbitrary rectangular element Ω_e by the transformation

$$\mathbf{x} = \mathbf{T}\hat{\mathbf{x}} + \mathbf{b} \quad \text{or} \quad \begin{bmatrix} x \\ y \end{bmatrix} = \begin{bmatrix} T_{11} & T_{12} \\ T_{21} & T_{22} \end{bmatrix} \begin{bmatrix} \hat{x} \\ \hat{y} \end{bmatrix} + \begin{bmatrix} b_1 \\ b_2 \end{bmatrix} \tag{3}$$

in which the matrix $\mathbf{T}$ is given by

$$2\mathbf{T} = \begin{bmatrix} x_j - x_i & y_j - y_i \\ x_l - x_i & y_l - y_i \end{bmatrix} \tag{4}$$

and $\mathbf{b}$ is the position vector of the centroid of the rectangle Ω_e.

Next, we set up linear local basis functions on $\hat{\Omega}$ that satisfy (41.8);

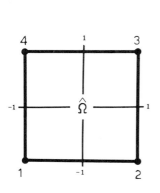
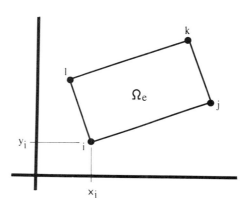

Figure 11.18

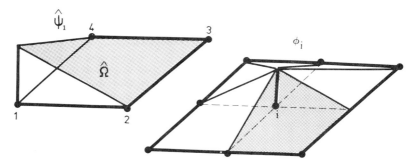

Figure 11.19

these are

$$\hat{\psi}_1(\hat{\mathbf{x}}) = \tfrac{1}{4}(1 - \hat{x})(1 - \hat{y}), \qquad \hat{\psi}_2(\hat{\mathbf{x}}) = \tfrac{1}{4}(1 + \hat{x})(1 - \hat{y})$$
$$\hat{\psi}_3(\hat{\mathbf{x}}) = \tfrac{1}{4}(1 + \hat{x})(1 + \hat{y}), \qquad \hat{\psi}_4(\hat{\mathbf{x}}) = \tfrac{1}{4}(1 - \hat{x})(1 + \hat{y}). \tag{5}$$

Then the functions $\psi_i^{(e)}$, $\psi_j^{(e)}$, $\psi_k^{(e)}$, $\psi_l^{(e)}$ are obtained in the same way as (42.3). As in the case of triangular elements, the positioning of the nodes and the choice of local basis functions ensures that the basis functions ϕ_i will be continuous across element boundaries, as shown in Figure 11.19.

Piecewise *biquadratic* approximations may be constructed by requiring that X_e be equal to $Q_2(\Omega_e)$. Now any member of Q_2 can be written as

$$f(x, y) = a_1 + a_2 x + a_3 y + a_4 x^2 + a_5 xy + a_6 y^2 + a_7 x^2 y$$
$$+ a_8 xy^2 + a_9 x^2 y^2,$$

so that nine values of the function determine it uniquely. We therefore place additional nodes in the finite element mesh and on the master element as shown in Figure 11.20.

The local basis functions $\hat{\psi}_i$ and $\psi_i^{(e)}$ are biquadratic polynomials that satisfy (41.8), and the basis functions ϕ_i formed by patching together the function ψ_i^e associated with node i are piecewise biquadratic polynomials

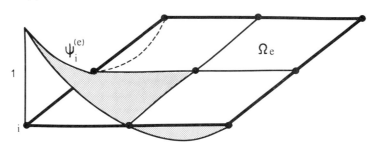

Figure 11.20

that are continuous across inter-element boundaries. We omit details of the basis functions, which are readily available in texts on the finite element method.

This concludes the discussion on elements for second-order problems in two dimensions. We now work through a simple example involving rectangular elements.

Example

Consider the problem

$$-\nabla^2 u = 2 - (x^2 + y^2) \qquad \text{in } \Omega = (0, 1) \times (0, 1),$$
$$u = 0 \qquad \text{on } \Gamma_1, \tag{6}$$
$$\partial u / \partial v = 0 \qquad \text{on } \Gamma_2,$$

where Γ_1 and Γ_2 are the parts of the boundary shown in Figure 11.21(a). The corresponding VBVP is: find $u \in V$ such that

$$\int_\Omega \nabla u \cdot \nabla v \, dx = \int_\Omega [2 - (x^2 + y^2)] v \, dx \qquad \text{for all } v \in V,$$

where $V = \{v \in H^1(\Omega) : v = 0 \text{ on } \Gamma_1\}$, and the approximate problem is: find $u_h \in V_h \subset V$ such that

$$\int_\Omega \nabla u_h \cdot \nabla v_h \, dx = \int_\Omega [2 - (x^2 + y^2)] v_h \, dx \qquad \text{for all } v_h \in V_h.$$

We divide the domain into four square elements and choose for X_h the space of piecewise *bilinear* functions, so that nodes are required at the corners of elements only (see Figure 11.21(b)).

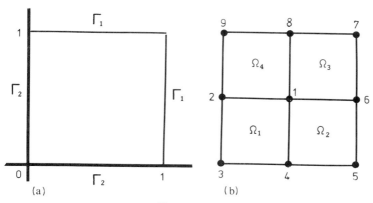

Figure 11.21

Next, we construct V_h: from (41.9) we require that

$$V_h = \text{span } \{\phi_i \in X_h : \phi_i(\mathbf{x}) = 0 \quad \text{on } \Gamma_1\} = \text{span } \{\phi_1, \phi_2, \phi_3, \phi_4\},$$

the functions ϕ_i being piecewise-bilinear functions; the restriction of ϕ_i to Ω_e is of course $\psi_i^{(e)}$.

Now we require

$$K_{ij} = \int\int_\Omega \nabla\phi_i \cdot \nabla\phi_j \, dx \, dy = \sum_{e=1}^4 \underbrace{\int\int_{\Omega_e} \nabla\psi_i^{(e)} \cdot \nabla\psi_j^{(e)} \, dx \, dy}_{K_{ij}^{(e)}}. \tag{7}$$

Since all elements have the same geometry, the amount of computational work can be reduced considerably by observing that many of the integrals have the same value. Indeed, if nodes i and j both belong to Ω_e then a moment's thought will convince us that

$$K_{ij}^{(e)} = \int\int_{\Omega_e} \nabla\psi_i^{(e)} \cdot \nabla\psi_j^{(e)} \, dx \, dy = \begin{cases} K_{11}^{(1)} & i = j, \\ K_{12}^{(1)} & \text{if} \quad i, j \text{ are adjacent,} \\ K_{13}^{(1)} & i, j \text{ are diagonally} \\ & \quad \text{opposite,} \end{cases} \tag{8}$$

so in fact we only have to evaluate three integrals. We have

$$K_{11}^{(1)} = \int\int_{\Omega_1} \nabla\psi_1^{(1)} \cdot \nabla\psi_1^{(1)} \, dx \, dy = \int\int_{\Omega_1} \left[\left(\frac{\partial\psi_1^{(1)}}{\partial x}\right)^2 + \left(\frac{\partial\psi_1^{(1)}}{\partial y}\right)^2 \right] dx \, dy$$

$$= \int_{-1}^1\int_{-1}^1 \left[\left(\frac{\partial\hat\psi_1}{\partial\hat x}\frac{\partial\hat x}{\partial x} + \frac{\partial\hat\psi_1}{\partial\hat y}\frac{\partial\hat y}{\partial x}\right)^2 + \left(\frac{\partial\hat\psi_1}{\partial\hat x}\frac{\partial\hat x}{\partial y} + \frac{\partial\hat\psi_1}{\partial\hat y}\frac{\partial\hat y}{\partial y}\right)^2 \right] |J| \, d\hat x \, d\hat y \tag{9}$$

using the chain rule and the rule $dx \, dy = |J| \, d\hat x \, d\hat y$ for changing variables in area integrals: here $J = \det \mathbf{T}$ where $\mathbf{T}$ is given by (4). For Ω_1 nodes i, j, k, l are nodes 1, 2, 3, 4, so that, from (4),

$$\mathbf{T} = \frac{1}{4}\begin{bmatrix} -1 & 0 \\ 0 & -1 \end{bmatrix} \quad \text{and} \quad \mathbf{T}^{-1} = \begin{bmatrix} -4 & 0 \\ 0 & -4 \end{bmatrix}. \tag{10}$$

By inverting (3) we obtain $\hat x$ and $\hat y$ in terms of x and y, and find that $\partial\hat x/\partial x = \partial\hat y/\partial y = -4$, $\partial\hat x/\partial y = \partial\hat y/\partial x = 0$. Also, $J = 1/16$. With all of this available and with the use of (5) we can now evaluate (9):

$$K_{11}^{(1)} = \int_{-1}^1\int_{-1}^1 [\{\tfrac14(\hat y - 1)(-4)\}^2 + \{\tfrac14(\hat x - 1)(-4)\}^2]\tfrac{1}{16} \, d\hat x \, d\hat y = \tfrac23.$$

Similarly, $K_{12}^{(1)} = -\frac{1}{6}$, $K_{13}^{(1)} = -\frac{1}{3}$. Using (7) and (8) we get

$$
6\mathbf{K} = \begin{bmatrix} 4 & -1 & -2 & -1 \\ & 4 & -1 & -2 \\ \text{sym} & & 4 & 1 \\ & & & 4 \end{bmatrix} + \begin{bmatrix} 4 & 0 & 0 & -1 \\ & 0 & 0 & 0 \\ \text{sym} & & 0 & 0 \\ & & & 4 \end{bmatrix}
$$

$$
+ \begin{bmatrix} 4 & 0 & 0 & 0 \\ & 0 & 0 & 0 \\ \text{sym} & & 0 & 0 \\ & & & 0 \end{bmatrix} + \begin{bmatrix} 4 & -1 & 0 & 0 \\ & 4 & 0 & 0 \\ \text{sym} & & 0 & 0 \\ & & & 0 \end{bmatrix}
$$

$$
= \begin{bmatrix} 16 & -2 & -2 & -2 \\ & 8 & -1 & -2 \\ \text{sym} & & 4 & -1 \\ & & & 8 \end{bmatrix}.
$$

The next task is to evaluate $F_i = \iint_\Omega f \phi_i \, dx \, dy$. As in the one-dimensional case, we replace $f(x, y)$ by its *interpolate* $\tilde{f}_h(x, y)$ which is given by

$$
\tilde{f}_h(x, y) = \sum_{i=1}^{9} f_i \phi_i(x, y), \qquad f_i = f(x_i, y_i).
$$

Then

$$
F_j = \iint_\Omega \left(\sum_{i=1}^{9} f_i \phi_i \right) \phi_j \, dx \, dy = \sum_{e=1}^{4} \underbrace{\iint_\Omega \left(\sum_{i=1}^{9} f_i \psi_i^{(e)} \right) \psi_j^{(e)} \, dx \, dy}_{F_j^{(e)}}.
$$

Thus we need integrals of the form $\iint_\Omega \psi_i^{(e)} \psi_j^{(e)} \, dx \, dy$, and once again we can save ourselves a great deal of effort by noting that

$$
\iint_{\Omega_e} \psi_i^{(e)} \psi_j^{(e)} \, dx \, dy = \begin{cases} \displaystyle\iint_{\Omega_e} \psi_1^{(1)} \psi_1^{(1)} \, dx \, dy & \text{if } i = j, & = 1/24, \\[2ex] \displaystyle\iint_{\Omega_e} \psi_1^{(1)} \psi_2^{(1)} \, dx \, dy & \begin{array}{l}\text{if } i \text{ and } j \text{ are} \\ \text{adjacent nodes,}\end{array} & = 1/72, \\[2ex] \displaystyle\iint_{\Omega_1} \psi_1^{(1)} \psi_3^{(1)} \, dx \, dy & \begin{array}{l}\text{if } i \text{ and } j \text{ are} \\ \text{diagonally} \\ \text{opposite nodes,}\end{array} & = 1/144. \end{cases}
$$

Hence

$$\mathbf{F} = \sum_{e=1}^{4} \mathbf{F}^{(e)} = \begin{bmatrix} \frac{1}{24}(f_1) + \frac{1}{72}(f_2 + f_4) + \frac{1}{144}(f_3) \\ \frac{1}{24}(f_2) + \frac{1}{72}(f_1 + f_3) + \frac{1}{144}(f_4) \\ \frac{1}{24}(f_3) + \frac{1}{72}(f_2 + f_4) + \frac{1}{144}(f_1) \\ \frac{1}{24}(f_4) + \frac{1}{72}(f_1 + f_3) + \frac{1}{144}(f_2) \end{bmatrix} = \begin{bmatrix} 0.0938 \\ 0.0972 \\ 0.1007 \\ 0.0972 \end{bmatrix}.$$

Finally, we solve $\mathbf{Ka} = \mathbf{F}$ to get

$$\mathbf{a} = \begin{bmatrix} 0.1585 \\ 0.2568 \\ 0.4213 \\ 0.2568 \end{bmatrix}$$

which are the values of u_h at nodes 1 to 4. The approximate solution is

$$u_h(\mathbf{x}) = 0.1585\phi_1(\mathbf{x}) + 0.2568(\phi_2(\mathbf{x}) + \phi_4(\mathbf{x})) + 0.4213\phi_3(\mathbf{x}),$$

which is shown in Figure 11.22 together with the exact solution to this problem,

$$u(x, y) = \tfrac{1}{2}(x^2 - 1)(y^2 - 1).$$

In spite of the relatively crude approximation we see that the approximate solution compares favourably with the exact solution.   □

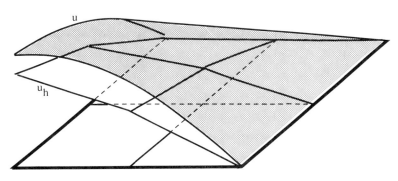

Figure 11.22

§44. Affine families of elements

In this section we start to set up the machinery that is vital to a proper development of error estimates for finite element approximations.

Affine-equivalent elements. We start here with a situation in which

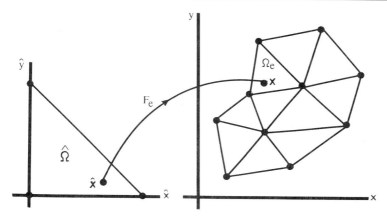

Figure 11.23

a domain Ω has been partitioned into E finite elements, all elements being of the same geometrical type (e.g. all triangles) and having the same degree of approximation (e.g. all 3-noded triangles). We regard such a finite element mesh as having been generated simply by setting up a single *master element* $\hat{\Omega}$, say, and by mapping or transforming $\hat{\Omega}$ into each one of the elements Ω_e in turn.

The basic idea is very simple, as we have already seen: first, define the *master element* $\hat{\Omega}$, this element being of the same geometrical type as the elements that make up Ω. Next, define an *affine transformation*, that is, a transformation which maps straight lines into straight lines, by

$$F_e : \hat{\Omega} \to \Omega_e, \qquad F_e(\hat{\mathbf{x}}) \equiv \mathbf{T}_e \hat{\mathbf{x}} + \mathbf{b}_e = \mathbf{x}, \tag{1}$$

so that F_e maps each point $\hat{\mathbf{x}}$ of $\hat{\Omega}$ to a point $\mathbf{x}$ of Ω_e. Here $\mathbf{T}_e$ is an invertible $n \times n$ matrix and $\mathbf{b}_e$ is a translation vector. We also require of F_e that it maps each nodal point $\hat{\mathbf{x}}_i$ of $\hat{\Omega}$ to a nodal point $\mathbf{x}_i$ of Ω_e:

$$F_e(\hat{\mathbf{x}}_i) = \mathbf{x}_i, \qquad i = 1, \ldots, N. \tag{2}$$

Once a set of affine transformations has been constructed in this way for each element, we need to focus attention only on the master element $\hat{\Omega}$ and the family of transformations $F_1, F_2, \ldots, F_E$, since *these provide a complete description of the mesh.*

When two elements $\hat{\Omega}$ and Ω_e are related to each other by a transformation of the type (1), (2), they are said to be *affine-equivalent*. Also, a set of finite elements $\Omega_1, \ldots, \Omega_E$ is called an *affine family* if all elements are affine-equivalent to a single reference element $\hat{\Omega}$.

It should be clear from the discussion in Section 43 that affine maps of the form (1), (2) exist in R, and in R^2 from one triangle to another, and

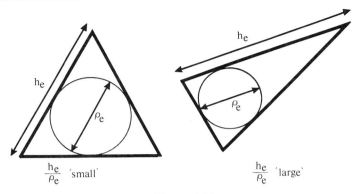

Figure 11.24

from one rectangle to another. Indeed, (42.2) and (43.3) are examples of affine maps. Similar results hold in R^3 for tetrahedra and 3-rectangles or "bricks". We are thus assured that affine maps are always available for the elements with which we are concerned.

The relative size and shape of an arbitrary element Ω_e are quantified in a natural way by defining the constants

$$h_e = \text{diam}\,(\Omega_e) = \max\,\{|\mathbf{x} - \mathbf{y}|, \mathbf{x}, \mathbf{y} \in \Omega_e\} \tag{3}$$

and

$$\rho_e = \sup\,\{\text{diameters of all spheres contained in } \Omega_e\}.$$

When dealing with the reference element $\hat{\Omega}$ we denote the corresponding constants by $\hat{h}$ and $\hat{\rho}$. These quantities are illustrated in Figure 11.24: whereas h_e gives some idea of the "size" of Ω_e, the ratio h_e/ρ_e tells us how "thin" the element is.

We now give some useful properties of the affine transformation (1).

Lemma 1. *Let $F_e : \hat{\Omega} \to \Omega_e$ be the affine map from $\hat{\Omega}$ to Ω_e defined by (1), for $\hat{\Omega}, \Omega_e \subset R^n$. If the matrix norm $\|\mathbf{T}_e\|$ is given by*

$$\|\mathbf{T}_e\| = \sup_{\hat{\mathbf{x}} \neq \mathbf{0}} \frac{\|T_e\hat{\mathbf{x}}\|}{\|\hat{\mathbf{x}}\|}$$

with $\|\hat{\mathbf{x}}\| = (\sum_{i=1}^n \hat{x}_i \hat{x}_i)^{1/2}$ for any $\hat{\mathbf{x}} \in R^n$, then

$$\|\mathbf{T}_e\| \leq h_e/\hat{\rho}, \qquad \|\mathbf{T}_e^{-1}\| \leq \hat{h}/\rho_e.$$

Proof. Let $\hat{\mathbf{z}} = \hat{\rho}\hat{\mathbf{x}}/\|\hat{\mathbf{x}}\|$; then $\|\hat{\mathbf{z}}\| = \hat{\rho}$. Hence

$$\|\mathbf{T}_e\| = \sup \frac{\|\mathbf{T}_e\hat{\mathbf{x}}\|}{\|\hat{\mathbf{x}}\|} = \sup \frac{\|(\|\hat{\mathbf{x}}\|/\hat{\rho})\mathbf{T}_e\hat{\mathbf{z}}\|}{\|\hat{\mathbf{x}}\|}, \qquad \|\hat{\mathbf{x}}\| \neq 0.$$

Now pick any two points $\hat{\mathbf{x}}$ and $\hat{\mathbf{y}}$ in $\hat{\Omega}$ that lie on the sphere of diameter $\hat{\rho}$: then $\|\hat{\mathbf{x}} - \hat{\mathbf{y}}\| = \hat{\rho}$ and so

$$\|\mathbf{T}_e\| = \hat{\rho}^{-1} \sup \|\mathbf{T}_e(\hat{\mathbf{x}} - \hat{\mathbf{y}})\| = \hat{\rho}^{-1} \sup \|(\mathbf{T}_e\hat{\mathbf{x}} + \mathbf{b}_e) - (\mathbf{T}_e\hat{\mathbf{y}} + \mathbf{b}_e)\|$$
$$= \hat{\rho}^{-1} \sup \|\mathbf{x} - \mathbf{y}\| \leqslant h_e/\hat{\rho}.$$

The second inequality follows similarly (see Exercise 44.1).   ■

Mappings of functions. Suppose that we are given a continuous function v defined on Ω_e; making use of the affine map (1), we can set up an operator K_e from $C(\Omega_e)$ to $C(\hat{\Omega})$ that maps v to a function $\hat{v}$ in $C(\hat{\Omega})$, the function $\hat{v}$ being defined by

$$K_e : C(\Omega_e) \to C(\hat{\Omega}), \qquad K_e v = \hat{v}, \qquad \hat{v}(\hat{\mathbf{x}}) = v(\mathbf{x}), \tag{5}$$

where $\mathbf{x} = F_e(\hat{\mathbf{x}})$ (Figure 11.25).

The operator K_e is invertible with inverse K_e^{-1}, so that

$$K_e^{-1} : C(\hat{\Omega}) \to C(\Omega_e), \qquad K_e^{-1}\hat{v} = v. \tag{6}$$

Now suppose that $\{\hat{\psi}_i\}_{i=1}^N$ is a set of *local basis functions* defined on $\hat{\Omega}$ with the usual property that $\hat{\psi}_i(\hat{\mathbf{x}}_j) = \delta_{ij}$ for nodal points $\hat{\mathbf{x}}_j$. The function $\hat{\psi}_i$ is a polynomial of degree k, say, which we can map to $C(\Omega_e)$ using (6):

$$K_e^{-1}\hat{\psi}_i = \psi_i^{(e)}. \tag{7}$$

Here $\{\psi_i^{(e)}\}_{i=1}^N$ is the corresponding set of polynomial *local basis functions* defined on Ω_e; these functions satisfy the required identity $\psi_i^{(e)}(\mathbf{x}_j) = \delta_{ij}$ since (5) implies that $\hat{\psi}_i(\hat{\mathbf{x}}_j) = \psi_i^{(e)}(\mathbf{x}_j)$ (we have in fact carried out this transformation in Sections 42 and 43).

As usual, $\{\hat{\psi}_i\}$ spans a space $\hat{X}$ (of polynomials, in our case) and so we can construct *a projection operator* $\hat{\Pi}$ which maps any $\hat{v} \in C(\hat{\Omega})$ to its *interpolate* $\hat{\Pi}\hat{v}$ in $\hat{X}$:

$$\hat{\Pi} : C(\hat{\Omega}) \to \hat{X}, \qquad \hat{\Pi}\hat{v} = \sum_{j=1}^N \hat{v}(\hat{\mathbf{x}}_j)\hat{\psi}_j. \tag{8}$$

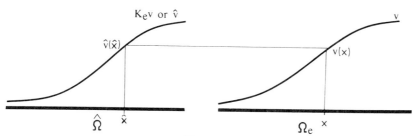

Figure 11.25

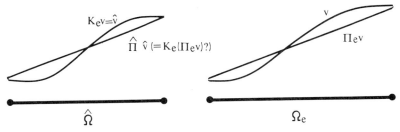

Figure 11.26

Similarly, we define the projection operator Π_e by

$$\Pi_e : C(\Omega_e) \to X_e, \qquad \Pi_e v = \sum_{i=1}^{N} v(\mathbf{x}_i)\psi_i^{(e)}. \tag{9}$$

Here $\Pi_e v$ is the interpolate of v in X_e. We come now to a crucial question about such interpolations: given a function v in $C(\Omega_e)$ and its image $K_e v$ or $\hat{v}$ in $C(\hat{\Omega})$, are $\hat{\Pi}(K_e v)$ and $K_e(\Pi_e v)$ the same functions? That is, if we map v to $\hat{v}$ and then interpolate in $\hat{\Omega}$, is this the same as first interpolating v and then mapping it? A glance at the sketch in Figure 11.26 (for linear interpolations) would seem to indicate that this is plausible; we now prove the assertion.

Theorem 1.   *Let $\hat{\Omega}$ and Ω_e be affine-equivalent finite elements. Then the interpolation operators $\hat{\Pi}$ and Π_e are such that*

$$\hat{\Pi}(K_e v) = K_e(\Pi_e v) \quad or \quad \hat{\Pi}\hat{v} = \widehat{(\Pi_e v)}.$$

Proof.   We have

$$\Pi_e v = \sum_{i=1}^{N} v(\mathbf{x}_i)\psi_i^{(e)}$$

by virtue of (9). Hence

$$K_e(\Pi_e v) = K_e\left(\sum_{i=1}^{N} \hat{v}(\hat{\mathbf{x}}_i)\psi_i^{(e)} \right)$$

$$= \sum_{i=1}^{N} \hat{v}(\hat{\mathbf{x}}_i) K_e \psi_i^{(e)} \quad (K_e \text{ is a linear operator})$$

$$= \sum_{i=1}^{N} \hat{v}(\hat{\mathbf{x}}_i)\hat{\psi}_i$$

which is precisely $\hat{\Pi}\hat{v}$.   ∎

§45. Local interpolation error estimates

Recall from our discussion of the convergence of Galerkin approxima-
tions, in Section 39, that the error $\|u - u_h\|$, measured in some
appropriate norm, can be bounded above by the *interpolation error*
$\|u - \bar{u}_h\|$ where $\bar{u}_h$ is the interpolate of u in V_h. Our task of estimating the
Galerkin error consequently reduces to one of estimating the interpola-
tion error. We go one step further towards obtaining such an estimate by
deriving in this section an estimate of the interpolation error $\|v - \Pi_e v\|$
for functions defined on a *single* finite element Ω_e. Once this estimate has
been found, we can derive from it an estimate pertaining to functions
defined over the entire domain Ω.

As before, the finite-dimensional space X_e spanned by local basis
functions $\psi_i^{(e)}$ contains polynomials of degree $\leqslant k$, for some $k \geqslant 1$. In
other words, either $X_e = P_k(\Omega_e)$ or (as in the case of rectangular elements
in R^2) $X_e = Q_l(\Omega_e)$ with l large enough so that $P_k(\Omega_e) \subset Q_l(\Omega_e)$. We will
show eventually that an interpolation error estimate in the H^m-norm can
be derived for a function v which is smooth enough to be in $H^{k+1}(\Omega_e)$,
and so we consider the following situation: we have two spaces $H^{k+1}(\Omega_e)$
and $H^m(\Omega_e)$ with $k + 1 \geqslant m$, and a projection operator Π_e which maps
members of $H^{k+1}(\Omega_e)$ to $H^m(\Omega_e)$, the images $\Pi_e v$ all lying in X_e:

$$\Pi_e : H^{k+1}(\Omega_e) \to H^m(\Omega_e), \quad \text{Range }(\Pi_e) = X_e. \tag{1}$$

The projection operator Π_e is defined by (44.9), and since $P_k(\Omega_e) \subset X_e$ by
assumption it has the property that

$$\Pi_e v = v \qquad \text{for any } v \in P_k(\Omega_e). \tag{2}$$

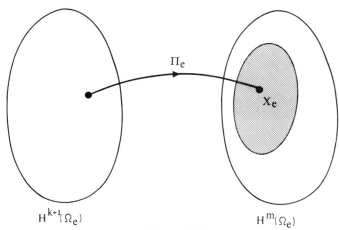

$$H^{k+1}(\Omega_e) \qquad \qquad H^m(\Omega_e)$$

Figure 11.27

Similarly,

$$\hat{\Pi}\hat{v} = \hat{v} \qquad \text{for any } \hat{v} \in P_k(\hat{\Omega}). \tag{3}$$

The main result in this section will be the following: for $v \in H^{k+1}(\Omega_e)$ and Π_e satisfying the above properties, *the interpolation error in the H^m-norm can be estimated by*

$$\|v - \Pi_e v\|_{m,\Omega_e} \leqslant C h_e^{k+1-m} |v|_{k+1,\Omega_e}$$

where h_e is defined in (44.3) and $|\cdot|_{s,\Omega_e}$ denotes the Sobolev *semi-norm*:

$$|v|^2_{s,\Omega_e} = \sum_{|\alpha|=s} \int_{\Omega_e} (D^\alpha v(\mathbf{x}))^2 \, dx$$

(note that the Sobolev norm $\|\cdot\|_{s,\Omega_e}$ is given by $\|v\|^2_{s,\Omega_e} = \sum_{l=1}^s |v|^2_{l,\Omega_e}$). Here and henceforth we will denote the norm on $H^s(\Omega)$ by $\|\cdot\|_{s,\Omega}$ rather than the more cumbersome $\|\cdot\|_{H^s(\Omega)}$. We start our development by recording an important result which will be required later.

Theorem 1. *For all $v \in H^{k+1}(\Omega)$ there is a constant C, depending only on the geometry of Ω, such that*

$$\inf_{p \in P_k(\Omega)} \|v + p\|_{k+1,\Omega} \leqslant C |v|_{k+1,\Omega}. \tag{4}$$

Proof. We use the Poincaré inequality, equation (26.10): replacing u by $v + p$ and noting that $D^\alpha p = 0$ for $|\alpha| = k + 1$, we have

$$\|v + p\|^2_{k+1} \leqslant C \left[|v|^2_{k+1} + \sum_{|\alpha| < k+1} \left\{ \int_\Omega D^\alpha(v + p) \, dx \right\}^2 \right]$$
$$\text{for } v \in H^{k+1}(\Omega), \qquad p \in P_k(\Omega). \tag{5}$$

Now construct a polynomial $\bar{p}$ in $P_k(\Omega)$ which has the property that

$$\int_\Omega D^\alpha(v + \bar{p}) \, dx = 0 \quad \text{for} \quad |\alpha| \leqslant k. \tag{6}$$

This can always be done: set $|\alpha| = k$, then $D^\alpha \bar{p}$ equals the coefficient of $\mathbf{x}^\alpha$, which can be solved for using (6). Having solved for all coefficients of terms of order k, set $|\alpha| = k - 1$, and use (6) to solve for coefficients of terms of order $k - 1$. Proceeding in this way, we find $\bar{p}$ for any given v.

With $p = \bar{p}$ in (5), we have

$$\inf_{p \in P_k(\Omega)} \|v + p\|^2_{k+1} \leqslant \|v + \bar{p}\|^2_{k+1} \leqslant C |v|^2_{k+1},$$

from which (4) follows. ∎

Next, we need to know how the semi-norms of functions v and $\hat{v}$ are related.

Theorem 2. *Let Ω_e and $\hat{\Omega}$ be two affine-equivalent open subsets of R^n. Then for any functions $v \in H^s(\Omega_e)$ and $\hat{v} = K_e v \in H^s(\hat{\Omega})$,*

$$|\hat{v}|_{s,\hat{\Omega}} \leqslant \|\mathbf{T}_e\|^s \, |\det \mathbf{T}_e|^{-1/2} \, |v|_{s,\Omega_e} \tag{7}$$

and

$$|v|_{s,\Omega_e} \leqslant \|\mathbf{T}_e^{-1}\|^s \, |\det \mathbf{T}_e|^{1/2} \, |\hat{v}|_{s,\hat{\Omega}} \tag{8}$$

where $\mathbf{T}_e$ is the matrix occurring in the affine map (44.1).

Proof. We prove (7); (8) is proved in a similar fashion. We have

$$|\hat{v}|^2_{s,\hat{\Omega}} = \sum_{|\alpha|=s} \int_{\hat{\Omega}} [D^\alpha \hat{v}(\hat{\mathbf{x}})]^2 \, d\hat{\mathbf{x}}$$

$$= \sum_{|\alpha|=s} \int_{\Omega_e} [D^\alpha \hat{v}(\hat{x})]^2 \, |\det \mathbf{T}_e|^{-1} \, dx \tag{9}$$

(using the result from multivariable calculus that if $\hat{x}_i = f_i(x_j)$ then $d\hat{x} \equiv d\hat{x}_1 \, d\hat{x}_2 \cdots d\hat{x}_n = |\det (\partial f_i/\partial x_j)| \, dx_1 \, dx_2 \cdots dx_n$).

Now by an application of the chain rule we have (see Exercise 45.5), for fixed $\mathbf{x}$ and $\hat{\mathbf{x}}$,

$$|D^\alpha \hat{v}(\hat{\mathbf{x}})| \leqslant \|\mathbf{T}_e\|^s \, |D^\alpha v(\mathbf{x})|, \qquad |\alpha| = s \tag{10}$$

(since $\hat{\mathbf{x}}$ and $\mathbf{x}$ are fixed, $D^\alpha \hat{v}(\hat{\mathbf{x}})$ and $D^\alpha v(\mathbf{x})$ are simply real numbers). Hence (9) becomes

$$|\hat{v}|^2_{s,\hat{\Omega}} \leqslant \sum_{|\alpha|=s} \int_{\Omega_e} [D^\alpha v(\mathbf{x})]^2 \, \|\mathbf{T}_e\|^{2s} (\det \mathbf{T}_e)^{-1} \, dx$$

from which (7) follows, since $\|\mathbf{T}_e\|$ and $(\det \mathbf{T}_e)$ are constant. ∎

We come now to the interpolation error estimate for the semi-norm $|v - \Pi_e v|_{m,\Omega_e}$.

Theorem 3. *Let k and m be non-negative integers such that*

$$H^{k+1}(\hat{\Omega}) \subset C(\hat{\Omega}), \qquad H^{k+1}(\hat{\Omega}) \subset H^m(\hat{\Omega}),$$

and

$$P_k(\hat{\Omega}) \subset \hat{X} \subset H^m(\hat{\Omega}).$$

Let Π_e and $\hat{\Pi}$ be the operators defined in (44.9) *and* (44.8). *Then for any* *affine equivalent element Ω_e and all functions $v \in H^{k+1}(\Omega_e)$,*

$$|v - \Pi_e v|_{m,\Omega_e} \leqslant \hat{C}(h_e^{k+1}/\rho_e^m) |v|_{k+1,\Omega_e}, \tag{11}$$

where h_e and ρ_e are defined in (44.3) and $\hat{C}$ is a constant depending on $\hat{\Omega}$ and $\hat{\Pi}$.

Proof.   We have, for all $\hat{v} \in H^{k+1}(\hat{\Omega})$ and all $\hat{p} \in P_k(\hat{\Omega})$,

$$\begin{aligned}
|\hat{v} - \hat{\Pi}\hat{v}|_{m,\hat{\Omega}} &\leqslant \|\hat{v} - \hat{\Pi}\hat{v}\|_{m,\hat{\Omega}} = \|\hat{v} - \hat{\Pi}\hat{v} + \hat{p} - \hat{\Pi}\hat{p}\|_{m,\hat{\Omega}} \quad \text{(using (3))}\\
&= \|I(\hat{v} + \hat{p}) - \hat{\Pi}(\hat{v} + \hat{p})\|_{m,\hat{\Omega}}\\
&\leqslant \|I(\hat{v} + \hat{p})\|_{m,\hat{\Omega}} + \|\hat{\Pi}(\hat{v} + \hat{p})\|_{m,\hat{\Omega}}\\
&\leqslant (\underbrace{\|I\| + \|\hat{\Pi}\|}_{\hat{C}}) \|\hat{v} + \hat{p}\|_{k+1,\hat{\Omega}}
\end{aligned}$$

[I and $\hat{\Pi}$ are bounded operators from $H^{k+1}(\hat{\Omega})$ to $H^m(\hat{\Omega})$ (see Exercise 45.1).] But this means that

$$|\hat{v} - \hat{\Pi}\hat{v}|_{m,\hat{\Omega}} \leqslant \hat{C} \inf_{\hat{p}} \|\hat{v} + \hat{p}\|_{k+1,\hat{\Omega}} \leqslant C\hat{C} |\hat{v}|_{k+1,\hat{\Omega}} \quad \text{(using (4)).} \tag{12}$$

Now from Theorem 44.1 we have $\hat{\Pi}(K_e v) = K_e(\Pi_e v)$, so that

$$\hat{v} - \hat{\Pi}\hat{v} = K_e v - \hat{\Pi}(K_e v) = K_e(v - \Pi_e v); \tag{13}$$

consequently, using (8) (replace v by $v - \Pi_e v$ and set $s = m$) and (13) we obtain

$$\begin{aligned}
|v - \Pi_e v|_{m,\Omega_e} &\leqslant \|\mathbf{T}_e^{-1}\|^m |\det \mathbf{T}_e|^{1/2} |K_e(v - \Pi_e v)|_{m,\hat{\Omega}}\\
&= \|\mathbf{T}_e^{-1}\|^m |\det \mathbf{T}_e|^{1/2} |\hat{v} - \hat{\Pi}\hat{v}|_{m,\hat{\Omega}}. \tag{14}
\end{aligned}$$

From (7), with $s = k + 1$,

$$|\hat{v}|_{k+1,\hat{\Omega}} \leqslant \|\mathbf{T}_e\|^{k+1} |\det \mathbf{T}_e|^{-1/2} |v|_{k+1,\Omega_e}. \tag{15}$$

Finally, substituting (12) in (14), then (15) in that result we obtain

$$|v - \Pi_e v|_{m,\hat{\Omega}} \leqslant \hat{C} \|\mathbf{T}_e^{-1}\|^m \|\mathbf{T}_e\|^{k+1} |v|_{k+1,\Omega_e},$$

and from (44.4), we obtain (11). ∎

Remarks.   1. Since we wish to evaluate $|v - \Pi_e v|_{m,\Omega_e}$, it follows that both v and $\Pi_e v$ must be in $H^m(\Omega_e)$ for this term to make sense. Equivalently, $\hat{v}$ and $\hat{\Pi}\hat{v}$ must be in $H^m(\hat{\Omega})$. This accounts for the inclusions $H^{k+1}(\hat{\Omega}) \subset H^m(\hat{\Omega})$ and $\hat{X} \subset H^m(\hat{\Omega})$. Note that $v \in H^{k+1}(\Omega_e)$ implies $\hat{v} \in H^{k+1}(\hat{\Omega})$. The inclusion $H^{k+1}(\hat{\Omega}) \subset H^m(\hat{\Omega})$ of course holds if $m \leqslant k + 1$.

2. In evaluating the interpolant $\Pi_e v$ of v, it is necessary to know the nodal values of v. This in turn requires that v be continuous, so that we

must have $v \in H^{k+1}(\Omega_e) \subset C(\Omega_e)$ or equivalently, $\hat{v} \in H^{k+1}(\hat{\Omega}) \subset C(\hat{\Omega})$. By the Sobolev embedding theorem, this inclusion holds if $k + 1 > n/2$ for a problem in R^n.

The two parameters h_e and ρ_e appearing in (11) may be reduced to one if we restrict ourselves to finite elements for which the ratio h_e/ρ_e is bounded above, so that elements are not allowed to become too "flat". For this purpose we introduce the notion of a *regular* family of finite elements: a family $\{\Omega_1, \ldots, \Omega_E\}$ of finite elements is said to be regular if

(i) there exists a constant σ such that $h_e/\rho_e \leq \sigma$ for all elements;
(ii) the diameters h_e approach zero.

In the case of regular families we can express the error estimate of Theorem 3 in terms of a *norm*. We record this result in the following

Corollary to Theorem 3. *Let the conditions of Theorem 3 hold, and let $\{\Omega_1, \ldots, \Omega_E\}$ be a regular family of finite elements. Then there is a constant C such that, for any element Ω_e in the family, and all functions $v \in H^{k+1}(\Omega_e)$,*

$$\|v - \Pi_e v\|_{m,\Omega_e} \leq Ch_e^{k+1-m} |v|_{k+1,\Omega_e}. \tag{16}$$

In the Exercises you are asked to deduce this result, and in particular to show that it depends on the property (ii) above of regular families of finite elements. ∎

We end this section with a simple example illustrating the estimate (16); other examples are given in Exercise 45.2.

Example

Let Ω_e be the three-noded triangle in R^2. The space X_e spanned by the local interpolation functions is $P_1(\Omega_e)$, so that $k = 1$. Assuming that v is smooth enough to belong to $H^2(\Omega_e)$, (16) gives

$$\|v - \Pi_e v\|_{m,\Omega_e} \leq Ch^{2-m} |v|_{2,\Omega_e}. \tag{17}$$

We confirm that the conditions of Theorem 3 hold: $H^{k+1}(\hat{\Omega}) = H^2(\hat{\Omega}) \subset C(\hat{\Omega})$ by the Sobolev embedding theorem (Theorem 26.2). Secondly, the estimate (17) holds for all m such that $m \leq k + 1$, that is, $0 \leq m \leq 2$.

§46. Error estimates for second-order problems

Having established properties of finite element interpolations over individual elements, we turn now to the question of interpolation of a

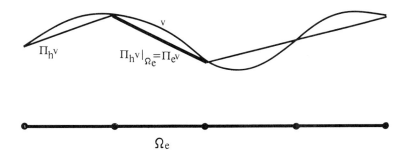

Figure 11.28

function defined on the entire domain Ω. Specifically, we have a function $v \in C(\Omega)$, and we set up its interpolant $\tilde{v}_h$ in the finite element space X_h according to

$$\tilde{v}_h(\mathbf{x}) = \sum_{i=1}^{n} v(\mathbf{x}_i)\phi_i(\mathbf{x}),$$

where ϕ_i are the basis functions that span X_h. As in Section 44, we define a *projection operator* Π_h which maps v to its interpolant $\tilde{v}_h$ or $\Pi_h v$:

$$\Pi_h : C(\Omega) \to X_h, \qquad \Pi_h v = \sum_{i=1}^{N} v(\mathbf{x}_i)\phi_i. \tag{1}$$

From the way in which the functions ϕ_i are constructed from local basis functions $\psi_i^{(e)}$, it should be clear that the restriction of $\Pi_h v$ to any element Ω_e is in fact $\Pi_e v$:

$$\Pi_h v \mid_{\Omega_e} = \Pi_e v, \qquad \Pi_h v \mid_{\Omega_e} = \Pi_e v. \tag{2}$$

Since we are primarily interested in this section in obtaining error estimates for second-order problems we must, according to (39.12), estimate $\|u - v_h\|_1$ for any $v_h \in V_h$. We choose for convenience $v_h = \Pi_h u$, and so we seek an estimate of the *interpolation error* $\|u - \Pi_h u\|_{1,\Omega}$ (recall that $m = 1$ for second-order problems). In the same way as $\|u - \Pi_e u\|_{m,\Omega_e}$ is estimated in terms of the parameter h_e (equation (45.16)), we must decide on a suitable parameter to use for the global estimate. For this purpose, suppose that we are dealing with a *regular* family of finite elements, and set

$$h = \max_{1 \leq e \leq E} \{h_e\}. \tag{3}$$

The constant h is called the *mesh parameter*, and is a measure of how refined the mesh is: the smaller h is, the larger the number of elements for a given domain Ω. Hence, if we can obtain an interpolation error estimate of the form

$$\|u - \Pi_h u\|_{1,\Omega} \leqslant Ch^\beta |u|_{k+1,\Omega}, \tag{4}$$

then we are assured of convergence as $h \to 0$, provided that $\beta > 0$.

The mesh parameter provides a natural way of quantifying the dimension of the spaces X_h or V_h that occur in Galerkin approximations. Recall from Section 39 that we discussed the notion of a *family* of problems, parametrized by a real parameter h. The idea is that for each value of h the approximate solution is sought in a finite-dimensional space V_h, with the hope that the error $\|u - u_h\|$ approaches zero as $h \to 0$. At the time we thought of h as being, for example, $1/(\dim V_h)$. As mentioned above, though, the mesh parameter gives a measure of how fine the subdivision of Ω is: the smaller h is, the finer the subdivision. Furthermore, the smaller h is, the larger the number of elements and nodal points will be, and hence the larger the dimension of V_h will be. Consequently we may use h, as defined in (3), as a measure of the size of the subspace V_h relative to V, and we take as our goal the task of showing that

$$\lim \|u - u_h\| \to 0 \quad \text{as} \quad h \to 0 \tag{5}$$

for selected classes of elements. First we obtain a global *interpolation* error estimate.

Theorem 1.   *Assume that all the conditions of Theorem 45.3 and its corollary hold. Then there exists a constant c independent of h such that, for any $v \in H^{k+1}(\Omega)$,*

$$\|v - \Pi_h v\|_{m,\Omega} \leqslant ch^{k+1-m} |v|_{k+1,\Omega} \quad \text{for } m = 0 \text{ or } m = 1. \tag{6}$$

Proof.   When $m = 1$, then $\hat{X} \subset H^1(\hat{\Omega})$ and $X_h \subset C(\bar{\Omega})$ imply that $X_h \subset H^1(\Omega)$ (see Exercise 41.1). Hence $\Pi_h u \in H^1(\Omega)$ with $\Pi_h u|_{\Omega_e} = \Pi_e u$ and we thus have, applying the corollary of Theorem 45.3 with $m = 0$ or 1,

$$\|u - \Pi_h u\|_{m,\Omega} = \left(\sum_{e=1}^{E} \|u - \Pi_e u\|_{m,\Omega_e}^2 \right)^{1/2}$$

$$\leqslant \left(\sum_{e=1}^{E} C^2 h_e^{2(k+1-m)} |u|_{k+1,\Omega_e}^2 \right)^{1/2}$$

$$\leqslant Ch^{k+1-m} \left(\sum_{e=1}^{E} |u|_{k+1,\Omega_e}^2 \right)^{1/2} = Ch^{k+1-m} |u|_{k+1,\Omega}. \quad \blacksquare$$

Finally, we come to the error estimate for second-order problems.

Theorem 2. *Consider the* VBVP *of finding* $u \in V$ *such that*

$$B(u, v) = l(v) \quad \text{for all} \quad v \in V \subset H^1(\Omega)$$

where $B(. , .)$ *is continuous and* V*-elliptic and* $l(.)$ *is continuous on* V. *If* u_h *is the finite element approximation of the solution in* V_h, *then there exists a constant* C *independent of* h *such that*

$$\|u - u_h\|_{1,\Omega} \leqslant Ch^k |u|_{k+1,\Omega}.$$

Proof. From (39.12) with $v_h = \Pi_h u$ and (6) with $m = 1$ we obtain

$$\|u - u_h\|_{1,\Omega} \leqslant (k/\alpha) \|u - \Pi_h u\|_{1,\Omega} \leqslant Ch^k |u|_{k+1,\Omega}$$

with $C = ck/\alpha$. ∎

It may happen in practice that the solution u is not smooth enough to belong to $H^{k+1}(\Omega)$. For example, if we know from the theory of elliptic BVPs (Section 33) that u is in $H^2(\Omega)$, then the use of quadratic six-noded triangles for a problem in R^2 will mean that $k = 2$ or $k + 1 = 3$, and the semi-norm $|v|_{3,\Omega}$ in (6) cannot necessarily be evaluated. We overcome this problem by going back to Section 45, and by noting that the entire theory developed there still holds if we replace $k + 1$ by r, and hence also k by $r - 1$, where $r \leqslant k + 1$ is any positive integer. Specifically, we do this in Theorems 45.1 and 45.3, and in the corollary to Theorem 45.3. Of course, r must be such that $H^r(\hat{\Omega}) \subset C(\hat{\Omega})$ (that is, $r > n/2$ and $r \geqslant m$). The estimate (45.16) then reads, for $v \in H^r(\Omega_e)$,

$$\|v - \Pi_e v\|_{m,\Omega_e} \leqslant Ch_e^\mu |v|_{r,\Omega_e}$$

where $\mu = k + 1 - m$ if $r \geqslant k + 1$ (since in this case $v \in H^{k+1}(\Omega_e)$ also) and $\mu = r - m$ if $r < k + 1$. Coming to the global estimate (6), we may alter this accordingly so that, for $v \in H^r(\Omega)$,

$$\|u - u_h\|_{1,\Omega} \leqslant Ch^\alpha |u|_{r,\Omega} \tag{7}$$

where $\alpha = \min (k, r - 1)$.

We make one more improvement to the error estimate (7). As it stands, it involves the *unknown* quantity $|u|_{r,\Omega}$ on the right-hand side. This dependence on u is easily removed, however, if we know that the solution depends continuously on the data; for example, from Theorem 33.1 we know that if the original PDE is of the form $Au = f$ with $f \in H^s(\Omega)$ and with Ω having a *smooth* boundary, then this theorem assures us that the solution u lies in $H^{s+2}(\Omega)$, and that

$$\|u\|_{s+2} \leqslant C_1 \|f\|_s, \tag{8}$$

for some constant $C_1 > 0$. Hence we may set $r = s + 2$, and since

$$|u|_r = |u|_{s+2} \leqslant \|u\|_{s+2} \leqslant C_1 \|f\|_s, \tag{9}$$

we may remove dependence on $|u|_r$ in (7). But (7) applies only to a polygonal domain; fortunately, (8) often holds even when Γ is not smooth, for many problems (for example, if the domain is convex with Lipschitz domain, (8) holds for (36.5) for $s = 0$) ([3], p. 267).

Corollary to Theorem 2.   *Let the conditions of Theorem 2 hold, and let the data f be given in $H^s(\Omega)$, $s \geq 0$. Furthermore, assume that (8) holds. Then a constant C exists such that, as $h \to 0$*

$$\|u - u_h\|_{1,\Omega} \leq Ch^\beta \|f\|_s \tag{10}$$

where $\beta = \min(k, s+1)$.   ■

According to the theorem and its corollary, since the order of convergence β is governed by the lesser of k and $s + 1$, when $s \leq k - 1$ then convergence is governed by the smoothness of f. For example, if f is only in $L_2(\Omega) = H^0(\Omega)$, then it suffices to use elements that contain only polynomials of degree ≤ 1 (such as two-noded elements in R, three-noded triangles and four-noded rectangles in R^2).

We conclude with a simple example.

Example

Consider the problem

$$-\nabla^2 u = f \qquad \text{in } \Omega \subset R^n,$$
$$u = 0 \qquad \text{on } \Gamma.$$

The corresponding VBVP is: find $u \in H_0^1(\Omega)$ such that

$$\int_\Omega \nabla u \cdot \nabla u \, dx = \int_\Omega fv \, dx \qquad \text{for all } v \in H_0^1(\Omega),$$

and we know from Section 36 that this problem has a unique solution. Similarly, the VBVP corresponding to the approximate solution is: find $u_h \in V_h$ such that

$$\int_\Omega \nabla u_h \cdot \nabla v_h \, dx = \int_\Omega fv_h \, dx \qquad \text{for all } v_h \in V_h,$$

and this problem also has a unique solution. Here V_h consists of those piecewise polynomial functions in X_h that satisfy the boundary condition, so that $V_h \subset H_0^1(\Omega)$. If $f \in H^s(\Omega)$, then the error is estimated by

$$\|u - u_h\|_{1,\Omega} \leq ch^\beta \|f\|_{s,\Omega}$$

where $\beta = \min(k, s+1)$. Thus if linear ($k = 1$) elements are used, the error is of order h^1 since $s + 1$ will not be less than 1.   □

Bibliographical remarks

The basic ideas set out in Sections 41 to 43 may be found in most books on finite elements, though the style and emphasis vary considerably from one book to another. The subject first gained prominence through its use in the solution of problems in solid and structural mechanics, and the vast majority of texts, though covering most of the essential ideas, are directed at those whose interests lie in mechanics. For some insight into the "real" applications of the method the book of Zienkiewicz [50] is recommended.

Good elementary expositions of the basic theory may be found in the texts by Becker, Carey and Oden [6] and Reddy [36] while the text by Bathe [5], though also very useful, is of interest mainly to those with an engineering background. We have not given any coverage to computer programming aspects of the method: for comprehensive accounts the books by Bathe [5] and by Hinton and Owen [20] are good references.

The final three sections of this chapter have dealt entirely with the analysis of convergence of the finite element method, and have drawn heavily on the works by Ciarlet [11], Oden and Reddy [34] and Oden and Carey [33], all of which can be consulted for further elucidation and for extensions of the theory to higher-order problems. The text by Strang and Fix [48] is also a very useful source.

We have not given any coverage to problems posed on domains with *curved* boundaries. Using the methods outlined for elements with straight boundaries, it is possible to deal with elements having curved boundaries, the affine map F_e from the master element $\hat{\Omega}$ to the element now being replaced by a nonlinear map. Various such maps can be used, the most popular being the *isoparametric* map, which is of the form

$$\mathbf{x} = \sum_{i=1}^{N} \mathbf{x}_i \hat{\psi}_i(\hat{\mathbf{x}});$$

that is, the basis functions $\hat{\psi}_i$ are used to approximate the curved boundary (Figure 11.29). Full details of isoparametric elements may be found, for example, in Becker, Carey and Oden [6], Bathe [5] and Reddy [36].

The interpolation theory for curved elements indicates that, provided that the map F_e is "almost affine", in that it differs from an affine map by a perturbation, then the interpolation error estimate is essentially unchanged. This extension of the basic theory is discussed fully by Ciarlet [11]. Other extensions, for example, those that take account of errors due to the use of numerical integration, may be found in Ciarlet [11], Oden and Reddy [34] and Strang and Fix [48]. Finally, Grisvard [18] discusses extensions of the theory which accommodate regularity results for

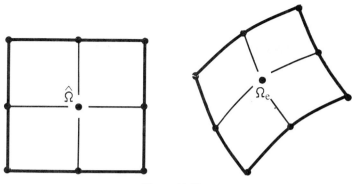

Figure 11.29

polygonal domains (recall that the results of Chapter 8 are for smooth domains).

Exercises

41.1 Assume that the space X_e spanned by local basis functions belongs to $H^1(\Omega_e)$, and that $X_h \subset C(\Omega)$. Show that $X_h \subset H^1(\Omega)$. [Take any $v \in X_h$: apply Green's theorem to $\int_{\Omega_e} (\partial v/\partial x_i)\psi \, dx$ where $\psi \in C_0^\infty(\Omega_e)$; then sum over all elements.]

42.1 Rework the Example in Section 42 using the quadratic local basis functions (42.5)–(42.6).

42.2 Let X_h be the space spanned by piecewise linear functions, that is, $X_e = P_1(\Omega_e)$, where $\Omega_e \subset \Omega \subset R$. Let f be any function defined on Ω, and assume that f can be differentiated as many times as desired. Let $\tilde{f}_h$ be the interpolate of u in X_h. The purpose of this Exercise is to show that the *interpolation error* $\tilde{e} = f - \tilde{f}_h$ satisfies the *error bound*

$$\|e\|_\infty = \max_{0 \leqslant x \leqslant l} |f(x) - \tilde{f}_h(x)| \leqslant \frac{h^2}{8} \max_{0 \leqslant x \leqslant l} |f''(x)| \qquad (1)$$

where h is the length of an element. Expand $\tilde{e}(x)$ in a Taylor series about any point $\bar{x}$ in Ω_e, that is

$$\tilde{e}(x) = \tilde{e}(\bar{x}) + \tilde{e}'(\bar{x})(x - \bar{x}) + \tfrac{1}{2}\tilde{e}''(z)(x - \bar{x})^2$$

where z is a point between x and $\bar{x}$. Select $\bar{x}$ to be the point at which $\tilde{e}$ is a maximum (note that $\tilde{e} = 0$ at nodal points); then derive the result

$$|\bar{e}(\bar{x})| = \tfrac{1}{2} |\bar{e}''(z)| (x_i - \bar{x})^2$$

where x_i is one of the nodes of Ω_e. Assuming that x_i is the node nearer to $\bar{x}$, obtain (1). Note that (1) tells us that the error converges to zero at a rate proportional to h^2: the more elements there are, the smaller h will be, and consequently the smaller the error will be.

42.3 Use Exercise 42.2 to estimate the error $\|f - \tilde{f}_h\|_\infty$ if f is the function $f(x) = x \sin \pi x$ on the domain $\Omega = (0, 1)$. Compute the actual error using two, three and four elements.

42.4 Suppose that we wish to solve a *fourth-order* BVP on the interval $[0, 1]$, so that the inclusion $X_h \subset H^2(0, 1)$ is required. On a master element with two nodes sketch the local basis functions

$$^0\hat{\psi}_1(\hat{x}) = \tfrac{1}{4}(2 - 3\hat{x} + \hat{x}^3), \qquad ^0\hat{\psi}_2(\hat{x}) = \tfrac{1}{4}(2 + 3\hat{x} - \hat{x}^3),$$

$$^1\hat{\psi}_1(\hat{x}) = \tfrac{1}{4}(1 - \hat{x} - \hat{x}^2 + \hat{x}^3), \qquad ^1\hat{\psi}_2(\hat{x}) = \tfrac{1}{4}(-1 - \hat{x} + \hat{x}^2 + \hat{x}^3).$$

Construct *Hermite basis functions* $^0\phi_i$ and $^1\phi_i$, obtained by mapping the functions $^0\hat{\psi}_i$ and $^1\hat{\psi}_i$ onto the finite element mesh in the usual way, and by patching together local basis functions associated with node i. Verify that

$$^0\phi_i(x_j) = \frac{d(^1\phi_i)}{dx}(x_j) = \begin{cases} 1 & \text{if } i = j, \\ 0 & \text{if } i \neq j, \end{cases} \qquad \frac{d(^0\phi_i)}{dx}(x_j) = {}^1\phi_i(x_j) = 0.$$

Show that the space $X_h = \text{span} \{^0\phi_i, {}^1\phi_i\}$ is a subspace of $H^2(0, 1)$, and that the interpolate $\tilde{v}_h \in X_h$ of $v \in H^2(0, 1)$ is

$$\tilde{v}_h(x) = \sum_{i=1}^{2} (v_i {}^0\phi_i(x) + v_i' {}^1\phi_i(x))$$

where

$$v_i = v(x_i), \qquad v_i' = v'(x_i).$$

43.1 Show that the basis functions ϕ_i obtained by patching together quadratic local basis functions $\psi_i^{(e)}$ on triangular elements are continuous. Sketch a typical basis function ϕ_i.

43.2 Derive explicit formulae for *biquadratic* basis functions $\hat{\psi}_i$ $(i = 1, \ldots, 9)$ on a square master element $\hat{\Omega}$.

43.3 Rework the Example in Section 43 using the mesh shown in Figure 11.30.

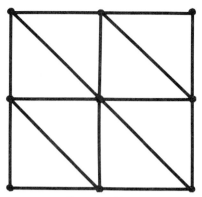

Figure 11.30

44.1 Complete the proof of Lemma 44.1 by showing that $\|T_e^{-1}\| \leqslant \hat{h}/\rho_e$.

45.1 Show that $I : H^{k+1}(\hat{\Omega}) \to H^m(\hat{\Omega})$ and $\hat{\Pi} : H^{k+1}(\hat{\Omega}) \to H^m(\hat{\Omega})$ are bounded operators, where $\hat{\Pi}$ is defined by (44.8) [Theorem 26.2 will be useful when dealing with $\hat{\Pi}$].

45.2 Complete Table 11.1.

Largest k for which $P_k(\Omega_e) \subset X_e$	$k = 1$	?	?
$\|u - \tilde{u}_h\|_{m,\Omega_e}$	$O(h_e^{2-m})$ $0 \leq m \leq 2$	?	?
Regularity of u	$H^2(\Omega_e)$	?	?
	?		

Table 11.1

45.3 Consider a *regular* family of triangular finite elements, that is, one for which

$$h_e/\rho_e \leq \sigma, \tag{i}$$

for some $\sigma > 0$. Show that the condition (i) is satisfied if the smallest angle θ_e in an element is bounded below by some constant, that is

$$\theta_e \geq \theta_0 \quad \text{for some } \theta_0.$$

This is known as *Zlámal's condition*; it ensures that elements are not too severely distorted.

45.4 Derive the estimate (45.16): explain where in the derivation the condition $h_e \to 0$ is required. Also show that the constant C in (45.16) is proportional to σ^a in (i), for some positive number a, and explain how this affects the error estimate.

45.5 The purpose of this Exercise is to derive the relation (45.10) for functions defined on domains in R^2. We start by defining the *Fréchet derivative* $\mathcal{D}v$ of a function v to be the linear map

$$\mathcal{D}v : R^2 \to R, \qquad \mathcal{D}v(\mathbf{a}) = \sum_{i=1}^{2} \frac{\partial v}{\partial x_i} a_i.$$

The second Fréchet derivative is defined to be the bilinear map

$$\mathcal{D}^2 v : R^2 \times R^2 \to R, \qquad \mathcal{D}^2 v(\mathbf{a}, \mathbf{b}) = \sum_{i,j=1}^{2} \frac{\partial^2 v}{\partial x_i \, \partial x_j} (a_i, b_j)$$

and higher derivatives are defined similarly. Generally $\mathcal{D}^k v$ is an operator from $R^2 \times \cdots \times R^2$ (k times) to R, and is linear in each slot. The space L^k of all kth Fréchet derivatives is a normed space with norm

$$\|\mathcal{D}^k v\| = \sup |\mathcal{D}^k v(\mathbf{a}^{(1)}, \dots, \mathbf{a}^{(k)})|, \qquad \|\mathbf{a}^{(k)}\| \leq 1. \tag{i}$$

Clearly then, we have

$$|D^\alpha v(\mathbf{x})| \leq \|\mathcal{D}^k v\| \quad \text{for} \quad |\alpha| = k; \tag{ii}$$

for example, if $\alpha = (1, 1)$ then

$$|D^\alpha v(\mathbf{x})| = |\partial^2 v/\partial x \, \partial y| = |\mathcal{D}^2 v(\mathbf{e}_1, \mathbf{e}_2)| \leq \sup_{\substack{\|\mathbf{a}\| \leq 1 \\ \|\mathbf{b}\| \leq 1}} |\mathcal{D}^2 v(\mathbf{a}, \mathbf{b})| \,.$$

To derive (45.10), show that

$$\mathscr{D}^k \hat{v}(\mathbf{a}^{(1)}, \ldots, \mathbf{a}^{(k)}) = \mathscr{D}^k v(\mathbf{T}_e \mathbf{a}^{(1)}, \ldots, \mathbf{T}_e \mathbf{a}^{(k)})$$

and then use (i) and (ii). [Carry out the calculation first for $k = 1$ and $k = 2$, then proceed to the general case.]

46.1 The theory of Sections 45 and 46 does not enable us to obtain error estimates in the L_2-norm for second-order problems, mainly because of the central rôle played by the inequality (39.12). It is possible to obtain L_2 estimates, though, using a method known as the *Aubin–Nitsche method*. This method is outlined here. Consider the auxiliary VBVP of finding $w \in V \subset H^1(\Omega)$ such that

$$B(w, v) = (u - u_h, v)_{L_2} \quad \text{for all } v \in V,$$

and let $\bar{w}_h$ be the interpolate of w in V_h. Show that

$$B(w - \bar{w}_h, u - u_h) = \|u - u_h\|_{L_2}^2$$

and use the continuity of B and the results of Sections 45 and 46 to obtain the estimate

$$\|u - u_h\|_{L_2}^2 \leqslant Ch^{\beta + \gamma} \|u\|_{H^r} \|w\|_{H^p}$$

where

$$\beta = \min(k, r - 1) \quad \text{and} \quad \gamma = \min(k, p - 1).$$

Finally use Theorem 33.1 (remember that $Aw = u - u_h$) to obtain the estimate

$$\|u - u_h\|_{L_2} \leqslant Ch^\nu \|u\|_{H^r}$$

where

$$\nu = \min(2k, k + 1, r).$$

46.2 Use the result of Exercise 46.1 to obtain L_2-estimates of the error for the problem

$$-\nabla^2 u = f \quad \text{in } \Omega \subset R^2,$$
$$u = 0 \quad \text{on } \Gamma,$$

assuming $f \in L_2(\Omega)$ and using linear or bilinear elements.

46.3 Suppose that we have to solve a fourth-order BVP defined on $\Omega = (0, 1)$, and assume that we intend using the Hermite basis functions described in Exercise 42.4. Verify that the theory developed in Sections 44 to 46 remains essentially unchanged except that, for example, we must specify $H^{k+1}(\hat{\Omega}) \subset C^1(\hat{\Omega})$ in Theorem 45.3. Derive an estimate of the error in finite-element approximations of the problem

$$\frac{d^4u}{dx^4} + ku = f \quad \text{in } (0, 1)$$

$$u(0) = u(1) = 0,$$

$$u'(0) = u'(1) = 0,$$

assuming that $f \in L_2(0, 1)$, and using the cubic Hermite functions in Exercise 42.4.

References

The list below contains all texts and articles referred to in the Bibliographical Remarks at the ends of chapters. This list is not exhaustive: there are many other books beside those referred to here, which would be useful to readers. However, I have confined the extent of the list to those texts with which I am most familiar, and which I have found particularly helpful. The omission of texts of similar scope to those listed here in no way implies their inferiority of quality. For more extensive bibliographies on functional analysis the reader is referred the list of references given by Roman [40], [41], while Oden and Reddy [35] have compiled a useful bibliography for variational methods, including variational boundary-value problems.

1. Adams, R. A., *Sobolev Spaces*. Academic Press (New York) 1975.
2. Apostol, T. M., *Mathematical Analysis: A Modern Approach to Advanced Calculus*. Addison-Wesley (Reading, Mass.) 1957.
3. Babuška, I. and Aziz, A. K., "Survey Lectures on the Mathematical Foundations of the Finite Element Method", in *The Mathematical Foundations of the Finite Element Method with Applications to Partial Differential Equations* (ed. A. K. Aziz). Academic Press (New York) 1972.
4. Baiocchi, C. and Capelo, A., *Variational and Quasi-Variational Inequalities*. Wiley (New York) 1984.
5. Bathe, K-J., *Finite Element Procedures in Engineering Analysis*. Prentice-Hall (Englewood Cliffs, New Jersey) 1982.
6. Becker, E. B., Carey, G. F. and Oden, J. T., *Finite Elements*, Volume 1: *An Introduction*. Prentice-Hall (Englewood Cliffs, New Jersey) 1981.
7. Binmore, K. G., *Mathematical Analysis: A Straightforward Approach*. Cambridge University Press (Cambridge) 1977.
8. Binmore, K. G., *The Foundations of Analysis: A Straightforward Introduction*. Book 2: *Topological Ideas*. Cambridge University Press (Cambridge) 1981.

9. Blanchard, P. and Brüning, E., *Direkte Methoden der Variationsrechnung: Ein Lehrbuch.* Springer-Verlag (Wien) 1982.
10. Carey, G. F. and Oden, J. T., *Finite Elements,* Vol. 2: *A Second Course.* Prentice-Hall (Englewood Cliffs, New Jersey) 1983.
11. Ciarlet, P. G., *The Finite Element Method for Elliptic Problems.* North-Holland (Amsterdam) 1978.
12. Courant, R. and Hilbert, D., *Methods of Mathematical Physics,* Vol. 2: *Partial Differential Equations.* Wiley-Interscience (New York) 1962.
13. Duvaut, G. and Lions, J. L., *Inequalities in Mechanics and Physics.* Springer-Verlag (Berlin) 1976.
14. Fučik, S. and Kufner, A., *Nonlinear Differential Equations.* Elsevier (Amsterdam) 1980.
15. Gilbarg, D. and Trudinger, N. S., *Elliptic Partial Differential Equations of Second Order.* Springer-Verlag (Berlin) 1977.
16. Glowinski, R., *Numerical Methods for Nonlinear Variational Problems.* Springer-Verlag (Berlin) 1984.
17. Goffman, C. and Pedrick, G., *First Course in Functional Analysis.* Prentice-Hall (Englewood Cliffs, New Jersey) 1965.
18. Grisvard, P., *Elliptic Problems in Nonsmooth Domains.* Pitman (London) 1985.
19. Halmos, P., *Finite Dimensional Vector Spaces.* Van Nostrand Reinhold (New York) 1958.
20. Hinton, E. and Owen, D. R. J., *Finite Element Programming.* Academic Press (London) 1977.
21. Hoffman, K. and Kunze, R., *Linear Algebra.* Addison-Wesley (Reading, Mass.) 1973.
22. Kellogg, B., "Properties of Solutions of Elliptic Boundary Value Problems", in *The Mathematical Foundations of the Finite Element Method with Applications to Partial Differential Equations* (ed. A. K. Aziz). Academic Press (New York) 1982.
23. Kolmogorov, A. N. and Fomin, S. V., *Elements of the Theory of Functions and Functional Analysis,* Volume 1, *Metric and Normed Spaces.* Graylock Press (Rochester, New York) 1957.
24. Kolmogorov, A. N. and Fomin, S. V., *Elements of the Theory of Functions and Functional Analysis,* Volume 2, *Measure, Lebesgue Integrals and Hilbert Space.* Academic Press (New York) 1961.
25. Kreyszig, E., *Introductory Functional Analysis with Applications.* Wiley (New York) 1978.
26. Kufner, A. and Kadlec, J., *Fourier Series.* Iliffe Books (London) 1971.
27. Lions, J. L. and Magenes, E., *Non-Homogeneous Boundary-Value Problems and Applications,* Volume 1. Springer-Verlag (New York) 1972.
28. Mikhlin, S. G., *Variational Methods in Mathematical Physics.* Pergamon (Oxford) 1964.
29. Naylor, A. W. and Sell, G. R., *Linear Operator Theory in Engineering and Science.* Springer-Verlag (Berlin) 1982.
30. Nečas, J., *Les Méthodes Directes en Théorie des Equations Elliptiques.* Masson (Paris) 1967.

31. Noble, B., *Applied Linear Algebra*. Prentice-Hall (Englewood Cliffs, New Jersey) 1969.
32. Oden, J. T., *Applied Functional Analysis: An Introductory Treatment for Students of Mechanics and Engineering Science*. Prentice-Hall (Englewood Cliffs, New Jersey) 1979.
33. Oden, J. T. and Carey, G. F., *Finite Elements*, Volume 4: *Mathematical Aspects*. Prentice-Hall (Englewood Cliffs, New Jersey) 1982.
34. Oden, J. T. and Reddy, J. N., *An Introduction to the Mathematical Theory of Finite Elements*. Wiley (New York) 1976.
35. Oden, J. T. and Reddy, J. N., *Variational Methods in Theoretical Mechanics*. 2nd Edition. Springer-Verlag (Berlin) 1983.
36. Reddy, J. N., *An Introduction to the Finite Element Method*. McGraw-Hill (New York) 1984.
37. Reddy, J. N., *Energy and Variational Methods in Applied Mechanics*. Wiley (New York) 1984.
38. Reed, M. and Simon, B., *Methods of Modern Mathematical Physics* I: *Functional Analysis*. Academic Press (New York) 1980.
39. Rektorys, K., *Variational Methods in Mathematics, Science and Engineering*, 2nd Edition. D. Reidel (Dordrecht) 1980.
40. Roman, P., *Some Modern Mathematics for Physicists and Other Outsiders*, Volume 1: *Algebra, Topology and Measure Theory*. Pergamon (Oxford) 1975.
41. Roman, P., *Some Modern Mathematics for Physicists and Other Outsiders*, Volume 2: *Functional Analysis with Applications*. Pergamon (Oxford) 1975.
42. Schwartz, L., *Théorie des Distributions*. Hermann (Paris) 1950.
43. Schwartz, L., *Mathematics for the Physical Sciences*. Hermann (Paris) 1966.
44. Showalter, R. E., *Hilbert Space Methods for Partial Differential Equations*. Pitman (Boston) 1977.
45. Simmons, G. F., *Introduction to Topology and Modern Analysis*. McGraw-Hill (New York) 1963.
46. Smirnov, V. I., *A Course of Higher Mathematics*, Volume 5: *Integration and Functional Analysis*. Pergamon (Oxford) 1964.
47. Strang, G., *Linear Algebra and its Applications*. Academic Press (New York) 1976.
48. Strang, G. and Fix, G. J., *An Analysis of the Finite Element Method*. Prentice-Hall (Englewood Cliffs, New Jersey) 1973.
49. Vichnevetsky, R., *Computer Methods for Partial Differential Equations*, Volume 1: *Elliptic Equations and the Finite Element Method*, Prentice-Hall (Englewood Cliffs, New Jersey) 1981.
50. Zienkiewicz, O. C., *The Finite Element Method*. 3rd Edition. McGraw-Hill (London) 1977.

Solutions to exercises

Chapter 1

1.1 $A = \{-2, 3\}$, $B = \{-3, -2, -1, 0, 1, 2, 3\}$. $A \cup B = B$; $A \cap B = A$; $A \cap Z^+ = 3$, $A - Z^+ = \{-2\}$.

1.2 $A \cup C = \{1, 2, 9\}$ so $B \times (A \cup C) = \{(7, 1), (7, 2), (7, 9), (8, 1), (8, 2), (8, 9)\}$; $A \cap C = \{1\}$ so $(A \cap C) \times B = \{(1, 7), (1, 8)\}$.

1.3 Let $x \in A \cap (B \cup C)$. Then $x \in A$ and $x \in B$ or C, i.e. $x \in A$ and $x \in B$, or $x \in A$ and $x \in C$. Hence $x \in (A \cap B) \cup (A \cap C)$. The second identity is proved in a similar way.

2.1 Consider the table

$$
\begin{array}{lllll}
1/1 & 1/2 & 1/3 & 1/4 & 1/5 \quad \ldots \\
2/1 & 2/2 & 2/3 & 2/4 & 2/5 \quad \ldots \\
3/1 & 3/2 & 3/3 & 3/4 & 3/5 \quad \ldots \\
4/1 & \ldots
\end{array}
$$

The rationals can be listed by writing down the numbers in the above table in the order shown, omitting those already listed (e.g. omit $2/2 = 1$). This gives a listing of all rationals whose numerator and denominator add up to 2, then 3, and so on. In this way all positive rationals are covered. Multiply by -1 to get negative rationals.

2.2 Assume $\sqrt{2} = p/q$, p and q having no common divisor (if they have

306

a common divisor, this can be cancelled out). Thus $p^2/q^2 = 2$ or $p^2 = 2q^2$. Hence p^2 is even $\Rightarrow p$ is even. Then p^2 is divisible by 4, hence q is even $\Rightarrow p$ and q have a common divisor, 2: contradiction.

2.3 (i) $[a, b]$; (ii) R; (iii) $[0, 1]$.

2.4 (i) is closed; (ii) open; (iii) neither open nor closed.

2.5 Assume I is closed. Let $x \in I'$; since $x \notin I$, the distance from x to I is finite. Hence we can set up a neighbourhood of radius $\varepsilon < d$ about x which lies entirely in I'. Hence I' is open. Conversely, assume I' is open. We always have $I \subset \bar{I}$, so we want to show that $\bar{I} \subset I$. Let $x \in \bar{I}$ and assume $x \notin I$. Then x is in I'. Since I' is open, there is a neighbourhood N of x with $N \cap I = \varnothing$, which is a contradiction. Thus $x \in I$ and so $\bar{I} \in I$.

2.6 (i) $\max A = 1 = \sup A$, $\min A$ is undefined but $\inf A = 0$. (ii) $\max A$, $\min A$ undefined; $\sup A = 1$, $\inf A = -1$.

2.7 (a) Let $p = \sup I$; then $x \leqslant p$ for any $x \in I$. Let $J = \{\alpha x; x \in I\}$; since $\alpha > 0$, $\alpha x \leqslant \alpha p$. Hence J is bounded above by αp. Let the supremum of J be q (we must prove that $q = \alpha p$). Since αp is an upper bound for J and q is the least upper bound, $q \leqslant \alpha p$. But for any $y \in J$ we have $y \leqslant q \Rightarrow \alpha^{-1} y \leqslant \alpha^{-1} q$. But $I = \{\alpha^{-1} y : y \in J\}$, hence $\alpha^{-1} q$ is an upper bound for I. Thus $p \leqslant \alpha^{-1} q$ or $\alpha p \leqslant q$. Since $q \leqslant \alpha p$ also, we have $\alpha p = q$.

3.1 (a) Ω is neither open nor closed. (b) Open. Set of limit points is
$$\Omega \cup \{\mathbf{x}: x^2 + y^2 + z^2 = a^2, z > 0\} \cup \{\mathbf{x}: x^2 + y^2 < a^2, z = 0\}.$$

3.2 (a) $\sqrt{2}$; (b) $2a$.

Chapter 2

4.1 (a) not continuous at $x = \pm 1$; (b) continuous on $(-\infty, 0]$.

4.2 Suppose that $|x - y| < \delta$. Then
$$
\begin{aligned}
|p(x) - p(y)| &= |a_1(x - y) + a_2(x^2 - y^2) + \cdots + a_n(x^n - y^n)| \\
&\leqslant |a_1| \, |x - y| + \cdots + |a_n| \, |x^n - y^n| \\
&= |x - y| \, [|a_1| + |a_2| \, |x + y| \\
&\quad + \cdots + |a_n| \, |x^{n-1} + x^{n-2} y + \cdots + y^{n-1}|] < \delta \cdot C
\end{aligned}
$$

since term in square brackets is bounded above. Set $\delta = \varepsilon/C$ for any given ε.

4.3 For $0 \leqslant y \leqslant x$ we have $\sqrt{y} \leqslant \sqrt{x} \Rightarrow 2y \leqslant 2\sqrt{(xy)} \Rightarrow x - 2\sqrt{(xy)} + y \leqslant x - y$ or $(\sqrt{x} - \sqrt{y})^2 \leqslant x - y$. Hence $\sqrt{x} - \sqrt{y} \leqslant (x - y)^{1/2}$, so if $|x - y| < \delta$ then $|u(x) - u(y)| < \delta^{1/2}$. For given ε set $\delta = \varepsilon^2$.

4.4 $|f(\mathbf{x}) - f(\bar{\mathbf{x}})| = |(x^2 + 2y) - (\bar{x}^2 + 2\bar{y})| = |(x^2 - \bar{x}^2) + 2(y - \bar{y})| \leqslant |x^2 - \bar{x}^2| + 2|y - \bar{y}|$.
   Suppose that $|\mathbf{x} - \bar{\mathbf{x}}| < \delta$, i.e. $(x - \bar{x})^2 + (y - \bar{y})^2 < \delta^2$. Then $|x^2 - \bar{x}^2| = |x - \bar{x}||x + \bar{x}| < \delta \cdot C$. Also, $|y - \bar{y}|^2 < \delta^2$ so $|y - \bar{y}| < \delta$. Hence $|f(\mathbf{x}) - f(\bar{\mathbf{x}})| < (C + 2)\delta$. Set $\delta = \varepsilon/(C + 2)$.

4.5 $|f(x) - f(y)| = |x^{-1} - y^{-1}| = |y - x|/|xy|$. But $x > a$, $y > a$, so $xy > a^2$ or $1/xy < 1/a^2$. Hence $|f(x) - f(y)| < a^{-2}|x - y|$.
   Suppose $|x - y| < \delta$: then for any $\varepsilon > 0$ set $\delta = a^2\varepsilon$.

4.6 We have $|f(x_0) - f(x)| < \varepsilon$ whenever $|x_0 - x| < \delta$, i.e. for $x \in (x_0 - \delta, x_0 + \delta)$. Pick any such x: either $0 < f(x_0) - f(x) < \varepsilon$ in which case $f(x) > f(x_0) - \varepsilon$, or $0 < f(x) - f(x_0) < \varepsilon$ in which case $f(x) < f(x_0) + \varepsilon$. For the first case choose ε smaller than $f(x_0)$ so that $f(x)$ is positive. For the second case $f(x) > f(x_0) > 0$.

4.7 Assume that $f(a) < 0$, $f(b) > 0$. Since $f(a) < 0$, there is an interval $[a, a + h]$ in which $f(x) < 0$. This interval has a least upper bound c, say, and $f(c) \leqslant 0$. We cannot have $f(c) < 0$ since we would then be able to find an interval about c for which $f(x) < 0$, which would imply that c is not a l.u.b. Hence $f(c) = 0$. A similar argument applies if $f(a) > 0$ and $f(b) < 0$.

4.8 (a) $u \in C(-1, 1)$; (b) $u \in C^\infty([0, \pi] \times [0, 1])$; (c) $u \in C^1[0, 1]$.

6.1 $\int_0^1 |u(x)|^p \, dx = \int_0^1 x^{-p/a} \, dx = (a/(a - p))[x^{(a-p)/a}]_0^1 < \infty$ if $a > p$.
   $\sup |u(x)| = \sup |x^{-1/a}|$ which is not bounded, hence $u \notin L_\infty(0, 1)$.

Chapter 3

7.1 (a) Linear space; (b) not a linear space; (c) linear space: if u and v are solutions, then $a(\alpha u + \beta v)'' + b(\alpha u + \beta v)' + c(\alpha u + \beta v) = \alpha[au'' + bu' + cu] + \beta[av'' + bv' + cv] = \alpha.0 + \beta.0 = 0$, hence $\alpha u + \beta v$ is a solution; (d) not a linear space: $a(\alpha u + \beta v)'' + b(\alpha u + \beta v)' + c(\alpha u + \beta v) + d \neq \alpha[au'' + bu' + cu + d] + \beta[av'' + bv' + cv + d]$.

7.2   (a) Subspace; (b) not a subspace: $\mathbf{0} \notin V$.

7.3   (a) Subspace; (b) not a subspace: for $u,\ v \in V,\ \alpha u(a) + \beta v(a) = \alpha + \beta \neq 1$; (c) subspace.

7.4   Suppose that $U = V \oplus W$, and let $u = v_1 + w_1 = v_2 + w_2$ for $v_1, v_2 \in V$ and $w_1, w_2 \in W$. Then $v_1 - v_2 = w_1 - w_2$. But $v_1 - v_2 \in V$ and $w_1 - w_2 \in W$, so that $v_1 - v_2 = w_1 - w_2 = 0$, or $v_1 = v_2$, $w_1 = w_2$. Conversely, suppose that $u = v + w$ for $v \in V$, $w \in W$, with v and w uniquely defined. If $V \cap W \neq \{0\}$ then there exists $z \in V \cap W$ with $z \neq 0$. Hence we can write $u = (v + z) + (w - z)$ so that the decomposition of u is not unique, a contradiction.

7.5   For any $u \in C[0, 1]$, $u(x) = v(x) + w(x)$ where $v(x) = \frac{1}{2}(u(x) + u(-x))$ and $w(x) = \frac{1}{2}(u(x) - u(-x))$. Thus $v \in V$ and $w \in W$. Also, $V \cap W = \{v : v \text{ is even and odd}\} = \{0\}$.

7.6 $\alpha\beta \leq \text{area } A + \text{area } B$, hence $\alpha\beta \leq \alpha^p/p + \beta^q/q$ since $A = \int_0^\alpha x^{p-1}\,dx = \alpha^p/p$, etc. The proof now follows easily from the hints given.

8.1 $(u, v)_0 = \int_0^1 x^3(1 - \frac{3}{2}x^2)\,dx = 0$; $(u, v)_1 = (u, v)_0 + (u', v')_0 = 0 + \int_0^1 (3x^2)(-3x)\,dx \neq 0$: not orthogonal.

9.1 $\|u - v\|^2 = (u - v, u - v) = \|u\|^2 + \|v\|^2 - 2(u, v) \geq \|u\|^2 + \|v\|^2 - 2|(u, v)| \geq \|u\|^2 + \|v\|^2 - 2\|u\|\,\|v\| = (\|u\| - \|v\|)^2$. Take square roots.

9.2 $\|u + v\|^2 + \|u - v\|^2 = (u + v, u + v) + (u - v, u - v)$. Expand and rearrange.

9.3   Verify that $(.\,,.)$ defined by $(u, v) = \int_a^b u'v'\,dx$ is an inner product on U.

9.4 $(u, w) = (v, w) \Rightarrow (u - v, w) = 0$ for all w. Set $w = u - v : 0 = (u - v, u - v) = \|u - v\|^2 \Rightarrow u = v$.

9.5 $\|\alpha u + (1 - \alpha)v\| \leq \alpha\|u\| + (1 - \alpha)\|v\| \leq \alpha + (1 - \alpha) = 1$ since $\|u\| \leq 1$, $\|v\| \leq 1$.

9.7 $\int |u^r v^r| \leq [\int |u|^{r(p/r)}]^{r/p}[\int |v|^{r(q/r)}]^{r/q}$. Take rth roots of both sides.

9.8   If $v = \alpha u$ then $\|u + v\| = (u + \alpha u, u + \alpha u)^{1/2} = (1 + \alpha)\|u\|$. But

$\|u\| + \|v\| = (1 + \alpha) \|u\|$. Conversely, assume that $\|u + v\| = \|u\| + \|v\|$. Then $\|u + v\|^2 = \|u\|^2 + \|v\|^2 + 2(u, v) = (\|u\| + \|v\|)^2 = \|u\|^2 + \|v\|^2 + 2\|u\| \|v\|$. Hence $\|u\| \|v\| = (u, v)$ or $(\hat{u}, \hat{v}) = 1$ where $\hat{u} = u/\|u\|$, $\hat{v} = v/\|v\|$. Suppose $\hat{v} \neq \hat{u}$: then $\hat{v} = \hat{u} + w$, and $1 = (\hat{u}, \hat{u} + w) = 1 + (\hat{u}, w) \Rightarrow (\hat{u}, w) = 0$. Also, $\|\hat{v}\|^2 = 1 = 1 + \|w\|^2 + 2(\hat{u}, w)$, i.e. $\|w\| = 0 \Rightarrow w = 0$. Hence $\hat{v} = \hat{u}$ or $v = \alpha u$ for some α.

Chapter 4

10.1 (i) -1, $1/2$, $-1/6$, $1/24, \ldots$; (ii) $1, 0, 1, 0, 1 \ldots$; (iii) -3, $6/7$, $9/13$, $12/19, \ldots$

10.2 (a) Converges to $-3/2$; (b) not convergent; (c) converges to 1.

10.3 $\left| \dfrac{3n + 2}{n - 1} - 3 \right| = \left| \dfrac{5}{n - 1} \right| < 0.001$. Assume $n > 1$, so that $5 < 0.001(n - 1) \Rightarrow n > 5001$. Take $n = 5001$.

10.4 Suppose $\{u_n\}$ is monotone increasing, with sup $= m$. For any $\varepsilon > 0$ there exists N such that $|u_n - m| < \varepsilon$ for $n > N$, so $u_n \to m$. The same reasoning applies if $\{u_n\}$ is monotone decreasing.

10.5 $|(u_n, v_n) - (u, v)| = |(u_n - u, v_n - v) + (u, v_n - v) + (v, u_n - u)| \leq \|u_n - u\| \|v_n - v\| + \|u\| \|v_n - v\| + \|v\| \|u_n - u\| \to 0$ as $n \to \infty$ since $u_n \to u$, $v_n \to v$. Hence $(u_n, v_n) \to (u, v)$. Set $v_n = v$ (i.e. the sequence $v, v, \ldots$) to get $(u_n, v) \to (u, v)$. Finally, $|(u_n, v) - (u, v)| \leq |(u_n - u, v)| \leq \|u_n - u\| \|v\|$, hence $(u_n, v) \to (u, v)$. Set $v_n = u_n$: $\|u_n\|^2 \to \|u\|^2$, i.e. $|\|u_n\|^2 - \|u\|^2| < \varepsilon$ for $n > N$. Thus $|\|u_n\| - \|u\|| \, |\|u_n\| + \|u\|| < \varepsilon$ or $|\|u_n\| - \|u\|| < \varepsilon$ (u_n and u are bounded), i.e. $\|u_n\| \to \|u\|$.

11.1 (a) $(-1, 1]$; (b) $(-\infty, \infty)$.

11.2 (a) $u_n(x) \to 0$ pointwise. But $\|u_n - u\|_{L_2}^2 = \int_{1/n}^{2/n} n^2 \, dx = n \to \infty$ as $n \to \infty$; (b) $u_n(x) \to 0$ pointwise since $u_n(x) = n^{3/2}x/\exp(n^2 x^2) = n^{3/2}x/[1 + n^2 x^2 + \frac{1}{2}n^4 x^4 + \cdots] \to 0$ as $n \to \infty$. But $\|u_n - u\|_{L_2}^2 = \int_{-n}^{n} y^2 \exp(-2y^2) \, dy$ (setting $y = nx$) $= -\frac{1}{2}([y \exp(-2y^2)]_{-n}^{n} - \int_{-n}^{n} \exp(-2y^2) \, dy) = -\frac{1}{2}(0 + \sqrt{(\pi/2)})$ as $n \to \infty$.

11.3 Sup $|u_n(x)| = \frac{1}{2}$ at $x = 1/n$. Thus in $[0, 1]$, $u_n(x) \to 0$ pointwise but $\|u_n - u\|_\infty = \frac{1}{2}$, so convergence is not uniform. But convergence is uniform in $(a, 1]$ $(a > 0)$: $\sup |u_n(x)| = na/(1 + n^2 a^2)$ at $x = a$ for $n > 1/a$ (check this by sketching $u_n(x)$!) and $\sup |u_n(x)| \to 0$ as $n \to \infty$.

11.4 $\sup |u_n(x) - u(x)| < \varepsilon$ for $n > N$. Hence

$$\int_a^b |u_n(x) - u(x)|^p \, dx \leqslant$$

$$(\sup |u_n(x) - u(x)|)^p \cdot (b - a) < (b - a)\varepsilon^p,$$

so $\|u_n - u\|_{L_p} < (b - a)^{1/p} \varepsilon = \varepsilon'$, say, when $n > N$.

12.1 $$\|u_n - u_m\|_{L_2}^2 = \frac{n}{n + 2} - \frac{2mn}{mn + m + n} + \frac{m}{m + 2}$$

$$= 2 \frac{(m - n)^2}{(m + 2)(n + 2)(mn + m + n)}.$$

Numerator $(m - n)^2 \leqslant (m + n)^2$. Now show that $\|u_n - u_m\|_{L_2}^2 \to 0$ as $n, m \to \infty$.

12.2 $\|u_n - u_m\|_{L_1} = \int_0^1 |x^n - x^m| \, dx = \frac{1}{n + 1} - \frac{1}{m + 1}$ (taking $m > n$)

$$= \frac{m - n}{(n + 1)(m + 1)}$$

$$\leqslant \frac{m}{(n + 1)(m + 1)} \to 0 \text{ as } n, m \to \infty,$$

hence $\{u_n\}$ is a Cauchy sequence.

12.3 $\{u_n\}$ is Cauchy, so $\sup |u_n(x) - u_m(x)| < \varepsilon$ for $m, n > N$. For any x_0, $|u_n(x_0) - u_m(x_0)| < \varepsilon$, so $\{u_n(x_0)\}$ is a Cauchy sequence of real numbers. R is complete, so $u_n(x_0) \to u(x_0)$, say, which defines a function $u(x)$. The rest of the proof follows easily from the hints given.

12.4 Let $\{\mathbf{x}^k\}$ be a Cauchy sequence in R^n: $\|\mathbf{x}_k - \mathbf{x}_l\| < \varepsilon$ for $k, l > N$, i.e. $\sum_i (x_{ki} - x_{li})^p < \varepsilon^p$. Hence $|x_{ki} - x_{li}|^p < \varepsilon^p$ for each i. But R is complete so $x_{ki} \to x_i$, say. Hence $\mathbf{x}_k \to \mathbf{x}$ in R^n, so R^n is complete.

12.5 Assume $\{u_n\}$ convergent: $\|u_n - u\| < \varepsilon$ for $n > N$. Also $\|u_m - u\| < \varepsilon'$ for $m > N'$. Hence $\|u_n - u_m\| = \|(u_n - u) + (u_m - u)\| \leqslant \|u_n - u\| + \|u_m - u\| < \varepsilon + \varepsilon'$ for $n, m > N$ (assume $N > N'$).

12.6 $\|u_n - u_m\|^2 = \int_{1/2}^{1/2 + 1/n} [n(x - \frac{1}{2}) - m(x - \frac{1}{2})]^2 \, dx +$
$$\int_{1/2 + 1/n}^{1/2 + 1/m} [1 - m(x - \frac{1}{2})]^2 \, dx.$$

Show that this $\to 0$ as m, $n \to \infty$, so that $\{u_n\}$ is Cauchy. Also,

$$\|u_n - u\|^2 = \int_{1/2}^{1/2+1/n} [n(x - \tfrac{1}{2}) - 1]^2 \, dx \to 0 \quad \text{as} \quad n \to \infty$$

(check this). So $u_n \to u$ in L_2.

13.1 Let $v(x)$ be defined by

$$v(x) = \begin{cases} -1, & -1 \leqslant x < -\varepsilon, \\ 1/\varepsilon, & -\varepsilon \leqslant x \leqslant \varepsilon, \\ +1, & \varepsilon < x \leqslant 1 \end{cases}$$

(sketch this). Obviously $v \in C[-1, 1]$. We have

$$\|u - v\|_{L_2}^2 = \int_{-\varepsilon}^{0} (-1 - \varepsilon^{-1})^2 \, dx + \int_{0}^{\varepsilon} (1 - \varepsilon^{-1})^2 \, dx = \varepsilon^3/3 - \varepsilon^2 + \varepsilon.$$

Hence v can be made arbitrarily close to u by choosing ε small enough.

13.2 $\|u - v\|_\infty = \sup |1 - v(x)|$ where $|v(x)| < 1$ and $v(0) = 0$. Hence $\|u - v\|_\infty = 1$; neighbourhoods of u of radius less than 1 do not contain members of V, so u is not a limit point.

13.4 $v \in \bar{B}(u_0, r) \Rightarrow \|u_0 - v\|_\infty \leqslant r$,   i.e. $\sup |\sin 2\pi x - \cos 2\pi r| \leqslant r$. $\sup |u_0 - v| = \sqrt{2}$ (at $x = \tfrac{3}{8}$) so we require $r \geqslant \tfrac{3}{8}$.

13.5 See solution to Exercise 2.5.

13.6 Assume that V is complete, and let v be a limit point of V. Then each open ball $B(v, 1/n)$, $n = 1, 2, \ldots$, contains a point v_n, say, in V. The sequence $\{u_n\}$ is convergent, hence Cauchy, in V. Since V is complete, $v \in V$. Hence V contains all its limit points, and is closed. Conversely, assume that V is closed, and let $\{v_n\}$ be a Cauchy sequence in V. Then $\{v_n\}$ is a Cauchy sequence in U, and so converges to v in U. From Theorem 13.1, v is in V also, so V is complete.

13.7 U dense in $V \Rightarrow$ for any $v \in V$ there is a $u \in U$ such that $\|u - v\| < \varepsilon$. Similarly, for any $w \in W$ there is $v \in V$ such that $\|v - w\| < \varepsilon$. Hence $\|u - w\| \leqslant \|u - v\| + \|v - w\| < 2\varepsilon$, so U is dense in W.

14.1   Suppose that there are two points v_0, v_0' such that $\|u - v_0\| = \|u - v_0'\| = d$. Then $w \equiv (v_0 + v_0')/2$ is in V hence, using the parallelogram law, it can be shown that $d^2 \leq \|u_0 - w\|^2 < \frac{1}{2} \|u_0 - v_0\|^2 + \frac{1}{2} \|u_0 - v_0'\|^2 = d^2$, a contradiction.

14.2   Consider $\{u_n\} \subset V^\perp$ with limit u_0 in U. We must show that $u_0 \in V^\perp$ also. By definition $(u_n, v) = 0$ for any $v \in V$; thus

$$0 = \lim_{n \to \infty} (u_n, v) = \left(\lim_{n \to \infty} u_n, v \right) = (u_0, v) = 0 \Rightarrow u_0 \in V^\perp.$$

14.3   Theorem 14.1(ii), which requires completeness of H, is used in Lemma 14.1.

14.4   Let $u \in U$ and $w \in V^\perp$. Then $u \in V$ also, so $(u, w) = 0$. u is arbitrary, hence $w \in U^\perp \Rightarrow V^\perp \subset U^\perp$.

Chapter 5

15.1   (i) $R(M) =$ points on the upper unit semi-circle, $N(M) = \emptyset$; (ii) $R(K) = [0, \infty)$, $N(K) = \{0\}$; (iii) $R(f) = (0, \infty)$, $N(f) = \emptyset$.

15.2   (i) One-to-one, not surjective; (ii) one-to-one, surjective (T is a reflection about a line at $45°$ through the origin).

15.3   (i) $ST(\mathbf{x}) = S(x, -y) = (-2y, x)$; $TS(\mathbf{x}) = T(2y, x) = (2y, -x)$; (ii) $ST(x) = S(\sin x) = \sin^2 x - 1$, $TS(x) = T(x^2 - 1) = \sin (x^2 - 1)$.

15.4 $S^{-1}: V \to U$ and $T^{-1}: W \to V$ exist. Clearly $TS: U \to W$ is one-to-one onto W, so $(TS)^{-1}$ exists. Furthermore, $(TS)u = w \Rightarrow u = (TS)^{-1}w$. But $(TS)u = T(Su) = w$, so $Su = T^{-1}w$ and $u = S^{-1}T^{-1}w$. Hence $(TS)^{-1} = S^{-1}T^{-1}$.

16.1   (i) linear; (ii) linear; (iii) nonlinear.

16.2   Let $Tu_1 = v_1$, $Tu_2 = v_2$. Then $T(\alpha u_1 + \beta u_2) = \alpha v_1 + \beta v_2$ by the linearity of T. Hence $T^{-1}(\alpha v_1 + \beta v_2) = \alpha u_1 + \beta u_2$. But $\alpha T^{-1}v_1 = \alpha u_1$, $\alpha T^{-1}v_2 = \alpha u_2 \Rightarrow T^{-1}(\alpha u_1 + \beta u_2) = \alpha T^{-1}v_1 + \beta T^{-1}v_2$.

16.3   For $u \neq 0$, $\|T\| = \sup (\|Tu\|/\|u\|) = \sup \|T(u/\|u\|)\|$ (T is linear) $= \sup \|Tu\|$, $\|u\| = 1$. To prove the second result, consider $\|Tu\| \leq \|T\| \|u\|$. For every $\varepsilon > 0$, there is a u_0 such that $\|Tu_0\| > (\|T\| - \varepsilon) \|u_0\|$. If $\|u\| \leq 1$ then $\|Au\| \leq \|A\| \|u\| \leq \|A\| \Rightarrow$

$\sup \|Au\| \leqslant \|A\|$, $\|u\| \leqslant 1$. But if we put $u_1 = u_0/\|u_0\|$ then $\|Au_1\| = \|u_0\|^{-1} \|Au_0\| > \|A\| - \varepsilon$, so for $\|u\| \leqslant 1$, $\sup \|Au\| \geqslant \|Au_1\| > \|A\| - \varepsilon$, or $\sup \|Au\| \geqslant \|A\|$.

16.4 $\displaystyle \|\mathbf{Ax}\|_\infty = \max_{1 \leqslant i \leqslant n} \left| \sum_{j=1}^n A_{ij} x_j \right| \leqslant \max_{1 \leqslant i \leqslant n} \sum_{j=1}^n |A_{ij}| \, |x_j|$

$\displaystyle \leqslant \max_{1 \leqslant i \leqslant n} \sum_{j=1}^n |A_{ij}| \max_{1 \leqslant j \leqslant n} |x_j| = \max_{1 \leqslant i \leqslant n} \sum_{j=1}^n |A_{ij}| \cdot \|\mathbf{x}\|_\infty.$

Hence

$$\|\mathbf{A}\| = \sup \frac{\|\mathbf{Ax}\|_\infty}{\|\mathbf{x}\|_\infty} \leqslant \max_{1 \leqslant i \leqslant n} \sum_{j=1}^n |A_{ij}|.$$

Suppose maximum occurs for $i = k$. Then for $\mathbf{x}$ such that $x_j = +1$ if $A_{kj} \geqslant 0$, $x_j = -1$ if $A_{kj} < 0$ we have

$$\|\mathbf{Ax}\|_\infty / \|\mathbf{x}\|_\infty = \sum_{j=1}^n |A_{ij}|.$$

Hence

$$\|\mathbf{A}\| = \max_{1 \leqslant i \leqslant n} \sum_{j=1}^n |A_{ij}|.$$

16.5 $\|Iu\| = \|u\|$; I is bounded. Consider $u(x) = \sin nx$: $\|u\|_V = 1$ but $\|Iu\|_W = 1 + n$ which cannot be bounded.

16.6 $\|ST(u)\| = \|S(Tu)\| \leqslant \|S\| \, \|Tu\| \leqslant \|S\| \, \|T\| \, \|u\|$.

16.7   Let $\{u_n\} \subset N(T)$ with limit u in U. Then $Tu_n = 0$. Thus $0 = \lim_{n \to \infty} Tu_n = T(\lim_{n \to \infty} u_n) = Tu \Rightarrow u \in N(T)$.

16.8 T is one-to-one since, if $Tu_1 = Tu_2 = v$, then $\|Tu_1 - Tu_2\| = 0 \geqslant K \|u_1 - u_2\|$, i.e. $u_1 = u_2$. Next, $\|T^{-1}v\| = \|u\| \leqslant K^{-1} \|Tu\| = K^{-1} \|v\|$.

$$u(x) = \int_0^x u'(s) \, dx \leqslant \sup_{0 \leqslant x \leqslant 1} |u'(x)| = \|Du\|.$$

Now take sup of both sides.

17.1 $(I - P)(I - P) = I^2 - PI - IP + P^2 = I - P$. $R(I - P) = N(P)$, $R(P) = N(I - P)$.

17.2 From Theorem 17.2, $\|Pu\| \le \|u\|$. Thus $\|P\| \le 1$. But for $u \in R(P)$ we have $Pu = u$, so $\|Pu\| = \|u\|$. Hence $\|P\| = 1$.

17.3 $P:[0, \infty) \to [0, \infty)$, $P(x) = 1$ for $x \ge 1$, $P(x) = 0$ for $0 \le x < 1$.

17.4 Let $u \in N(P)$. By definition $(u, v) = 0$ for $v \in R(P)$. Hence $N(P) \subset R(P)^\perp$. Let $u \in R(P)^\perp$. Then $(u, z) = 0$ for $z \in R(P)$. By Theorem 17.3, $u = v + w$ for $v \in R(P)$, $w \in N(P)$, so $Pu = Pv + Pw = Pv = v$. Also, $0 = (u, z) = (v, z) + (w, z) = (v, z)$, hence $v = 0$. Thus $Pu = 0 \Rightarrow u \in N(P)$.

17.5 T is a projection since T is linear and $T^2 u = Tv$ (where $v = u(x)$ if $|x| < 1$ and 0 otherwise) $= v = Tu$. $R(T) = \{u \in L_2(R) : u(x) = 0$ for $|x| \ge 1\}$, $N(T) = \{u \in L_2(R) : u(x) = 0$ for $|x| < 1\}$.

18.1 (i) $\mathbf{x}$ satisfies $\mathbf{Ax} = \mathbf{1}$ where $\mathbf{1} = (1, \ldots, 1)$; (ii) $\mathbf{x}$ satisfies $\mathbf{Ax} = \boldsymbol{\alpha} = (1, 0, \ldots, 0)$.

18.2 $u(x) = \dfrac{2}{e^3 - 1}(-e^{3-2x} + e^x) - 2x + 2$, $l(f) = \displaystyle\int_0^1 u(x)\,dx$.

$\dfrac{2}{e^3 - 1}(\tfrac{3}{2}e - e^3) = \displaystyle\int_0^1 g(x)2x\,dx$; so g satisfies $\displaystyle\int_0^1 (gf - u)\,dx = 0$.

18.4 Let $\{l_n\}$ be a Cauchy sequence in U'. Then for any $u \in U$, $|l_n(u) - l_m(u)| \le \|l_n - l_m\| \, \|u\| \to 0$ as $m, n \to 0$ so $\{l_n(u)\}$ is a Cauchy sequence in R, with limit $l(u)$, say. Complete the proof by showing that l is bounded and linear, with $\|l_n - l_m\| \to 0$ as $n \to \infty$. Hence $l_n \to l$ in U'.

18.5 In the use of the projection theorem.

18.6 To show uniqueness, proceed as follows: if there are two elements u_1, u_2 such that $(u_1, v) = (u_2, v) = f(v)$, then $(u_1 - u_2, v) = 0$. Set $v = u_1 - u_2 : \|u_1 - u_2\|^2 = 0$ or $u_1 = u_2$. $\|f\| = \sup(|f(v)|/\|v\|)$ (for $v \neq 0$) $= \sup(u, v)/\|v\|) \le \sup(\|u\| \, \|v\|/\|v\|) = \|u\|$. Also, $|f(u)| = (u, u) = \|u\|^2 \le \|f\| \, \|u\|$ so $\|f\| \ge \|u\|$. Hence $\|f\| = \|u\|$.

19.1 cf. Exercise 10.5.

19.2 $\quad |f(v)| = \left| \int_0^1 (-1 - 4x)v(x)\, dx \right| = |(-1 - 4x, v)_{L_2}|$

$\qquad \leqslant \|-1 - 4x\|_{L_2} \|v\|_{L_2}$

$\qquad \leqslant k\, \|v\|_{H^1}.\ |B(u, v)| \leqslant 2 \left| \int_0^1 u'v'\, dx \right|$

$\qquad \leqslant 2\, \|u'\|_{L_2} \|v'\|_{L_2} \leqslant 2\, \|u\|_{H^1} \|v\|_{H^1}$, hence continuous.

$$B(v, v) \geqslant \int_0^1 (v')^2\, dx.\ \text{Now } \|v'\|_{L_2}^2 \geqslant C^{-2} \|v\|_{L_2}^2$$

so $(C^{-2} + 1)\, \|v'\|_{L_2}^2 \geqslant C^{-2} \|v\|_{H^1}^2$. Hence we get H_0^1-ellipticity.

$$\int_0^1 (-1 - 4x)v\, dx = \int_0^1 (x + 1)u'v'\, dx = [(x + 1)u'v]_0^1$$

$$- \int_0^1 u' + (x + 1)u''\, dx$$

$$\Rightarrow \int_0^1 \{(x + 1)u'' + u' - (1 + 4x)\}v\, dx = 0.$$

Term in brackets is thus zero.

19.3 $\quad |\bar{B}(u, v)| \leqslant |B(u, v)| + |(u, cv)_{L_2}| \leqslant K\, \|u\|\, \|v\| + K'\, \|u\|\, \|v\|$

where $K' = \sup |c(x)|$. $\bar{B}(v, v) = B(v, v) + (v, cv)$

$\geqslant \alpha\, \|v\|^2 + \beta(v, v) \geqslant \alpha\, \|v\|^2$ where $\beta = \inf c(x)$.

Chapter 6

20.1 (a) Linearly dependent; (b) linearly independent.

20.2 $\quad \sum_{k=1}^n \alpha_k e^{ikx} = 0 \Rightarrow \sum_{k=1}^n \alpha_k \cos kx = 0$ and $\sum_{k=1}^n \alpha_k \sin kx = 0$ which holds only for all $\alpha_k = 0$. Hence $\{e^{ikx}\}$ is linearly independent.

20.3 If $u,\ v \in U$ then $(\alpha u + \beta v)'' - 2(\alpha u + \beta v)' + (\alpha u + \beta v) = \alpha(u'' - 2u' + u) + \beta(v'' - 2v' + v) = 0$, hence $\alpha u + \beta v \in U$. dim $U = 2$. Basis for U is $\{u_1(x) = e^x, u_2(x) = xe^x\}$.

20.4 Let dim $V = m$ with basis $\{v_1, \ldots, v_m\}$ and dim $W = n$ with basis $\{w_1, \ldots, w_n\}$. Every $u \in V \oplus W$ is of the form $u = v + w$ for some $v \in V$, $w \in W$. But

$$v = \sum_i \alpha_i v_i \quad \text{and} \quad w = \sum_j \beta_j w_j \quad \text{so} \quad u = \sum_i \alpha_i v_i + \sum_j \beta_j w_j.$$

Hence $B = \{v_1, \ldots, v_m, w_1, \ldots, w_n\}$ spans $V \oplus W$. It remains to show that B is linearly independent.

21.1 $\phi_0(x) = \sqrt{(\frac{1}{2})}$, $\phi_1(x) = \sqrt{(\frac{3}{2})}x$, $\phi_2(x) = \frac{1}{2}\sqrt{(\frac{5}{2})}(3x^2 - 1)$, $\phi_3(x) = \frac{1}{2}\sqrt{(\frac{7}{2})}(5x^3 - 3x)$.

21.2 $\boldsymbol{\phi}_1 = \frac{1}{2}(1, 0, 1)$, $\boldsymbol{\phi}_2 = \frac{1}{2}(1, 0, -1)$, $\boldsymbol{\phi}_3 = (0, 1, 0)$.

21.3 $A_{11} = \frac{1}{2}(e^2 - 1)$, $A_{12} = A_{21} = \frac{1}{2}(1 - e^{-2})$, $A_{22} = \frac{1}{6}(1 - e^{-6})$. $\det \mathbf{A} \neq 0$.

21.4 Consider $I: U_1 \to U_2 : \|Iu\|_2 = \|u\|_2 \leqslant k \|u\|_1$ (show this using Lemma 21.1; see also Theorem 22.1). Similarly, $\|u\|_1 \leqslant K \|u\|_2$ if we consider $I: U_2 \to U_1$.

22.1 $T_{12} = 2$, $T_{23} = 6$, others zero.

22.2 $T_{11} = 2\pi$, $T_{22} = \cos x$, others zero.

22.3 Assume

$$\mathbf{c} \in N(\mathbf{T}^t) : (\mathbf{c}, \mathbf{Ta}) = (\mathbf{c}, \mathbf{b}) \Rightarrow (\mathbf{T}^t\mathbf{c}, \mathbf{a}) = 0$$
$$= (\mathbf{c}, \mathbf{b}), \text{ hence } \mathbf{b} \in N(\mathbf{T}^t)^\perp.$$

Let $\mathbf{d} \in R(\mathbf{T})^\perp$. Then $(\mathbf{d}, \mathbf{Tu}) = 0 = (\mathbf{T}^t\mathbf{d}, \mathbf{u}) \Rightarrow \mathbf{d} \in N(\mathbf{T}^t)$. Conversely, if $\mathbf{d} \in N(\mathbf{T}^t)$ then if $\mathbf{Tu} = \mathbf{v}$ we have $(\mathbf{T}^T\mathbf{d}, \mathbf{u}) = 0 = (\mathbf{d}, \mathbf{v}) \Rightarrow \mathbf{d} \in R(\mathbf{T})^\perp$. Hence $N(\mathbf{T}^t) = R(\mathbf{T})^\perp \Rightarrow N(\mathbf{T}^t)^\perp = R(\mathbf{T})$. $N(\mathbf{T}^t) = \{(1, 1, -1)\}$. $\mathbf{b} = (\alpha, \beta, \alpha + \beta)$.

22.4 Let $B_1 = \{e_1, \ldots, e_n\}$ and $B_2 = \{f_1, \ldots, f_n\}$ be orthonormal bases of U and R^n, respectively. For any $u \in U$ we have $u = \sum u_i e_i$, $u_i = (u, e_i)$. Define the map $T: U \to R^n$ by $T(u) = (u_1, \ldots, u_n)$. Then T is an isomorphism (show this) and $\|u\|_U^2 = (u, u) = (\sum u_i e_i, \sum u_j e_j) = \sum u_i^2 = \|Tu\|_{R^n}^2$.

23.1 (i) $u(x) = \sqrt{(2\pi)}(1/\sqrt{(2\pi)})$; (ii) $u(x) = \sum_{k=1}^{\infty} \frac{2}{k}(1 - (-1)^k) \sin kx$.

23.2 $(u, v) = (\sum_i u_i \phi_i, \sum_j v_j \phi_j) = \sum_i \sum_j u_i v_j (\phi_i, \phi_j) = \sum_i u_i v_i$.

23.3 $0 \leqslant \|u - \sum_{i=1}^{N} (u, \phi_i)\phi_i\|^2 = \|u\|^2 - \sum_{i=1}^{N} (u, \phi_i)^2$, hence $\sum_{i=1}^{N} (u, \phi_i)^2 \leqslant \|u\|^2$. Since sum is bounded, we can let $N \to \infty$.

23.4 (a) Set $s_n = \sum_{i=1}^{n} (u, \phi_i)\phi_i$. $\|s_n - s_m\|^2 = \sum_{i=m+1}^{n} (u, \phi_i)^2 \Rightarrow s_n$ is Cauchy. But H is complete, hence s_n is convergent. So $\|Pu\|^2 = \|\sum_{k=1}^{\infty} (u, \phi_k)\phi_k\|^2 = \sum_{k=1}^{\infty} (u, \phi_k)^2 \leqslant \|u\|^2$, using the Bessel inequality. (b) From Exercise 10.5, $\lim_{n\to\infty} (v, s_n) = (v, \lim_{n\to\infty} s_n)$ so the sequence $\{s_n = \sum_{k=1}^{n} (v, \alpha_k x_k)\}$ of partial sums converges to $(v, \sum_{k=1}^{\infty} \alpha_k x_k)$. Now proceed as in Theorem 23.1.

23.5 See Exercise 21.1. $Pu = \sum_{k=0}^{3} (u, \phi_k)\phi_k = \sqrt{(\tfrac{2}{5})}\phi_0 + \tfrac{8}{35}\sqrt{(\tfrac{5}{2})}\phi_2$.

Chapter 7

24.1 $|\alpha| = 0 \Rightarrow \alpha = (0, 0)$, $(x^\alpha/\alpha!)D^\alpha f(0) = f(0)$. $|\alpha| = 1 \Rightarrow$

$$\frac{x^1 y^0}{1! \, 0!} D^{(1,0)}f(0) + \frac{x^0 y^1}{0! \, 1!} D^{(0,1)}f(0) = x \left.\frac{\partial f}{\partial x}\right|_0 + y \left.\frac{\partial f}{\partial y}\right|_0, \quad \text{etc.}$$

24.2 $\int_{-a}^{a} \delta(x)\phi(x)\,dx \leqslant e^{-1}\int_{-a}^{a} \delta(x)\,dx$ since $\sup \phi_a(x) = e^{-1}$. If δ were locally integrable then $\lim_{a\to 0} \int_{-a}^{a} \delta(x)\,dx = 0$. But left-hand side = $\phi(0) = e^{-1}$.

24.3 $f(x)\phi(x) \in C(\Omega)$. Assume $f \neq 0$, but $\int f\phi\,dx = 0$. In particular, if $f(x_0) \neq 0$ then $f(x) \neq 0$ for all $x \in (x_0 - h, \; x_0 + h)$, for some h. Choose arbitrary ϕ with compact support inside $(x_0 - h, \; x_0 + h)$; can always find ϕ such that $\int f\phi\,dx \neq 0$, a contradiction.

25.1 Consider $\Omega \subset R^2$, for example: for $|\alpha| = m$, $\displaystyle\int_\Omega (D^\alpha u)v\,dx = \int_\Omega \frac{\partial^m u}{\partial x^k \partial y^{m-k}} v\,dx$ where $0 \leqslant k \leqslant m$. Use Green's theorem repeatedly to get results.

25.2 $(\text{sgn})'(\phi) = -\text{sgn}\,(\phi') = -\displaystyle\int_{-1}^{0} -\phi'\,dx - \int_{0}^{1} \phi'\,dx = [\phi]_{-1}^{0} - [\phi]_{0}^{1}$
$$= 2\phi(0) = 2\delta(\phi).$$

25.3 $(\sin ax \cdot H(x))''(\phi) = (\sin ax \cdot H(x))(\phi'') = H(x)(\phi'' \sin ax) = \int_{0}^{1} \phi'' \sin ax\,dx = [\phi' \sin ax - a\phi \cos ax]_{0}^{1} - \int_{0}^{1} a^2\phi \sin ax\,dx = a\phi(0) - a^2 \sin ax \cdot H(x)(\phi)$.

25.4 $\quad f'(\phi) = -f(\phi') = -\displaystyle\int_{-1}^{0} x\phi'(x)\,\mathrm{d}x - \int_{0}^{1}(x+c)\phi'(x)\,\mathrm{d}x$

$\qquad = -[x\phi]_{-1}^{0} + \displaystyle\int_{-1}^{0}\phi(x)\,\mathrm{d}x - [(x+c)\phi]_{0}^{1} + \int_{0}^{1}\phi(x)\,\mathrm{d}x$

$\qquad = c\phi(0) + \displaystyle\int_{-1}^{1} 1\cdot\phi(x)\,\mathrm{d}x = c\delta(\phi) + 1(\phi).$

25.5 Set $A = \{\mathbf{x}: -1 < x < 0,\ -1 < y < 0\}$, $B = \{\mathbf{x}: 0 < x < 1,\ 0 < y < 1\}$, $C = A \cup B$, with boundaries ∂A, ∂B. Then

$$D^{(1,1)}f(\phi) = f(D^{(1,1)}\phi) = \iint_{C} xy\,\frac{\partial^2 \phi}{\partial x\,\partial y}\,\mathrm{d}x\,\mathrm{d}y$$

$$= \int_{\partial A} xy v_x\,\frac{\partial \phi}{\partial y}\,\mathrm{d}s + \int_{\partial B} xy v_x\,\frac{\partial \phi}{\partial y}\,\mathrm{d}s - \iint_{C} y\,\frac{\partial \phi}{\partial y}\,\mathrm{d}x\,\mathrm{d}y$$

$$= -\iint_{C} y\,\frac{\partial \phi}{\partial y}\,\mathrm{d}x\,\mathrm{d}y$$

$$= -\iint_{\partial A} y\phi v_y\,\mathrm{d}s - \iint_{\partial B} y\phi v_y\,\mathrm{d}s + \iint_{C}\phi\,\mathrm{d}x\,\mathrm{d}y$$

$$= \iint_{C}\phi\,\mathrm{d}x\,\mathrm{d}y.$$

$$= \iint_{\Omega} g(\mathbf{x})\phi(\mathbf{x})\,\mathrm{d}x\,\mathrm{d}y.$$

25.6 Solution of homogeneous equation is $u(x) = \mathrm{e}^{-x}$. Now

$\qquad (u' + u)(\phi) = -u(\phi') + u(\phi)$

$\qquad\qquad = -H(f\phi') + H(f\phi)$ (using $u = Hf$)

$\qquad\qquad = \int_{0}^{1}(-f\phi' + f\phi)\,\mathrm{d}x = \mathrm{e}^{-x}\delta(\phi) = \delta(\mathrm{e}^{-x}\phi) = \phi(0).$

Left-hand side $= f(0)\phi(0) + \int_{0}^{1}(f' + f)\phi\,\mathrm{d}x \Rightarrow f(x) = \mathrm{e}^{-x}$. Hence $u(x) = (c + H(x))\mathrm{e}^{-x}$.

26.1 (a) $u \in H^2(0, 3)$; (b) $u \in H^1((0, 1) \times (0, 2))$.

26.2 $u \perp v$ in $H^1(0, 2)$.

26.3 Show that $|(u, u - v)| \leqslant \|u\|\,\|u - v\|$ in L_2- and H^1-norms.

26.4 $D^{\alpha}u \in L_2(\Omega)$ for $|\alpha| = 2$ so $m = 2 > n/2 = 1$.

27.1 Consider $\{u_n\}$, $\{v_n\} \subset C^1(\bar{\Omega})$ such that $u_n \to u$ and $v_n \to u$ in the H^1-norm with u, $v \in H^1(\Omega)$ (H^1 is the closure of C^1). Then $D^\alpha u_n \to D^\alpha u$, $D^\alpha v_n \to D^\alpha v$ in L_2, for $|\alpha| \leq 1$. Also, $v_n \to v$ and $u_n \to u$ in $L_2(\Gamma)$. Thus if we start with

$$\int_\Omega \frac{\partial u_n}{\partial x_i} v_n \, dx = \int u_n v_n v_i \, ds - \int_\Omega u_n \frac{\partial v_n}{\partial x_i} \, dx \Rightarrow \left(\frac{\partial u_n}{\partial x_i}, v_n \right)_{L_2(\Omega)}$$

$$= (u_n, v_n v_i)_{L_2(\Gamma)} - \left(u_n, \frac{\partial v_n}{\partial x_i} \right)_{L_2(\Omega)},$$

then taking $\lim_{n \to \infty}$ and using Exercise 10.5 we get the desired result.

27.2 Assume

$$\Omega \subset R^2 : \int_\Omega (\nabla^2 u)(\nabla^2 v) = \int_\Omega \left(\frac{\partial^2 u}{\partial x^2} + \frac{\partial^2 u}{\partial y^2} \right) \left(\frac{\partial^2 v}{\partial x^2} + \frac{\partial^2 v}{\partial y^2} \right) dx.$$

Now

$$\int_\Omega \frac{\partial^2 u}{\partial x^2} \frac{\partial^2 v}{\partial x^2} \, dx = \int_\Gamma \left(\frac{\partial^2 u}{\partial x^2} \frac{\partial v}{\partial x} - \frac{\partial^3 u}{\partial x^3} v \right) v_x \, ds + \int_\Omega \frac{\partial^4 u}{\partial x^4} v \, dx.$$

Proceed in this manner, use $\partial/\partial v = v_1 \, \partial/\partial x + v_2 \, \partial/\partial y$.

28.1 Show that $(u, v) \equiv \int_\Omega \sum_{|\alpha|=m} D^\alpha u D^\alpha v \, dx$ is an inner product. In particular,

$$(u, u) = 0 \Rightarrow \int_\Omega \sum_{|\alpha|=m} (D^\alpha u)^2 \, dx = 0 \Rightarrow \int_\Omega (D^\alpha u)^2 \, dx = 0$$

for $|\alpha| = m$, hence $D^\alpha u = 0$ for $|\alpha| = m$. But $u \in H_0^m(\Omega)$, so $u = 0$.

28.2 $\|\nabla^2 v\|_{L_2}^2 = \int_\Omega \left(\frac{\partial^2 v}{\partial x^2} + \frac{\partial^2 v}{\partial y^2} \right)^2 dx = \int \left(\frac{\partial^2 v}{\partial x^2} \right)^2 + 2 \frac{\partial^2 v}{\partial x^2} \frac{\partial^2 v}{\partial y^2} + \left(\frac{\partial^2 v}{\partial y^2} \right)^2 dx$

and

$$|v|_{H^2}^2 = \int_\Omega \left(\frac{\partial^2 v}{\partial x^2} \right)^2 + 2 \left(\frac{\partial^2 v}{\partial x \, \partial y} \right)^2 + \left(\frac{\partial^2 v}{\partial y^2} \right)^2 dx.$$

But

$$\int \left(\frac{\partial^2 v}{\partial x \, \partial y} \right)^2 dx = - \int_\Omega \frac{\partial^3 v}{\partial x^2 \, \partial y} \frac{\partial v}{\partial x} \, dx = \int_\Omega \frac{\partial^2 v}{\partial y^2} \frac{\partial^2 v}{\partial x^2} \, dx$$

using Green's theorem.

28.3 Require sup $|\delta(v)|$ to be defined, i.e. v continuous. Hence $m > n/2$. For example, $\delta : H_0^1(\Omega) \to R$ is not defined for $\Omega \subset R^2$.

28.4 $u \in H_0^1(\Omega)^\perp \Rightarrow (u, v)_{H^1} = 0$ for all $v \in H_0^1(\Omega)$, i.e.

$$0 = \int_\Omega \left(uv + \sum_{|\alpha|=1} D^\alpha u D^\alpha v \right) dx.$$

Set $v = \phi \in C_0^\infty(\Omega) : 0 = \int (u - \nabla^2 u)\phi \, dx$ using Green's theorem $\Rightarrow \nabla^2 u = u$. Since C_0^∞ is dense in H_0^1, we can extend this result in the usual way. $u \in H_0^1(\Omega)^\perp$ for $\Omega = (0, 1) \Rightarrow u'' - u = 0$. Basis for $H_0^1(\Omega)^\perp$ is $\{e^x, e^{-x}\}$.

Chapter 8

29.1 (a) Second order, nonlinear, $\Omega = $ upper unit semi-circle. (b) Fourth-order, linear, $\Omega = $ triangle with vertices at $(0, 0)$, $(1, 0)$, $(0, 1)$.

30.1 Elliptic in $A = \{\mathbf{x} : x > 1, \ y > 1\} \cup \{\mathbf{x} : x < 1, \ y < 1\}$, strongly elliptic in any open subset of A.

30.2 $\sum_{|\alpha|=2} a_\alpha \xi^\alpha = -(1 + x^2)\xi^2 + 3\eta^2 + 2(1 + z^2)\zeta^2 = 0$ at any x_0 for any ξ such that $\xi^2 = [3\eta^2 + 2(1 + z_0^2)\zeta^2]/(1 + x_0^2)$.

31.1 $\dfrac{\partial^2 u}{\partial x_1^2} v_1^2 + 2 \dfrac{\partial^2 u}{\partial x_1 \partial x_2} v_1 v_2 + \dfrac{\partial^2 u}{\partial x_2^2} v_2^2 = g$. Have to check

$$\sum_{|\alpha|=2} b_\alpha \mathbf{a}^\alpha = v_1^2 a_1^2 + 2v_1 v_2 a_1 a_2 + v_2 a_2 = (v_1 a_1 + v_2 a_2)^2$$

$$= (v_1^2 + v_2^2)^2 \neq 0 \quad \text{if} \quad \mathbf{a} = \mathbf{v}.$$

31.2 Use $\partial/\partial v = v_1 \partial/\partial x_1 + v_2 \partial/\partial x_2$, $\partial/\partial \tau = -v_2 \partial/\partial x_1 + v_1 \partial/\partial x_2$: $\sum_{|\alpha|=3} b_\alpha \mathbf{v}^\alpha = -1$, so normal for any σ.

31.3 Consider point $\mathbf{x} \in \Gamma$ with $\mathbf{v} = (0, 1)$ and $\mathbf{\tau} = (\xi, 0)$: require that $\beta \neq 0$ or $\beta = 0$, $\alpha\gamma \neq 0$.

32.1 $[u'''v - u''v' + u'v'' - uv''']_0^1 = [-B_1 u S_1^* v - B_0 u S_0^* v + S_1 u B_1^* v + S_0 u B_0^* v]_0^1$.

32.2 $\displaystyle\int_\Gamma \left(v \dfrac{\partial u}{\partial v} - u \dfrac{\partial v}{\partial v} \right) ds = \int_\Gamma (-S_0^* v B_0 u + S_0 u B_0^* v) \, ds.$

32.3   cf. Section 25 on derivatives of distributions.

33.2 $N(A) = \{u : u(x) = ax + b\} = N(A^*)$. Solution exists if $\int_0^1 f(x)\, dx = \int_0^1 xf(x)\, dx = 0$. Solution is unique if $\int_0^1 u(x)\, dx = \int_0^1 xu(x)\, dx = 0$.

33.3 $N(A) = \{u : u \text{ const.}\}$; unique solution if $\int_\Omega u(\mathbf{x})\, dx = 0$. $N(A^*) = \{u : u(\mathbf{x}) = \alpha_1 + \alpha_2(x - y)\}$; solution exists if $\int_\Omega f\, dx = \int_\Omega (x - y)f\, dx = 0$. If $\Omega = (-1, 1) \times (-1, 1)$ then $\int_\Omega f\, dx = 0$ if f is odd in x or y; $\int_\Omega (x - y)f(\mathbf{x})\, dx = 0$ if $f(x, y) = f(y, x)$.

33.4   (a) $\|Au\|_{s-2m} = \left\| \sum_{|\alpha| \leqslant 2m} a_\alpha D^\alpha u \right\|_{s-2m}$. But $\|a_\alpha D^\alpha u\|_{s-2m}^2$

$$= \int_\Omega \sum_{|\beta| \leqslant 2m} [D^\beta(a_\alpha D^\alpha u)]^2 \, dx$$

$$\leqslant C \sum_{2m \leqslant l \leqslant s} |u|_l^2 \leqslant C \|u\|_s^2.$$

(b) From (a), $A : N(A)^\perp \to R(A)$ is bounded. Hence, using the Banach theorem, $A^{-1} : R(A) \to N(A)^\perp$ is linear, bounded $\Rightarrow \|A^{-1}v\| \leqslant K \|v\|$ for all $v \in R(A)$, so setting $v = Au$ we have $\|u\| \leqslant K \|Au\|$ for $u \in N(A)^\perp$. If $\{v_n\}$ is a Cauchy sequence in $R(A)$ with limit v, then with $u_n = A^{-1}v_n$ we have $\|u_m - u_n\| \leqslant K \|v_m - v_n\| \to 0$ as $m, n \to \infty$, so $\{u_m\}$ is a Cauchy sequence in $N(A)^\perp$. $N(A)^\perp$ is closed, so $u_m \to u$ in $N(A)^\perp$. Since A is continuous, $v_n = Au_n \Rightarrow v = Au$. Hence $v \in R(A) \Rightarrow R(A)$ is closed.

Chapter 9

35.1 $\int_0^1 (pu'v' + ruv)\, dx + p(1)u(1)v(1) = \int_0^1 fv\, dx$; $V = \{v \in H^1(0, 1) : v(0) = 0\}$.

35.2   Let angle between $\boldsymbol{\tau}$ and $\mathbf{v}$ be β. VBVP is

$$\int_\Omega \nabla u \cdot \nabla v\, dx = \int_\Omega fv\, dx + \int v\nabla u \cdot \mathbf{v}\, ds, \ v \in H^1(\Omega).$$

But $\boldsymbol{\tau} = \cos \beta \mathbf{v} + \sin \beta \mathbf{s}$ ($\mathbf{s} = \text{tangent} = (-v_2, v_1)$),   or $\boldsymbol{\tau} = (v_1 \cos \beta - v_2 \sin \beta, \ v_2 \cos \beta + v_1 \sin \beta) \Rightarrow \mathbf{v} = (\tau_1 \cos \beta + \tau_2 \sin \beta, -\tau_1 \sin \beta + \tau_2 \cos \beta)$.   Boundary   term   is $\int_\Gamma v(g \cos \beta - \nabla u \cdot \boldsymbol{\mu} \sin \beta)\, ds$ where $\boldsymbol{\mu}$ is normal to $\boldsymbol{\tau}$.

35.3   Use the fact that $\partial v / \partial x = v_1 \partial v / \partial v - v_2 \partial v / \partial s$, $\partial v / \partial y = v_2 \partial v / \partial v + v_1 \partial v / \partial s$, and that $\partial^2 v / \partial v^2 = v_1^2 \partial^2 v / \partial x^2 + 2v_1 v_2 \partial^2 v / \partial x\, \partial y + v_2^2 \partial^2 v / \partial y^2$.

36.1 $\int_\Omega v^2 \, dx = \int_\Gamma v^2 x_1 v_x \, ds - \int_\Omega x_1 \, \partial(v^2)/\partial x_1 \, dx \le 2K \left| \int_\Omega v \, \partial v/\partial x_1 \, dx \right|$

$\le 2K \|v\|_{L_2} |v|_{H^1}$ where $K = \sup |x_1|$.

36.2 $\int_\Omega \left(\sum_{ij} a_{ij} \frac{\partial u}{\partial x_i} \frac{\partial v}{\partial x_j} + buv \right) dx = \int_\Omega fv \, dx, \; v \in H_0^1(\Omega).$

$B(v, v) \ge \int_\Omega \mu \sum_i \frac{\partial v}{\partial x_i} \frac{\partial v}{\partial x_i} dx$ using strong ellipticity. Complete by using the Poincaré–Friedrichs inequality.

36.3 Get $B(v, v) = \int_\Omega \left[\sigma(\nabla^2 v)^2 + (1-\sigma)\left\{ \left(\frac{\partial^2 v}{\partial x^2}\right)^2 + 2\left(\frac{\partial^2 v}{\partial x \, \partial y}\right)^2 \right. \right.$

$\left. \left. + \left(\frac{\partial^2 v}{\partial y^2}\right)^2 \right\} \right] dx \ge (1-\sigma) \int_\Omega \sum_{|\alpha|=2} (D^\alpha v)^2 \, dx$

$\ge C^{-1}(1-\sigma) \|v\|_{H_2}^2$ if $0 \le \sigma < 1.$

36.4 From Exercise 35.1, $B(u, v) = \int_0^1 (pu'v' + ruv) \, dx + p(1)u(1)v(1)$. V-ellipticity: $B(v, v) \ge \alpha \|v\|_{H^1}^2$, $\alpha = \min(p_0, r_0)$; derive this using an extension of the Poincaré–Friedrichs inequality, which states that $\int_a^b u^2 \, dx \le c_1 \int_a^b (u')^2 \, dx + c_2[u^2(a) + u^2(b)]$ (see Rektorys [39], Chapter 18). Continuity: $B(u, v) \le \int_0^1 (p_1 u'v' + r_1 uv) \, dx + p_1 u(1)v(1) \le k \int_0^1 (u'v' + uv) \, dx + p_1 \|u\|_\infty \|v\|_\infty$ $(k = \max(p_1, r_1))$ $\le k(u, v)_{H^1} + p_1 K \|u\|_{H^1} \|v\|_{H^1}$ (Sobolev embedding theorem) $\le (k + p_1 K) \|u\|_{H^1} \|v\|_{H^1}.$

36.5 VBVP is: $\int_0^1 u''v'' \, dx + [h_1 v(1) - g_1 v'(1) - h_0 v(0) + g_0 v'(0)] = \int_0^1 fv \, dx$, $v \in H^2(0, 1)$, so $P = P_1(0, 1)$. Hence $Q = \{v \in H^2(0, 1): \int_0^1 v \, dx = \int_0^1 xv \, dx = 0.\}$ Q-ellipticity is tricky, but see Rektorys [39], Chapter 35. A unique solution exists if and only if $0 = l(p) = \int_0^1 fp \, dx + [g_1 p'(1) - h_1 p(1) + h_0 p(0) - g_0 p'(0)]$ for all $p \in P_1(0, 1)$.

37.1 $DJ(u)v = \lim_{\theta \to 0} \dfrac{J(u + \theta v) - J(u)}{\theta}$ by definition. Set $f(\theta) = J(u + \theta v)$ for any given u, v. Then

$$DJ(u)v = \lim_{\theta \to 0} \frac{f(\theta) - f(0)}{\theta} = f'(0) = \frac{d}{d\theta} (DJ(u)v)|_{\theta=0}.$$

37.2 Since J is convex, $J(v + \theta(u - v)) - J(v) \le \theta(J(u) - J(v))$. Divide by θ, let $\theta \to 0$: $DJ(v)(u - v) \le J(u) - J(v)$. Similarly, $DJ(u)(v - u) \le J(v) - J(u)$. Use linearity of $DJ(u)$ to get result.

37.3 $\dfrac{\partial J}{\partial x_i} = \lim\limits_{\theta \to 0} \dfrac{J(\mathbf{x} + \theta(0, \ldots, y_i, \ldots, 0)) - J(\mathbf{x})}{\theta y_i}$ (y_i in ith slot) so

$\dfrac{\partial J}{\partial x_i} y_i = \lim\limits_{\theta \to 0} \dfrac{J(\mathbf{x} + \theta(0, \ldots, y_i, \ldots, 0)) - J(\mathbf{x})}{\theta}$.

Sum over i to get result.

37.4 $J(\theta u + (1 - \theta)v) = \frac{1}{2}\{\theta^2 B(u, u) + (1 - \theta)^2 B(v, v)$
$- 2\theta(1 - \theta)B(u, v)\} - \theta l(u) - (1 - \theta)l(v)$.
$B(u - v, u - v) > 0$ since B is V-elliptic, so $2B(u, v) <$
$B(u, u) + B(v, v)$. Use this to obtain strict convexity of J.

Chapter 10

38.3 u_n satisfies $B(u, v) = l(v)$ or $(u_n, \phi_k)_B = l(\phi_k)$. Also,

$$u_n = \sum_{k=1}^{n} (u_n, \phi_k)_B \phi_k = \sum_{k=1}^{n} l(\phi_k)\phi_k.$$

Now $J(u) = -\frac{1}{2}\|u\|_B^2$ (show this); But $J(u_n) = \frac{1}{2}\|u_n - u\|_B^2$ $-\frac{1}{2}\|u\|_B^2$, hence $\|u_n - u\|_B \to 0$. The result $\|u_n - u\|_H \to 0$ follows from continuity of $B(.\,,.)$.

39.1 $\|u - u_h\|_B^2 = B(u - u_h, u - u_h) = \|u\|_B^2 - \|u_h\|_B^2 - 2B(u_h, u - u_h)$.
The last term is zero.

39.2 $B(u, v) = l(v)$ so $B(u, u) = l(u)$, hence $J(u) = -\frac{1}{2}B(u, u)$.

39.3 Straightforward; see Section 17.

39.4 $\bar{u}_h(x) = \dfrac{\sqrt{2}}{2}(-\phi_1(x) - \phi_2(x) + \phi_3(x)) = \dfrac{\sqrt{2}}{2}(x^2 + \frac{1}{2}x - 1)$.

40.1 Replace v by Av_h in Green's formula ($G(u, v) = 0$): (40.13) gives $(Av_h, f) = (Av_h, Au_h) \Rightarrow (v_h, A^*f) = (v_h, A^*Au_h)$. If $A = -\nabla^2$, then $A^* = -\nabla^2$.

40.2 (a) $\int_0^1 (-u_h'' + u_h - \sin x)v_h \, dx = 0$, $v_h \in V^h \subset L_2(0, 1)$, $u_h \in U^h \subset H^2(0, 1) \cap H_0^1(0, 1)$. Solve $\mathbf{M}^T\mathbf{a} = \mathbf{F}$ where

$$M_{ij} = \int_0^1 (-\phi_i'' + \phi_i)\psi_j \, dx, \quad F_j = \int_0^1 (\sin x)\psi_j \, dx \quad \text{(see (40.12))}.$$

(b) Least squares: solve $\mathbf{M}^T\mathbf{a} = \mathbf{F}$ where

$$M_{ij} = \int_0^1 (-\phi_i'' + \phi_i)(-\psi_j'' + \psi_j)\, dx \quad \text{and}$$

$$F_j = \int_0^1 (\sin x)(-\psi_j'' + \psi_j)\, dx.$$

Collocation: solve

$$\sum_{k=1}^N (-\phi_k''(x_i) + \phi_k(x_i))a_k = f(x_i), \qquad i = 1, \ldots, N.$$

(c) Solve $\mathbf{M}^T\mathbf{a} = \mathbf{F}$ where $M_{ij} = \int_0^1 \phi_i(-\psi_j'' + \psi_j)\, dx$ and $F_j = \int_0^1 f\psi_j\, dx$.

Chapter 11

41.1 Must show that a function v_i, say, exists such that

$$\int_\Omega v_i \phi\, dx = -\int_\Omega v\, \partial\phi/\partial x_i\, dx.$$

For each Ω_e,

$$\int_{\Omega_e} (\partial v/\partial x_i)\phi\, dx = \int_{\Gamma_e} v v_i \phi\, ds - \int_{\Omega_e} v\, \partial\phi/\partial x_i\, dx$$

$$\text{since} \quad v\,|_{\Omega_e} \in H^1(\Omega).$$

Sum over all elements:

$$\int_\Omega (\partial v/\partial x_i)\phi\, dx = -\int_\Omega v\, \partial\phi/\partial x_i\, dx + \sum_e \int_{\Gamma_e} v v_i \phi\, ds.$$

Boundary term vanishes.

42.2 $\bar{e}_{max}$ exists at point $\bar{x}$ where $\bar{e}' = 0$; then $\bar{e}(x_i) = 0 = \bar{e}(\bar{x}) + \frac{1}{2}\bar{e}''(z)(x_i - \bar{x})^2 \Rightarrow |\bar{e}(\bar{x})| = \frac{1}{2}|\bar{e}''(z)|\,(x_i - \bar{x})^2$. If i is node nearer to $\bar{x}$, then $|x_i - \bar{x}| \leqslant \frac{1}{2}h$. Hence $|\bar{e}(\bar{x})| \leqslant \frac{1}{8}h^2\,|\bar{e}''(x)|$. Maximize over all elements to get result.

42.3 $f''(x) = 2\pi \cos \pi x - \pi^2 x \sin \pi x$. $|\text{Max. value}| = |2\pi|$ at $x = 0$, 1. Hence $\|\bar{e}\|_\infty \leqslant \frac{1}{8}h^2 \cdot 2\pi = \pi h^2/4$. See whether $\log\|\bar{e}\|_\infty \approx 2h + \text{const}$.

44.1 $\|\mathbf{T}_e^{-1}\| = \sup \|\mathbf{T}_e^{-1}\mathbf{y}\|/\|\mathbf{y}\|$, $\mathbf{y} \neq 0$. Set $\mathbf{z} = \rho_e\mathbf{y}/\|\mathbf{y}\|$: then $\|\mathbf{T}_e^{-1}\| = \sup \|\rho_e^{-1}\mathbf{T}_e^{-1}\mathbf{z}\|$. Pick $\mathbf{x}$, $\mathbf{y}$ in Ω_e such that $\|\mathbf{x} - \mathbf{y}\| = \rho_e$: $\|\mathbf{T}_e^{-1}\| = \rho_e^{-1} \sup \|\mathbf{T}_e^{-1}(\mathbf{x} - \mathbf{b} + \mathbf{b} - \mathbf{y})\| = \rho_e^{-1} \sup \|\mathbf{x} - \mathbf{y}\| = \hat{h}/\rho_e$.

45.1 $\|Iv\|_{m,\hat{\Omega}} = \|v\|_{m,\hat{\Omega}} \leq \|v\|_{k+1,\hat{\Omega}}$ since $m \leq k+1$.

$$\|\hat{\Pi}v\|_{m,\hat{\Omega}} = \left\| \sum_i \hat{v}(\hat{\mathbf{x}}_i)\hat{\psi}_i \right\|_{m,\hat{\Omega}} \leq \sum_i |\hat{v}(\hat{\mathbf{x}}_i)| \, \|\hat{\psi}_i\|_{m,\hat{\Omega}}$$

$$\leq C \sup |\hat{v}(\hat{\mathbf{x}}_i)|$$

(C is independent of $\hat{v}$). Use Sobolev embedding theorem: $\hat{v} \in H^{k+1}(\hat{\Omega}) \subset C(\hat{\Omega})$ by assumption so $\sup |\hat{v}(\mathbf{x}_i)| \leq \|\hat{v}\|_{k+1,\hat{\Omega}}$.

45.2

$k = 2$	$k = 3$
$0(h_e^{3-m})$, $0 \leq m \leq 3$	$0(h_e^{4-m})$, $0 \leq m \leq 4$
$u \in H^3(\Omega_e)$	$u \in H^4(\Omega_e)$

45.3 Let triangle have angles α, β, γ with $\theta_e = \alpha \leq \beta \leq \gamma$. Let the sides opposite α, β, γ be a, b, c, respectively. Then $a \leq b \leq c$ and $h_e = c$. The largest circle inscribed in the triangle touches all sides. Draw a sketch and show that $h_e = (\rho_e/2)(\cot \alpha/2 + \cot \beta/2)$. Now $\alpha < \pi/2$, $\beta < \pi/2$, so $\cot \beta/2 \leq \cot \alpha/2$. Hence $h_e/\rho_e \leq \cot \alpha/2 \leq \sigma$ if we prescribe $\alpha \leq \theta_0$, so that $\sigma = \cot \theta_0/2$.

45.4 $\|v - \Pi_e v\|_m^2 = \sum_{l=0}^m |v - \Pi_e v|_l^2 \leq C^2 h_e^{2(k+1)}[\sigma^0 h_e^0 + \sigma^2 h_e^{-2} + \cdots +$

$\sigma^{2m}h_e^{-2m}] |v|_{k+1}^2 \leq C^2 c h^{2(k+1-m)}[h_e^{2m} + h_e^{2m-2} + \cdots + 1] |v|_{k+1}^2$

(where $c = \max (\sigma^0, \ldots, \sigma^{2m})$). Given $K > 0$ we can always find $\varepsilon > 0$ such that term in square brackets $< 1 + K$ provided $h_e < \varepsilon$, i.e. $\|v - \Pi_e v\|_m \leq c_1 h^{k+1-m} |v|_{k+1}$ as $h_e \to 0$.

45.5 $\mathscr{D}\hat{v}(a) = \sum_i \dfrac{\partial \hat{v}}{\partial \hat{x}_i} a_i = \sum_{i,j} \dfrac{\partial v}{\partial x_j} \dfrac{\partial x_j}{\partial \hat{x}_i} a_i = \sum \dfrac{\partial v}{\partial x_j} T_{ji} a_i = \mathscr{D}v(\mathbf{Ta})$.

Proceed in the same way for higher derivatives. Then for $k = 2$, for example,

$$|D^\alpha \hat{v}(\hat{\mathbf{x}})| \leq \|\mathscr{D}^2 \hat{v}\| = \sup |\mathscr{D}^2 \hat{v}(\mathbf{a}, \mathbf{b})| \qquad (\|\mathbf{a}\| \leq 1, \|\mathbf{b}\| \leq 1)$$

$$= \sup |\mathscr{D}^2 v(\mathbf{Ta}, \mathbf{Tb})|$$

$$= \sup \left| \mathscr{D}^2 v\left(\frac{\mathbf{Ta}}{\|\mathbf{T}\|}, \frac{\mathbf{Tb}}{\|\mathbf{T}\|} \right) \right| \cdot \|\mathbf{T}\|^2 \qquad (*)$$

But $\|\mathbf{Ta}\| \leq \|\mathbf{T}\| \, \|\mathbf{a}\| \leq \|\mathbf{T}\|$, so $\|\mathbf{Ta}\|/\|\mathbf{T}\| \leq 1$, similarly $\|\mathbf{Tb}\|/\|\mathbf{T}\| \leq 1$. Hence $(*) = \|\mathscr{D}^2 v\| \cdot \|\mathbf{T}\|^2$.

46.1 $B(w, e) = \|e\|_{L_2}^2$ where $e = u - u_h$. Also, $B(\bar{w}_h, e) = 0$ so $B(w -$

$\bar{w}_h, e) = \|e\|^2_{L_2}$.   Hence $\|e\|^2_{L_2} \leqslant K \|w - \bar{w}_h\|_{1,\Omega} \|e\|_{1,\Omega} \leqslant KCh^\mu \cdot$
$\|w\|_{p,\Omega} \cdot h^\beta \|u\|_{r,\Omega}$ for $w \in H^p(\Omega)$, $u \in H^r(\Omega)$,   where $\mu = $
$\min(k, p-1)$ and $\beta = \min(k, r-1)$. Since $Aw = e$, we have
$w \in H^2(\Omega)$ and $\|w\|_{2m,\Omega} \leqslant c \|e\|_{L_2}$ so $\|e\|_{L_2} \leqslant C_1 h^\nu \|u\|_{r,\Omega}$.

46.2 $\|u - u_h\|_{L_2} \leqslant C_1 h^\nu \|u\|_{r,\Omega}$ where $\nu = \min(2, r)$ for linear or bilinear
elements.

46.3   Using the basis functions of Exercise 42.4 and making appropriate
changes (e.g. in Theorem 44.1 and Theorem 45.3 replace $C(\hat{\Omega})$ by
$C^1(\hat{\Omega})$), the estimate (45.16) remains valid. The VBVP is: find
$u \in H^2_0(0, 1)$ such that $\int_0^1 (u''v'' + k(x)uv)\,dx = \int_0^1 fv\,dx$ for all $v \in$
$H^2_0(0, 1)$. We obtain an error estimate from

$$\|u - u_h\|_{2,\Omega} \leqslant K \|u - \bar{u}_h\|_{2,\Omega} = K\left(\sum_e \|u - \bar{u}_h\|^2_{2,\Omega_e}\right)^{1/2}$$

$$\leqslant Kh_e^{k-1}\left(\sum_e |u|^2_{k+1,\Omega_e}\right)^{1/2} = Kh^{k-1} |u|_{k+1,\Omega},$$

provided that $P_k(\hat{\Omega}) \subset \hat{X} \subset H^2(\hat{\Omega})$ and $H^{k+1}(\hat{\Omega}) \subset C^1(\hat{\Omega})$ and $u \in$
$H^{k+1}(\Omega)$. Hence for a cubic Hermite basis we have $\|u - u_h\|_{2,\Omega} = $
$0(h^2)$.

Index

Page numbers referring to section headings are displayed in **bold** type.